The Lopsided Ape

The Lopsided Ape

Evolution of the Generative Mind

MICHAEL C. CORBALLIS
Department of Psychology
University of Auckland
Auckland, New Zealand

New York Oxford
OXFORD UNIVERSITY PRESS
1991

Oxford University Press

Oxford New York Toronto
Delhi Bombay Calcutta Madras Karachi
Petaling Jaya Singapore Hong Kong Tokyo
Nairobi Dar es Salaam Cape Town
Melbourne Auckland

and associated companies in
Berlin Ibadan

Published by Oxford University Press, Inc.,
200 Madison Avenue, New York, New York 10016

Library of Congress Cataloging-in-Publication Data
Corballis, Michael C.
The lopsided ape / Michael C. Corballis.
p. cm. Includes bibliographical references and index.
ISBN 0-19-506675-8
1. Laterality. 2. Human evolution.
3. Brain—Evolution. I. Title.
QP385.5.C67 1991 152.3′35—dc20
90-22905

2 4 6 8 9 7 5 3 1

Printed in the United States of America
on acid-free paper

Preface

In the 1950s, when I was an undergraduate, basic psychology seemed to be built very largely on work with animals. We spent a good deal of time discussing whether rats in mazes learned which turns to make or which places to go to, or discussing how the rates at which they pressed bars depended on the pattern of rewards they received. The dominant figure of the time was the behaviorist B. F. Skinner. I remember also the excitement of the ideas proposed by D. O. Hebb, in many ways the founder of modern neuroscience as it pertains to psychology. In all of this, it seemed to be taken for granted that the same principles of behavior applied to humans as to other animals, an idea that seemed at the time to have a vaguely subversive quality to those of us with conventional upbringings. Indeed, I am afraid it was probably the whiff of heresy that attracted me away from mathematics and to psychology.

But how things have changed! In my own department animals are slowly but surely disappearing, and mainstream psychology is now based primarily on research with humans. No longer are we comfortable with the assumption that results based on animals can be extrapolated to our own species. This change was due in part to the classic confrontations of the late 1950s and early 1960s between Skinner and Noam Chomsky, a linguist who revived the dualistic, Cartesian idea that there is an unbounded quality to the human mind that is not to be found in other animals. But change may have been in the wind anyway, since the behaviorism of the 1950s and 1960s had seemed to place a straitjacket on the study of the mind. A new sense of the special nature of humanity emerged, and such human-based topics as memory, attention, language, and consciousness moved to center stage.

This switch from an animal-based to a human-based science reflects an age-old tension in the way we view ourselves. For much of our history we have seen ourselves as closer to God (or gods) than to the animals, but there has always been a heretical and often suppressed sense of our animal nature. This was reinforced by the growth of secular science, which provided mechanistic accounts of the way our bodies worked, and indeed revealed close parallels between the workings of our own bodies and those of other animals. For some time, scholars such as Descartes were still able to preserve at least a part of the human mind for some higher realm. The ultimate challenge, of course, came from that mild but far-seeing Englishman, Charles Darwin. At first reticent over the implications of his theory of natural selection, he eventually made clear his view that we are indeed a part of the animal kingdom, and share a common ancestry, not so remote, with the apes.

Even in a secular age, it still seems reasonable to some to insist on a fundamental discontinuity between humans and other animals. In part, the issue has focused on the supposedly special nature of human language, and this is discussed in some detail in the following chapters. It has also been argued that humans are endowed with a special kind of consciousness, including self-consciousness, that is not possessed by other animals. This has proven a more intractable argument, partly because consciousness has proven peculiarly difficult to define. The notion of consciousness seems to depend on our own subjective sense of what it is to be conscious, and so long as this is so, other animals are placed at a distinct disadvantage .

This book focuses the question of human discontinuity on laterality. Although I was fearful that evidence on animal asymmetries would overtake me, it still seems to be the case that right-handedness and cerebral asymmetry are unique to humans—unique not so much in their presence as in their extent, pattern, and population bias. Moreover it is not laterality per se that is critical so much as the nature of functions that are lateralized, which themselves seem to capture much of the essence of what it is to be human. Handedness is related to our extraordinary ability to manufacture and manipulate, and cerebral asymmetry is most pronounced with respect to that putatively unique faculty, language.

In order to develop my themes, I have been obliged to cover material from a range of disciplines, some outside my own areas of expertise. I have therefore tried to avoid excessive jargon and to write in a way that is understandable to nonspecialists. Since my own expertise is in cognitive psychology and neuropsychology, I am aware that my attempts to do this are uneven. I am also aware that much of the material, especially that on human evolution, is highly controversial, and that my conclusions may be seen by some as biased or wrong. I have simply tried to make the best sense of the material that I could.

I have also tried to make each chapter as autonomous as possible, so that readers familiar with some material can skip chapters. No chapter is immune from my own interpretations, although I hope the coverage is sufficiently broad to serve also as a general review, even for those who disagree with my conclusions. The crux of my argument for human uniqueness is developed in Chapter 9, and knowledgeable readers avid for the message (or for blood) might proceed directly there. My theme also suggests a revision of popular notions of hemispheric duality, which is discussed in Chapter 10. However, I have tried to organize the material in such a way that the case is built up in logical sequence, so that readers with widely ranging backgrounds can follow it.

I have used notes at the end of each chapter rather than references in the text so as not to interrupt the flow of the text, and technical details are sometimes covered in the notes. I have nevertheless provided a fairly comprehensive list of references. In this, I was mindful of a colleague who once wrote to me to say how much he appreciated an earlier book I wrote—because he found the references useful. Anything to please.

I could not have finished this book had I not been granted Research and Study Leave from the University of Auckland in 1989. I am grateful to Professor Francesca Simion of the University of Padua, Italy, Dr. Alan Baddeley of the MRC Applied Psychology Unit in Cambridge, England, and Dr. Justine Sergent of the Montreal Neurological Institute, Montreal, Canada, for arranging for me to spend time in each of their institutions. All three visits were productive for me, and I am sorry that I was so obsessed with writing that I neglected to contribute much to their endeavors. Countless colleagues have contributed, but I should like to single out for special mention Marian Annett, Irving Biederman, Daniel Bub, Martha Farah, Philip Johnson-Laird, John Macnamara, Tony Marcel, Paula Marentette, Brenda Milner, Jenni Ogden, Isabelle Peretz, Gill Rhodes, Tim Shallice, Lynette Tippett, and Robert Zatorre—although I expect none of them is in complete agreement with me. I am also grateful to the students in my 1990 graduate seminar in "Mind and Brain," who discussed the chapters in class and made several useful suggestions.

Finally I thank my wife Barbara and sons Paul and Tim, who put up with me while we traveled to glorious places, while all I could think about was The Book.

Auckland
August 1990

M. C. C.

Contents

The Lopsided Ape

— 1 —

Are Humans Unique?

> What a piece of work is a man! How noble in reason! How infinite in faculty! In form, in moving, how express and admirable! In action how like an angel! In apprehension how like a god! The beauty of the world! The paragon of animals!

So said Hamlet, in an outburst of human (and male) chauvinism. Admiration of our own species is a common human foible; we humans have long attributed to ourselves special qualities of mind, spirit, or morality that are denied all other creatures. Religions seem to play a special role in this by bestowing uniquely upon us an immortal soul or by tracing our origins to some divine act of creation. Indeed, it may be argued that religions exist as much for the glorification of humans as for the worship of gods. To quote from the Psalms: "What is man, that thou art mindful of him? . . . For thou hast made him a little lower than the angels, and hast crowned him with glory and honour."

According to the American historian Lynn White, Jr., the Christian religion in particular is "the most anthropocentric religion the world has seen,"[1] standing in sharp contrast to ancient paganism and Asian religions, with the possible exception of Zoroastrianism. Much of the Christian attitude toward animals can be traced to the book of Genesis. The Garden of Eden was originally a paradise in which Adam was granted dominion over all living creatures. During this blissful era, humans lived harmoniously with other animals and were not carnivorous. But with the Fall, when humans rebelled against God, things changed. The earth degenerated and became less fertile, and pests such as thorns, thistles, gnats, and fleas appeared. The animals had to be forced into submission, and humans became carnivorous.

After the Flood, God gave humans a renewed but more repressive authority over the animals:

> The fear of you and the dread of you shall be upon every beast of the earth, and upon every fowl of the air, upon all that moveth upon the earth, and upon all the fishes of the sea; into your hand are they delivered. Every moving thing that liveth shall be meat for you. (Genesis, ix. 2–3)

This text indeed set the stage for much of Western history, with human dominance part of the divine plan. In eighteenth-century England, for example, it was widely preached that the domestication of animals was good for them, and the butchering of animals was held to be kind rather than cruel since it spared them the suffering of old age. Even the killing of animals for sport was justified, in the words of Thomas Fuller in 1642, by "man's charter of dominion over the creatures."[2] The killing of animals was also thought to have a civilizing influence on native peoples. At the Great Exhibition of 1851 in London, there was a booth displaying monkey skins from Africa. One observer noted the suffering that the animals must have undergone but added that "the work of catching these monkeys is civilizing the African."[3] Reports of how Eastern religions respected the lives of other animals were treated with contempt, as though they debased human ascendancy.

It would be wrong to blame our attitude toward animals wholly on the Christian religion. Similar attitudes can in fact be found in other religious doctrines. For example, in 1632 it was said of the American Indians that "they have it among them by tradition that God made one man and one woman and bade them live together, and get children, kill deer, beasts, birds, fish, and fowl and what they would at their pleasure."[4] Besides, the Old Testament is not unequivocal in its expressed attitude toward animals. In Proverbs (xii, 10) it is taught that a good man "regardeth the life of his beast," and in Hosea (ii, 18) it is implied that animals are members of God's covenant. St. Francis of Assisi, whom Lynn White describes as "the greatest spiritual revolutionary in Western history,"[5] tried valiantly to place animals on an equal footing with humans, but the tide was against him.

The truth is that our very survival has often depended on the exploitation of animals, which still serve as beasts of burden and as sources of food. In more recent times, animals have served the cause of human medicine. The term "guinea pig" is not confined to that pleasant and cooperative animal but applies even to the higher primates that most resemble ourselves and that can therefore be used to test the effectiveness of drugs or surgical procedures. Of course, not all exploitation of animals is in the cause of human survival. Animals are still hunted and kept in unnatural confinement for entirely frivolous reasons. Nevertheless, some degree of exploitation seems inevitable, and it is understandable that people should have sought justification for their cruelty in terms of divine principles. Religious attitudes toward animals are no doubt born of social practices rather than the reverse.

Negative attitudes toward animals are also sustained by the belief that animals embody the baser human instincts that need to be constantly repressed. In the sixteenth century, lust was said to make men "like . . . swine, goats, dogs and the most savage and brutish beasts in the world."[6] Words like "beastly," "brutish," and "swinish," as references to disagreeable human behavior, remain in common parlance. Bodily functions such as eating, copulation, or evacuation remind us of our animal natures, which may be why these functions are shrouded in ritual or secrecy. The idea of the beast within also persists in twentieth-century thought, as in Sigmund Freud's notion of the libidinous, murderous id that has constantly to be restrained by the rational ego and the censorious superego.

The exploitation of animals has never been completely unrestrained, however, and the very insistence of moral arguments for the killing of animals may owe something to the fact that people are naturally well disposed to many animals. Horses, dogs, and cats have always been the special objects of human affection; there is a proverb that runs "He cannot be a gentleman who loveth not a dog."[7] Organizations such as the Society for the Protection of Cruelty to Animals, or the more recent and extreme Animal Rights Movement, have ensured at least some measure of balance in the way we treat animals.

There has also been a long history of attempts to define the difference between humans and animals in other than theological terms. Aristotle regarded humans as the only political animals, while Thomas Willis noted that humans were the only creatures capable of laughter. Martin Luther in 1530 pronounced that an essential difference between humans and other creatures lay in the possession of private property. He should be so lucky. One eighteenth-century observer attached special importance to the nose: "Man is, I believe, the only animal that has a marked projection in the middle of the face."[8]

Aristotle also argued that humans were more beautifully formed than other animals, showing "a more exquisite symmetry of parts." This idea carries a special irony in the context of this book, since one of my aims is to evaluate the idea the special nature of human beings may lie, at least in part, in our asymmetry rather than in our symmetry. Other ideas that will be examined in this book also have historical antecedents. Benjamin Franklin seems to have been the first to suggest that it was tool making that distinguished us from other species. Our erect posture, unique among the mammals, has long been the focus of speculation, although again the argument often took on a theological slant; Plato, for example, noted that only humans looked up to heaven, while the beasts were doomed to gaze forever downward.[9]

There is no denying that humans are *different* from other species, just as other species are different from each other. The question that motivated the writing of this book, and that has motivated a good deal of discussion for as long as humans have pondered their own condition, is whether the difference between humans and other species somehow represents a special discontinuity. The most extreme position would be that this discontinuity is in effect a dichotomy, unmatched by any other difference between species. A more moderate position would be that the magnitude of the discontinuity is not unprecedented but is matched by, say, that between animals that can and cannot fly. Given the extraordinary similarity in genetic structure between ourselves and our closest relative, the chimpanzee,[10] it seems on the face of it unlikely that we are quite so special as we may like to think—unless we have indeed been touched by some divine wand. And yet, even in an objective sense, the scope of the human enterprise on our planet, and even beyond, seems to belie the biological similarity between ourselves and the other primates.

Descartes' Dichotomy

Although a concern with the question of human uniqueness can be traced back to ancient times, modern discussion really begins with the seventeenth-century French philosopher Rene Descartes (1596–1650), often regarded as the founder of modern philosophy. Descartes argued for a dichotomy between humans and other species, but this is perhaps less remarkable than the way in which he framed the issues, foreshadowing some of the debates of the present time. In maintaining that humans were indeed fundamentally different from other species, he seems in fact to have been guided at least in part by his religious principles: "After the error of those who deny God," he wrote, ". . . there is none that leads weak minds further from the

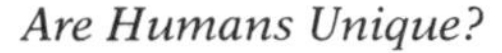

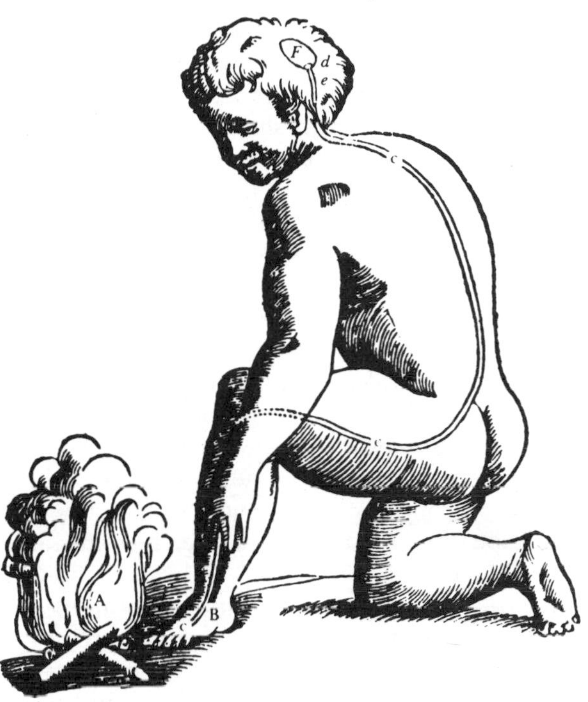

Figure 1.1. Descartes' view of the pathway from foot to brain that signals the presence of burning heat and causes the foot to be withdrawn (from Descartes, 1985).

straight path of virtue than that of imagining that the souls of beasts are of the same nature as ours."[11] However, Descartes also put forward rational arguments for human uniqueness that continue to exert an influence.

Descartes argued that animals, unlike humans, were mere machines, or automata. In this he was influenced in part by his fascination with clockwork models of animals, which were popular at the time, and in part by the growing knowledge of anatomy and physiology. He himself had made a substantial contribution to our knowledge of the workings of the eye by dissecting out the eye of an ox and observing the inverted image formed on the retina. He was also aware of the importance of nerve impulses in mediating actions and gave a purely mechanistic account of how reflexes work. Figure 1.1 shows his famous diagram of the reflex involved in moving a limb away from a source of burning heat.

Descartes also described the case of a girl who had had her arm removed at the elbow, yet still complained of pain in the absent hand. A limb perceived as though still present after amputation is known as a *phantom limb*. Descartes concluded that the "pain in the hand is not felt by the mind inasmuch as it is in the hand, but as it is in the brain." The phantom limb shows, in other words, that conscious perception depends on events in the brain rather than on the sensory input itself.

Such observations led Descartes to a mechanistic view of the functioning of the brain. Such a view is taken for granted in much of modern neuroscience, but in Descartes' time it was a revolutionary doctrine. It was also a dangerous one, since it threatened the idea that humans were different from the animals. However, Descartes was prudent enough to make an exception of the human mind. He accepted a mechanistic account of the behavior of animals and of human reflexes but considered that human rationality was on a different plane. Influenced again by mechanical toys, he argued that a perfect mechanical replica of an ape would in fact be indistinguishable from a real ape, but that no such replica of a human being could be mistaken for a real human.

In particular, there were two tests that the human replica would fail. First, a mechanical replica would be incapable of the creative use of language enjoyed even by "men of the lowest grade of intellect." Second, and more generally, the replica could do only what it was constructed to do, and would lack the human flexibility to respond intelligently to novel or unexpected situations. Human reason and will, Descartes seemed to say, transcend the workings of a mere mechanical device.

These arguments have a remarkably contemporary ring to them. Modern research on the ability of apes and other animals to use language has focused on the crucial question of whether that language has the creativity and flexibility of human language. Moreover, our modern age is also obsessed with mechanical devices—computers and robots—that simulate human thought or action, and the question again is whether the simulating device can be said to duplicate the human mind or human consciousness. It is sometimes said, for instance, that computers will never truly simulate the human mind because a computer can only do what it is programmed to do, and so cannot mimic the creativity and flexibility that humans possess. This was exactly the point that Descartes was making, although he was not, of course, able to dress it in computer jargon.

Since the human mind could not be reduced to mechanical principles, Descartes argued that humans must possess a nonmaterial soul that sets us apart from the other animals. The distinction between the nonmaterial mind and the physical brain and body is known as mind-body *dualism*. Descartes suggested that the soul could influence the workings of the brain through the pineal gland, a small structure that he thought important because of its location on the midline of the brain, just behind the forehead and roughly between the eyebrows. It was through the pineal gland that the soul received sensations and initiated voluntary action. So it is, according to Descartes, that humans, and only humans, possess consciousness and free will.

This may have been the most exalted status to have been achieved by the pineal gland—its finest hour, so to speak—although certain sects still hold that it possesses special powers. In theosophy and some Eastern religions it is known as the "third eye," and some parapsychologists still believe that it may be responsible for telepathy and other extrasensory powers. Nevertheless, Descartes' idea was greeted with skepticism even by his own contemporaries, and few now give it any credence whatever.

There have been other attempts to find some anatomical basis for human uniqueness and through which the soul might find expression. The nineteenth-century anatomist Richard Owen, for instance, maintained that certain brain structures, including the hippocampus minor, were present only in humans and were responsible for our special status. However, Thomas Henry Huxley, Charles Darwin's great friend and protagonist, showed conclusively that all apes possess these structures. As a consequence of the debate between Huxley and Owen, the hippocampus was the object of some befuddled ridicule among their contemporaries. In *The Water Babies* (1886), Charles Kingsley satirized Huxley and, incidentally, touched again on themes introduced by Descartes, as follows:

> You may think that there are other more important differences between you and an ape, such as being able to speak, and make machines, and know right from wrong, and say your prayers, and other little matters of that kind; but that is a child's fancy, my dear. Nothing is to be depended on but the great hippopotamus [*sic*] test.[12]

The hippocampus minor seems to have gone the way of the pineal gland, but the structure known simply as the hippocampus has often been implicated in modern theories of consciousness. This structure can be identified in lower mammals as well as in primates, but is absent in birds, which has led to the frivolous suggestion that its purpose in animals may be to prevent flight. It is still in fact the focus of considerable controversy and has been implicated in a number of mental functions, including conscious memory, the representation of space, and emotion and anxiety. Ironically, the very similarity of its function in humans and other apes might be taken to imply that consciousness is *not* unique to humans.[13]

Another modern idea is that the anatomical basis for a uniquely human consciousness might be found in the left side of the human brain, which appears to possess properties not found in the left brains of other mammals. These include a specialization for language, itself often taken to be a hallmark of human uniqueness. Some have argued that the left hemisphere of the human brain is also endowed with special qualities of consciousness that set us apart from other species. As the title of this book indicates, these themes are central to my concerns and will be discussed in more detail in later chapters.

Descartes' dualistic view, in which the nonmaterial soul was distinguished from the material brain, was something of a master stroke, for it appeased religious opinion while keeping the study of the brain and nervous system safe for future generations of scientists. Nevertheless, even at the time, there were those who doubted its plausibility. One was the young Princess Elizabeth of Palatine, cousin of King Charles II of England who founded the Royal Society. Elizabeth did not see how a nonmaterial entity, without physical properties or extension in space, could possibly influence the material brain, and she had some correspondence with Descartes about this. Nevertheless, Elizabeth may have been aware that, theologically, she was venturing into forbidden territory, and asked that her letters to Descartes be destroyed. Fortunately, they were not, but some were not recovered until some 200 years later.

In the eighteenth century, Descartes' compatriot J. O. de La Mettrie, in his book *L'Homme Machine* (1747),[14] did take the fateful step of arguing that "irritation" of the nerves would explain all behavior, whether animal or human, reflexive or intelligent. For this heresy he was attacked by the clergy and banished from France and later from Holland, to find refuge ultimately in the court of Frederick the Great of Prussia. However doubts about Cartesian dualism and the dichotomy between humans and other animals were probably quite widespread, if suppressed.

Darwin and Continuity

If Descartes laid the modern foundations for a dichotomy between humans and other animals, it was Charles Darwin who set the stage for continuity. The publication in 1859 of his famous treatise *The Origin of Species* posed a serious threat to notions of human uniqueness, and the repercussions are still with us.

The idea that humans evolved from other animals was itself not new, although it was generally regarded in Darwin's time as heretical. As early as the sixth century B.C., the Greek philosopher Anaximander argued that all animals, including humans, were descended from fishes. Perhaps evolution was in Darwin's genes, since his own grandfather, Erasmus Darwin, had speculated that the different forms of life developed gradually from a common ancestry. The French natural historians Buffon (1701–1788) and Cuvier (1769–1832) were also important influences; Darwin remarked of Buffon's ten-volume work, *Natural History*, that "whole pages are laughably like mine."[15] Although he argued that new species descended from old ones, and that humans and apes must share a common ancestor, Buffon was forced by the religious authorities to declare that his views did not contradict the biblical story of creation. Jean Baptiste Pierre Antoine de Monnet, Chevalier de Lamarck (1744–1829), otherwise known as Lamarck, a prominent French zoologist and a precursor of Darwin, also argued for evolution, but held that evolutionary change was brought about in part through the inheritance of characteristics acquired as a result of learning. Although revived in the late 1930s and 1940s by the Russian geneticist Trofim D. Lysenko, this theory has never held sway for long and is now generally discredited.

Darwin's contribution was twofold. First, he marshaled a huge array of facts to support the idea of evolution, so that it was no longer possible to ignore the likelihood that different organisms were linked by common physical descent. Second, he proposed the theory of natural selection as a simple and convincing explanation of how evolutionary changes occurred. Organisms vary, and those variations that are favored by the environment, so the theory goes, will tend to survive and propagate. Evolution works, in other words, through the survival of the fittest.

The theory of natural selection is in a way more heretical than the idea of evolution itself, because it implies that evolutionary change is without direction or purpose. Some theologians, for instance, were able to reconcile themselves to evolution by supposing that it leads to increasing perfection, with humans just a step away from the angels. But the idea that evolution is just nature's way of capitalizing on chance variations seems to border on the insulting, especially if applied to the evolution of humans. We are the product, not of divine creation, but of historical accident.

Darwin knew that his theory was likely to cause trouble, especially insofar as it implied that humans were descended from the apes. In *The Origin of Species* he largely avoided the issue, venturing only that "light will be thrown on the origin of man and his history."

Later, in *Descent of Man* (1871) and *The Expression of the Emotions in Man and Animals* (1872), he was more explicit. He was adamant that humans did not differ from other animals in kind, while conceding an immense difference in degree of mental powers. Yet he was somewhat equivocal, and in *Descent of Man* he was prudent enough to include references to God:

> The belief in God has often been advanced as not only the greatest, but the most complete of all the distinctions between man and the lower animals. It is however impossible . . . to maintain that this belief is innate or instinctive in man. On the other hand a belief in all-pervading spiritual agencies seems to be universal, and apparently follows from a considerable advance in man's reason, and from a still greater advance in his faculties of imagination, curiosity, and wonder. I am aware that the assumed instinctive belief in God has been used by many persons as an argument for His existence. But this is a rash argument, as we should be thus compelled to believe in the existence of many cruel and malignant spirits, only a little more powerful than man; for the belief in them is far more general than in a beneficent Deity. The idea of a universal and beneficent Creator does not seem to arise in the mind of man, until he has been elevated by long-continued culture.[16]

Another heretical aspect of the theory of evolution was its implied materialism. If humans are descended from animals, then it is difficult to sustain Descartes' view that humans uniquely possess a non-material mind or soul. A youthful experience had warned Darwin that he was on dangerous ground here. While he was an undergraduate in Edinburgh, his friend W. A. Browne had presented a paper giving a materialistic interpretation of life and mind to the Plinian Society. This was so controversial that it was decided to remove all references to the paper, including its announcement in the minutes of the previous meeting.

Darwin refrained from public comment on the materialistic implications of evolution, but in one of his notebooks he questioned why the idea that thought was a product of the brain should be more to be wondered at than the idea that gravity was a property of matter. A. R. Wallace, who shared with Darwin the credit for the theory of natural selection, could not accept a materialistic interpretation of the human mind. He maintained that consciousness was uniquely human, a divine gift rather than the product of evolution.

It is instructive to consider how Wallace reached this opinion, for it was based not on religious sentiment but on a logical argument. He observed that people of different races varied widely in their quality of life and intellect. In this he may be accused of cultural chauvinism, but not of racism, for he also believed that the brains of "savages" were as developed as those of the "civilized" Europeans. He based this conclusion, at least in part, on that distinctive human faculty that is so often the focus of arguments about human uniqueness, namely, language:

> Among the lowest savages with the least copious vocabularies the capacity of uttering a variety of distinct articulate sounds, and of applying to them an almost infinite amount of modulation and inflection, is not in any way inferior to that of the higher races.[17]

Wallace was also a rigid selectionist in that he believed that all evolutionary change was due to natural selection. However, this led to the paradox that people in primitive societies used only a tiny part of their intellectual and cultural potential. Plainly, this potential could not have been the outcome of natural selection, since it was not in fact used. Wallace therefore concluded that human development must have been guided not by natural selection but by some divine intelligence.[18]

Darwin himself was appalled by this argument and wrote to Wallace: "I hope you have not murdered too completely your own child and mine."[19] He recognized that not all evolutionary change is the immediate result of natural selection, and that features that have evolved for one purpose are often exploited for other quite different purposes. It is difficult to make any sense of human evolution without supposing that this is true of the human brain. That is, our powers of thought and reasoning may well have evolved in specific contexts, such as the use or manufacture of tools, but have been readily adapted to quite different activities, like playing chess or arguing about evolution. In modern evolutionary theory this is known as *exaptation*, and it is now thought to play at least as important a role in evolutionary change as adaptation in the Darwinian sense.[20]

It is also becoming apparent that the genes can undergo sudden, quite dramatic changes in structure. One way in which this occurs is through the influence of viruses. Viral infection can result in genetic material being transferred, or *transduced*, even from one *species* to another; for example, it has been suggested that development of the eye in the squid and in vertebrates was not a matter of independent, parallel evolution, but was actually a consequence of viral transduction—the "smuggling" of whole genes across species.[21] It is difficult to think benignly of viral infections, which are notoriously resistant to the body's immune system, or indeed to any form of "cure." Yet this very property may be testimony to the important role of viruses in evolutionary change.

However it is now clear, especially from the pioneering work of Nobel Prize winner Barbara McClintock, that wholesale changes in genetic structure may be brought about by mechanisms other than viral transduction. So-called jumping genes, transposons, and chromosomal rearrangements can also result in sudden and dramatic changes in genetic structure.[22] One consequence of this is that a change may be larger than that which is subsequently selected for. Genes lying close on the DNA string to a change that proves to have survival value may also be selected in terms of their sheer proximity. This has been termed *genetic hitch-hiking*.[23] Such genes are merely selected, rather than selected *for*—they have hitchhiked their way into destiny. Since the positioning of genes along a chromosome may bear no relation to function, the mechanism of hitchhiking adds considerably to the randomness and serendipity of evolution.

These newly discovered mechanisms imply that evolution need not be the slow, smooth process that Darwin envisaged. Rather, evolutionary change can occur in what has been called *punctuate* fashion[24] and can include characteristics that were not themselves selected in terms of survival value. It has been suggested, for example, that the organs of flight in insects were originally selected *for* as organs of thermal exchange, but it was the hitchhiking of other genes that serendipitously bestowed their uplifting properties.[25] Given examples such as this, it becomes not unreasonable to suppose that similar mechanisms may have produced a discontinuity between humans and other animals. Indeed, the extra but often unexploited human potential that so puzzled Wallace may have been bestowed not by God but by a hitchhiking gene.

Religious opposition to the Darwinian theory of evolution has, of course, persisted. In 1860, the year following publication of *The Origin of Species*, the famous debate between T. H. Huxley and Bishop Wilberforce took place, with victory generally conceded to Huxley and the theory of evolution against the religious opposition. Sixty-five years later, in 1925, history repeated itself with the famous Scopes trial in Dayton, Tennessee, representing a challenge to the law that made it illegal "to teach any theory that denies the story of the Divine Creation of man as taught in the Bible, and to teach instead that man has descended from a lower order of animals." In present-day America, the creationists continue to demand that the act of creation as presented in the Bible be given coverage equal to that of evolutionary theory in the teaching of biology.

Not everyone has seen the theory of evolution as a threat to human superiority. Ideas about "the great chain of being," with humans representing the pinnacle of biological development, had existed long before Darwin. Some of Darwin's contemporaries, notably the British philosopher Herbert Spencer, smugly took Darwin's theory as proof of the status of the English gentry as the highest flower on the evolutionary tree. Theologians, too, have tried to reconcile evolution with religious doctrine; as the American poet William Carruth Herbert (1859–1924) wrote, "Some call it Evolution, And others call it God."

One influential attempt at reconciliation was that of the paleontologist and Jesuit priest Pierre Teilhard de Chardin. His ideas were largely suppressed by the church during his lifetime but gained considerable influence with the posthumous publication of his book *The Phenomenon of Man* in 1959. Teilhard argued that evolution was characterized by the progressive dominance of spirit over matter, culminating in human consciousness. He saw the earth as consisting of four concentric layers—the lithosphere, the atmosphere, the biosphere, and the noosphere—the last being Teilhard's term for the outermost layer representing the psychic component. The evolution of the noosphere marked the beginning of a domination of spirit over matter. Its future development would see a progressive convergence of human lineages, a process already begun with human socialization, reaching ultimately a single point of spirit unencumbered by matter. This point is called Omega:

> The end of the world: the overthrow of equilibrium, detaching the mind, fulfilled at last, from its material matrix, so that it will henceforth rest with all its weight on God-Omega.[26]

The Phenomenon of Man was an international best-seller in the 1960s. Stephen Jay Gould writes that the Widener Library at Harvard University houses a whole tier of books devoted to Teilhard's ideas, and that two journals established to discuss his theories are still flourishing.[27]

Gould also points out that Teilhard's evolutionary theory makes little sense.[28] It implies that evolution is purposeful and is directed toward an ultimate goal. As we have seen, natural selection contradicts this belief. Evolution depends on random events, such that if the earth's history were to rerun its course, the probability that a species resembling ourselves would re-emerge is vanishingly small. In Gould's words, "the evolutionary tree [looks] more to me like a complexly ramifying bush than a bundle of parallel twigs growing upward in a definite direction."

Yet Gould himself, in many ways the heir to T. H. Huxley in his defense of Darwinian theory against reactionary forces, seems nevertheless to accept a fundamental discontinuity between humans and other animals:

> We are but a tiny twig on a tree that includes at least a million species of animals, but our one great evolutionary invention, consciousness—a natural product of evolution integrated with a bodily frame of no special merit—has transformed the surface of our planet. Gaze upon the land from an airplane window. Has any other species left so many visible signs of its relentless presence?[29]

In his reference to "a bodily frame of no special merit," Gould seems almost to accept a nonmaterial basis for consciousness. I do not suppose that he would want to claim that, but the passage does serve to illustrate how thoroughly we are steeped in mind-body dualism.

But as we have seen the idea that consciousness might have a purely mechanistic basis was recognized by Darwin himself and goes back at least to la Mettrie. Following publication of *The Origin of Species,* in fact, it became commonplace among scientists to emphasize the objective rather than the subjective nature of mind and consciousness. One enthusiastic evolutionist and materialist was the eminent French physician and anthropologist Paul Broca, who said that he would "rather be a transformed ape than a degenerate son of Adam."[30]

Broca is chiefly remembered for his discovery that a small region in the left frontal lobe of the human brain, now generally known as *Broca's area,* is of critical importance in the control of articulate speech. The remarkable thing about this discovery was that it implied an asymmetry of function in an organ that appeared to be anatomically symmetrical. This discrepancy between function and structure carries the implicit idea of some nonmaterial intervention in the left side of the brain—the pineal gland relocated—which may explain why cerebral functional asymmetry has been regarded with fascination for over a century. Broca himself remained a convinced and articulate materialist, however, and like Socrates was denounced for his heretical ideas and for the corruption of youth.

Many biologists also adopted an objective approach to the study of the mind. In his book *Mental Evolution in Animals,* published in 1888, the British biologist George John Romanes wrote as follows:

> Now throughout the present work we shall have to consider Mind as an object; and therefore it is well to remember that our only instrument of analysis is the observation of activities which we infer to be

> prompted by, or associated with, mental antecedents or accompaniments analogous to those of which we are directly conscious in our own subjective experience.[31]

To Romanes, the essence of consciousness was the ability to choose and to modify choice in the light of experience. By these criteria, humans are scarcely special, and visits to any number of modern laboratories will reveal rats, pigeons, insects, and even planaria making and learning choices. Romanes nevertheless felt qualified to construct an evolutionary tree of intellectual development. He identified some 50 levels, but only humans reached the top. Their nearest rivals, the apes and dogs, reached only to level 28, followed closely by monkeys and elephants at level 27. Through infancy and childhood, humans progressed through the various levels, so that a 12-week-old human, for instance, is at the level of a full-grown fish or frog.

Another influential figure of the late nineteenth century was Conwy Lloyd Morgan, whose 1894 book *Introduction to Comparative Psychology* also presented the case for an objective study of consciousness in animals. Morgan is famous for his so-called canon governing the interpretation of the behavior of nonhuman animals:

> *In no case may we interpret an action as the outcome of a higher psychical activity, if it can be interpreted as the outcome of one which stands lower in the psychological scale.*[32]

But while this admirable doctrine may prevent excesses in attributing intelligence to animals, it may equally (and conveniently) serve to widen artificially the gap between ourselves and our animal cousins, and so preserve our sense of superiority.

Despite their objective approaches, Romanes, Morgan, and other biologists of the time still regarded our human knowledge of our own minds as crucial to the exercise of attributing consciousness to animals. Objective data from another organism must still be assessed in relation to the laws of mind as revealed by subjective experience.

Scientific psychology, however, had been firmly entrenched in Cartesian dualism since Wilhelm Wundt set up the first laboratory of experimental psychology in Leipzig in 1879. To Wundt, psychology was the study of the human mind, to be explored through *introspection*, the examination of the inner workings of the mind. Psychology would therefore proceed in parallel with the physical sciences, discovering the laws of the nonmaterial mind as the physical sciences discovered the laws of physical nature. There was little place for animal psychology in such an approach, although Wundt later wrote a book in which he argued that animal consciousness might be estimated through comparisons of animal behavior with our own introspections.[33]

Behaviorism and the Denial of Mind

Things changed radically with the behaviorists, who emerged as the dominant force in North American psychology in the first half of the twentieth century. The father of behaviorism was John Broadus Watson, and in a famous article published in 1913 he wrote:

> Psychology, as the behaviorist views it, is a purely experimental branch of natural science. Its experimental goal is the prediction and control of behavior. Introspection forms no essential part of its methods, nor is the scientific value of its data dependent upon the readings with which they lend themselves to interpretation in terms of consciousness. The behaviorist, in his efforts to get a unitary scheme of animal response, recognizes no dividing line between man and brutes.[34]

Watson's credo led him to some farfetched and indeed unnecessary contortions, such as the idea that thinking could be reduced to subvocal movements of the larynx, but in the hands of later exponents, principally B. F. Skinner, behaviorism emerged as a powerful doctrine. It is still with us, nowadays more in applied than in mainstream psychology, although its purifying (not to say puritanical) influence can be seen even in such esoteric areas as artificial intelligence.

The dominance of behaviorism was such that, in North America at least, generations of students have been taught general principles of behavior based on experiments with rats and pigeons. The implication is that these principles apply equally to humans, and they have been carried, not without success, into human environments, such as the classroom or the therapy session, and even to society at large, as in Skinner's utopian novel *Walden Two*. The behaviorists did their best to rid psychology of all mentalistic terms, such as *mind, consciousness,* and *will,* and even today, discussion of such matters is often furtive.

Yet behaviorism may have forfeited some influence, at least within biology, precisely because it minimized the differences between animals and because it emphasized learning rather than instinct. The study of instinct and species-specific aspects of behavior was the domain of ethology, and for much of the present century, behaviorists and ethologists ignored each other, studying complementary aspects of behavior. Like the behaviorists, however, the so-called objectivists among the ethologists, such as Niko Tinbergen, deliberately ignored the subjective. But many ethologists retained an appreciation of the mental component and were guided by at least an informal reference to their own subjective states. It is ironic that behavioral psychologists should have tried so hard to eradicate the mind, while ethologists, whose main concern has been with animals, should have retained it.

The Cognitive Revolution

Over the last quarter of a century, the influence of behaviorism has waned, and it is no longer the dominant force in psychology. Among ethologists, too, the objectivist movement has lost some of its momentum, and there is growing interest in the nature of consciousness in animals. Perhaps the reason for this shift was simply that rigid behaviorist and objectivist doctrines seemed to rob psychology and biology of much of their interest and relevance. There were, however, some other identifiable influences at work, although whether they were the causes of change or the effects of some deeper shift in the zeitgeist is a matter for debate.

Chomsky and the Linguistic Revolution

One influential figure was the linguist Noam Chomsky, whose rather obscure and seemingly innocuous book *Syntactic Structures*, published in 1957, led to a revolution in our conceptions of language. Coincidentally, 1957 was also the year in which B.F. Skinner published his book *Verbal Behavior*, an ambitious attempt to reduce language to behaviorist principles. Two more different books on language can scarcely be imagined. Matters came to a head in 1959, when Chomsky wrote a blistering review of Skinner's book.

Since language features so insistently in discussions of human uniqueness, I have devoted a whole chapter (Chapter 5) to it, and Chomsky's ideas on language will be discussed in some detail there. For the present, it is sufficient to note that Chomsky effectively revived a Cartesian view of language as a uniquely human skill characterized by unlimited variety and flexibility. The unbounded nature of language was such that it could not be reduced to simple associations, linked through reinforcement (rewards and punishments), as Skinner had argued.

In the wake of Chomsky's trenchant critique, many psychologists adopted a Chomskian perspective on language rather than a Skinnerian one, and a new discipline called *psycholinguistics* was born. A whole generation of psychologists was weaned away from the Skinner box,[35] and the human laboratory flourished, with language a primary focus of inquiry. Mainstream psychology seemed to be about humans again, and not about rats or pigeons.

Computers and Artificial Intelligence

Another influence in the decline of behaviorism was the development of the digital computer, which provided an obvious metaphor for the human mind. Much of psychology has depended on the availability of suitable mechanical metaphors; it has always proven difficult to think about thinking without likening it to some mechanical device. Descartes, for instance, adopted a hydraulic model, and when this was abandoned, the Voltaic pile was for a time in vogue. The development of railroads, with switching mechanisms, provided another convenient metaphor, and ideas based on telegraphy (such as "circuits" and "relay stations") are still prominent. But the computer seemed so obviously capable of mimicking mental processes that scientists deliberately set about programming it to do so. So the science of artificial intelligence was born, and computers began to play chess, simulate human reasoning and expertise, write poetry and music, and provide psychotherapy.

The development of artificial intelligence was strongly influential in the growth of so-called cognitive psychology, which for the past two decades, at least, has supplanted behaviorism as the dominant influence in academic psychology. Cognitive psychologists freely borrow computer terminology in constructing theories of the mind, using concepts like storage, retrieval, scanning, filtering, coding, and transformation, in computer-based flow diagrams that are supposed to model the way the mind processes information.

Some enthusiasts have argued that computers can not only simulate the mind, they can also be said *possess* minds. That is, a computer in the process of playing chess or giving therapy may be said to actually be conscious. Needless to say, such claims also strike at the roots of our beliefs about human uniqueness and have led to considerable debate.[36] Although again haunted by the ghost of Descartes, the debate may still be premature, because for all their apparent sophistication, computers are still a very long way from mastering even the most elementary processes of the human mind. They have great difficulty with such apparently simple tasks as understanding ordinary sentences, or perceiving ordinary visual scenes, or even picking out the individual words from a sentence spoken in a normal voice. This, of course, says rather little for the level of mental processing required in chess or psychotherapy, where simulation has been reasonably successful.

It has long been held that only humans are capable of reason; in Shakespeare's *Julius Caesar*, for example, Mark Antony cries:

> Oh judgment thou art fled to brutish beasts
> And men have lost their reason!

Nevertheless, this comfortable sense of our unique rationality fails to set us apart from intelligent computers, which are in many respects the epitome of reason and logic. Ironically, then, when we assert our superiority to computers, we are inclined to stress our intuition, or our emotion, as attributes that no computer could simulate. As Sherry Turkle, in her book *The Second Self* (1984), puts it: "Where we once were rational animals, now we are feeling computers, emotional machines."[37] This leaves our sense of human uniqueness poised uneasily at the intersection of the worlds of animals and computers, combining the feelings of the one with the reason of the other.

Although the cognitive revolution ended the rule of behaviorism, subtle influences remain. Cognitive psychology remains fairly strictly materialistic, as evidenced by the profusion of computer-based terms; mentalistic terms are still largely avoided. Moreover, in artificial intelligence, the test of whether a simulation is successful is an essentially behavioral one. This is the so-called Turing test, named after the eminent British mathematician A. M. Turing, although again, the basic idea may be said to belong to Descartes.[38] Imagine a wall, behind which is a person, as well as a simulation of that person. Imagine also that it is possible to communicate with both of them by means, say, of a keyboard and a visual display unit. Your job is to try to find out which is the person and which is the simulation by asking appropriate questions and evaluating the answers. If you cannot tell which is which, then the simulation passes the test.

Continuity Reasserted

Although the cognitive revolution revived Cartesian ideas of human uniqueness, there are nevertheless forces in modern science and philosophy that tend to reassert the idea of continuity. One of these is the remarkable growth in recent years of neuroscience, a multidisciplinary field to do with the study of the brain and the nervous system. Neuroscience leads naturally to mechanistic theories of the human mind, and so to the continuity between humans and other species. Such is the influence of neuroscience that it has begun to have a direct impact on the philosophy of mind, to the point where a new term, *neurophilosophy*, is now in vogue.[39] Historically, however, the subdiscipline that has closest bearing on the relation between mind and brain is *neuropsychology*.

The Growth of Neuropsychology

In the late nineteenth and early twentieth centuries, there was considerable interest in brain function and its relation to mental processes. This was due in part to the observations of Paul Broca and others on the areas of the brain involved in language—this will be reviewed in Chapter 7. Another influential figure was the British neurologist John Hughlings Jackson, who foreshadowed the importance of the right side of the brain in nonverbal mental processes. Yet for a long time this work had curiously little impact on psychology. This was due at least in part to the narrowing influence of behaviorism. For all its materialism, behaviorism seemed to deny not only the mind but also the brain.

Things began to change after World War II. An important influence was the publication in 1949 of *Organization of Behavior* by the Canadian psychologist Donald O. Hebb, whose ideas stemmed in turn from his work with Karl S. Lashley. Hebb's book was a largely speculative account of how learning, perception, and thinking might be explained in neurophysiological terms. As such, it represented a thoroughly materialistic approach to the mind.

Two other prominent figures of the time helped revive interest in the psychological implications of brain impairments. Alexander R. Luria in the Soviet Union and Hans-Lukas Teuber in the United States. Between them, Hebb, Luria, and Teuber may be regarded as the founding fathers of the new discipline of neuropsychology. Two early converts, Oliver Zangwill in England and Henri Hécaen in France, helped establish neuropsychology as an international discipline.

An especially influential development in neuropsychology took place in the 1960s when two neurosurgeons in Los Angeles, Philip J. Vogel and Joseph E. Bogen, decided to sever the main commissures, or fiber tracts, connecting the two sides of the brain in a series of patients suffering from intractable epilepsy. By "splitting the brain" in this fashion, they argued, it should be possible to prevent the spread of epileptic seizures originating on one side. The idea of the operation was not new; it had been performed in the 1930s, but its effects had not been systematically explored or widely discussed. Besides having a beneficial effect on the patients' epilepsy, the new series led to dramatic discoveries, with profound implications for the relations between mind and brain.

By means of subtle techniques, which will be described in more detail in Chapter 7, it is possible to restrict input to one or the other cerebral hemisphere of a split-brained patient, and so to assess separately each hemisphere's mental capacities. The pioneering work on the split-brained patients was performed mainly by Roger W. Sperry, who received the Nobel Prize for his discoveries in 1981. This work confirmed that the left cerebral hemisphere was dominant for language, and also showed that the right hemisphere was capable of understanding language but not of producing speech. As anticipated by the eminent British neurologist John Hughlings Jackson, the right hemisphere also proved superior to the left in some nonverbal skills, including those involved in spatial perception.[40]

Perhaps the most intriguing aspect of the split-brained patient was the apparently independent mental life enjoyed by each hemisphere. Each remained oblivious to information presented to the other, and it was possible to arrange for the two hemispheres to perform conflicting tasks simultaneously, each operating with the opposite hand—the left hand knew not what the right hand was doing. Later observations suggested that the two hemispheres could even possess rather different personalities, not always in harmony with one another. As we shall see in Chapter 10, these observations revived the idea, popular in the latter part of the nineteenth century, that there are two minds in the one brain. Indeed, it is part of our contemporary folklore that the left hemisphere represents the rational, analytic aspect of our consciousness, while the right hemisphere represents the intuitive, holistic aspect.

Whether or not the notion of a "dual mind" can be extended to those of us with normal, intact brains, the separate mental lives of the two cerebral hemispheres of the split-brained patient have some implications for the understanding of consciousness. The split mind of the split-brained patient suggests a material basis for consciousness—the mind cannot traverse the physical gap between the separated hemispheres. Yet Sperry himself, the presiding architect of the split-brain studies, seemed to see in the results evidence for a transcendental view of consciousness. He argues that consciousness is an emergent property that can feed back on the neurophysiological process and so influence action.[41]

The split-brain studies are but one example of how mental processes may be affected by surgery or accident to the brain. Although neuropsychology provides far from a complete understanding of how mental processes are represented in neural tissue, few neuropsychologists doubt the materialistic assumption that mind is ultimately to be understood in terms of brain function.

Cognitive Ethology

In 1976, the ethologist D. R. Griffin published *The Question of Animal Awareness: Evolutionary Continuity of Mental Experience,* a second edition of which appeared in 1981. In some respects, this work is also part of the cognitive revolution, but it seeks to include animals as well as humans. Griffin was opposed to the objectivist, parsimonious interpretation of animal behavior that had dominated ethology (and psychology) under the influence of Morgan's famous canon. He collected a great many examples of animal behavior that appear to provide convincing evidence of conscious thought. For instance, although tool use has been regarded as an essentially human development, there are in fact many instances of animals using tools in apparently purposeful ways: Vultures throw stones at ostrich eggs, apparently in order to break them; sea otters carry stones around and use them as anvils upon which to pound shellfish; and even the lowly ant uses bits of leaf or wood as a sponge to soak up semiliquid foodstuffs so that they can be transported to their home colonies.

Much of Griffin's argument is based on an analysis of communication in animals. In particular, he is concerned with refuting Chomsky's assertion that the essential properties of human language are uniquely human. This matter will be treated more fully in Chapter 6.

Griffin's analysis is essentially a commonsense one, and he is least convincing when he tries to define subjective terms; in the second edition of his book, he complains of definitions and concepts being "quibbled to death" and places less reliance on them. Although Griffin advocates a new field that he calls *cognitive ethology,* his approach is actually little different from that of, say, Romanes, who similarly argued that we should draw on subjective experience in the interpretation of behavioral observations. Griffin does suggest, however, that developments in teaching language to apes might open new doors to the minds of other animals. Still, readers are advised not to hold their breaths while waiting for this to happen.

It is interesting to note the changing role of consciousness in the question of human uniqueness. So long as consciousness was conceived as subjective, and presumably involving a nonmaterial component, it seemed to serve as a criterion for distinguishing ourselves from the other animals. But as objective ways of studying conscious processes are developed, it appears more and more reasonable to suppose that consciousness is something that is possessed by at least some other animals besides ourselves.

Sociobiology

The publication of Edward O. Wilson's book *Sociobiology* in 1975 led to further attempts to reduce the behavior of humans and other animals to common principles. This book extends the application of evolutionary principles to the social behavior of animals. Most of it consists of carefully and extensively documented evidence from non-human animals, including insects. The final chapter, however, is titled "From Sociobiology to Sociology" and is a speculative account of the supposedly genetic basis of human social behavior. This, the most controversial aspect of sociobiology, was expanded by Wilson in 1978 with the publication of his book *On Human Nature*.

Wilson proposed that such human traits as aggression, xenophobia, social stratification, and even gender differences in behavior or ability were under genetic control. He argued that Darwinian theory would transform the human and social sciences just as it had earlier transformed much of biology. The implicit materialism, determinism, and continuity between humans and other species again awakened hostility, prompted in part no doubt by the challenge to our own uniqueness and self-determination. This time, protest came not so much from a religious establishment as from those who saw sociobiology as a threat to liberal ideals. As Stephen Jay Gould notes, biological determinism is too often invoked in order to preserve the status quo, be it the existence of a ruling class or discrimination against women.[42]

However, even from a strictly scientific point of view, Wilson undoubtedly overstated the case. One sociobiological strategy is to seek precursors of human behavior among "lower" animals, as evidence that these behaviors are a product of biological evolution. For instance, male mallard ducks appear to force females to copulate with them, suggesting a biological basis for human rape. The risk here is that of confusing analogy with homology. For instance, the wings of birds and insects evolved quite independently of each other (the common ancestor of the two groups did not possess wings) and are therefore analogies rather than homologies. Similarly, it is extremely unlikely that forced copulation in mallards and human rape, which is surely a pathological behavior that has little to do with sex or procreation, have a common genetic origin.[43]

In fact, in dealing with human behavior, it is very difficult indeed to distinguish cultural influences from biological or genetic ones. We are a slow-breeding species—slow, that is, relative to the pace of technological change and of the adaptational demands it places on us—and we do not permit genetic experiments upon ourselves. Moreover, cultural evolution, in which behavioral patterns are transmitted through instruction or imitation, often mimics biological evolution; both, after all, have to do with adaptation. Given the human capacity for language and flexible communication, cultural transmission is the more rapid and efficient means of adaptation, and the odds must favor it over genetic transmission as the likely source of most distinctively human behaviors.

More recently, Charles L. Lumsden and Wilson, in their 1983 book *Promethean Fire: Reflections on the Origin of Mind,* have argued for a compromise in terms of what they call *gene-culture coevolution.*[44] There are two sides to this thesis. First, human genes influence the structure of the human mind, which in turn influences the cultures we live in. Second, culture influences our behavior, and so can exert a selective influence on the genes. This coupling of genes and culture, with each influencing the other, was responsible for the very creation of the human mind. It also "drove the growth of the brain and the human intellect forward at a rate perhaps unprecedented for any organ in the history of life." Lumsden and Wilson propose in fact a "thousand year rule," which asserts that significant genetic changes in human evolution can occur in a mere 50 generations, or about a thousand years.

Lumsden and Wilson thus revive a sense of human uniqueness while insisting on strong genetic influences; they have their cake and eat it too. However, their book actually gives little evidence of the contingencies that might have shaped gene-culture coevolution as a mode of adaptation.[45] Despite the earlier promises of the sociobiologists, it seems that the issue of whether there is a discontinuity between humans and other species is still very much with us.

Perspective

In summary, the issues remain very much as defined by Descartes in the seventeenth century. However, opinion as to whether humans are or are not fundamentally different from other species has fluctuated. Probably the common view among laypersons is that humans *are* unique and superior to other creatures, but this view may be unconsciously motivated by our shabby treatment of animals. Darwinian theory strongly suggested the heretical idea that humans do not differ fundamentally from other species, and this view permeated behavioristic psychology, ethology, neuroscience, and sociobiology, at least in its earlier phase.

Nevertheless, the cognitive revolution of the past 30 years or so has revived in some quarters the notion of human uniqueness. In particular, Chomsky and other modern linguists have adopted the role of neo-Cartesians. The dominance of cognitive psychology over behaviorism meant that research on animals gave way to research on humans, and modern theories of cognition are presented with little concern for how they might apply to nonhumans. Humans were reinstated as the main object of inquiry in psychology.

It should not be thought, however, that the cognitive revolution, with its Cartesian overtones, is opposed to evolutionary theory. As we have seen, modern evolutionary theory differs from the traditional Darwinian view in that it allows for sudden and quite dramatic changes in genetic structure. The changes that can occur can be more extensive than the changes that are actually selected *for,* so that evolution may have a serendipitous quality not envisaged by Darwin. Far from denying an evolutionary basis for the discontinuity between humans and other animals, neo-Cartesians such as Chomsky argue explicitly that the discontinuity itself has a genetic basis.

For the most part, too, modern cognitive science eschews Cartesian dualism, retaining an objective approach to the study of the human mind. Whether their inclination is toward the study of brain function or the study of artificial intelligence, most cognitive scientists seek mechanistic interpretations of the mind. Even consciousness, it is implied, can be found in the electronic circuits of a computer or the neural circuits of a brain. However it then becomes natural to ask whether mental phenomena, including consciousness, might also apply to other animals. This is the domain of cognitive ethology. Even the modern behaviorists have grown steadily more "cognitive," studying such complex behaviors as choice, memory, and communication. Behaviorists are thus no longer isolated from ethologists, or psychologists from linguists, neurologists, or computer scientists. Even philosophers are rejoining the action.

As cognitive science grows ever more confident in the framing of objective, mechanistic theories of the mind, so the subjective aspect has diminished. Our own mental processes are in any event notoriously inaccessible to introspection (or subjective scrutiny), and even when they are apparently accessible, the information about them is often notoriously fallible. Introspection may have some heuristic value, in framing hypotheses or in guiding interpretation of experimental findings, but it no longer has any substantive role in cognitive theory itself. With the relentless march of progress in the neural and behavioral sciences, and in the construction of computer-based models of cognition, there is little room for mentalistic concepts.

Our sense that we are fundamentally different from other animals is nevertheless largely a subjective one, and is thus prey to our fears, prejudices, and need to justify the way we treat animals. Moreover, reliance on the subjective tends to lead us toward the conclusion that we are different simply because we are we: in our own minds, we find a consciousness that belongs to ourselves but does not belong to the family cat or dog. It clearly will not do to conclude from this that only humans are conscious. The family cat, feigning sleep on the hearth, might equally conclude that only cats are conscious. As the study of mental processes has been hijacked from the armchair to the laboratory, so the sense of human uniqueness has diminished. Whether it has vanished altogether is taken up in the following chapters.

Notes

1. White (1967, p. 1205).
2. Cited on p. 22 of Keith Thomas book *Man and the Natural World* This book provides an excellent review of attitudes towards animals, especially in the England of 1500–1800.
3. From E. Lankester, *The Uses of Animals in Relation to the Industry of Man* (n.d., p. 272), cited by Thomas (1984, p. 30).
4. From Thomas Morton's *The New English Canaan* and cited by Thomas (1984, p. 24).
5. White (1967, p. 1207).
6. From Bartholomew Batty's *The Christian Man's Closet* (1581), cited by Thomas (1984, p. 38).
7. Cited by Thomas (1984, p. 103).
8. From Uvedale Price's *Essays on the Picturesque* (1810), cited by Thomas (1984, p. 32).
9. From Plato's *Timaeus*, cited by Thomas (1984, p. 31).
10. Sarich and Wilson (1967).
11. From *Discourse and Essays* (1637). This is translated in Descartes (1985); the quotation is from p. 141 of Volume 1 of this work. For a more complete account of Descartes' life and ideas, see Chapter 1 of Walker's excellent book, *Animal Thought* (1983).
12. Kingsley (1889, p. 154).
13. This point was made to me by Morris Moscovitch.
14. La Mettrie (1747). For an English translation, see la Mettrie (1912).
15. Cited in Walker (1983, p. 35).
16. This quote is from the 1901 edition, p. 936.
17. Quoted by Eiseley (1959, p. 82).
18. From Gould's essay "Natural selection and the human brain," reprinted in his 1983 book *The Panda's Thumb*.
19. Quoted by Gould (1987, p. 121).
20. Gould and Vrba (1982).
21. Anderson (1970).
22. McClintock (1984).
23. Maynard Smith and Haigh (1974).
24. Gould and Eldredge (1977).

25. Kingsolver and Koehl (1985).
26. Teilhard de Chardin (1959, p. 288).
27. From Gould's essay "The Piltdown Conspiracy," which first appeared in the *Natural History Magazine* for August, 1980, but which is also included in Gould (1984). In this essay Gould discusses Teilhard's possible role in the famous Piltdown forgery.
28. In his essay "Our natural place," also printed in Gould (1984). The next two paragraphs, and the two quotations therein, are also drawn from this essay.
29. Gould (1984, p. 250).
30. Quoted in Sagan (1979).
31. Romanes (1888, pp. 15–16).
32. Morgan (1894, p. 53; his italics).
33. Wundt (1894).
34. Watson (1913, p. 158).
35. In which you put a rat or pigeon for training. Skinner also invented a box for holding babies.
36. See Searle (1980), and ensuing commentaries. Searle's article, which is critical of the idea that a computer could be said to be conscious, is also reprinted in *The Mind's I*, by Hofstadter and Dennett (1981), which contains further entertaining reflections on the issue.
37. Turkle (1984, p. 313).
38. See Turing (1950). In the original Turing test, the object was to distinguish a man from a woman rather than a computer from a person.
39. Churchland (1986).
40. See Sperry (1974, 1982).
41. Sperry (1985, 1988).
42. In his essay "Biological potentiality vs. biological determinism," first published in *Natural History Magazine* and reprinted in Gould (1980).
43. Ibid.
44. Lumsden and Wilson (1983). This book is a popularization of their earlier book, *Genes, Mind, and Culture: The Coevolutionary Process* (1981).
45. See the review "Genes on the brain" in Gould (1987).

— 2 —

Human Evolution

> The question is this: Is man an ape or an angel? My lord, I am on the side of the angels.

SO SAID BENJAMIN DISRAELI. It has become clear, however, that the rest of us surely share a common ancestry with the apes. We belong, in fact, to the order of creatures known as *primates*—in the biological rather than the ecclesiastical meaning of the term. The primates, in turn, are considered the highest order of the class known as *mammals*, so called because they suckle their young. Modern primates include lemurs, monkeys, anthropoid apes, and, of course, ourselves.

It is convenient to begin the story of human evolution not with the primates themselves but with an extinct species, the dinosaur. It might be argued, in fact, that we owe our exalted position on the planet to the extinction of that exotic creature.

From Dinosaurs to Apes

At one time, the dinosaurs and their cousins of the air (pterosaurs) and sea (ichthyosaurs, plesiosaurs, mosasaurs) were the lords of the earth, much as we humans are (or think we are) today. Some 65 million years ago, at the close of the period known as the Cretaceous, they all perished in one of several mass extinctions known to have occurred in the history of life. The reason for this is a matter of some controversy.

One theory has it that the earth was struck by a large asteroid perhaps 10 km in diameter. The impact would have produced a huge crater, and flung such quantities of dust into the atmosphere that the earth became dark. This, in turn, would have destroyed the photosynthetic plankton and the oceanic food sources based on them. It would also have destroyed living plants, although seeds would have remained dormant for the later reestablishment of plant life. These events would have been sufficient to wipe out the herbivorous dinosaurs. It is also thought that the dinosaurs had been in decline for some time, and the wayward asteroid might simply have delivered the coup de grace.[1]

During the reign of the dinosaurs, mammals were relatively insignificant, rat-like creatures, but they survived whatever it was that destroyed the dinosaurs. Included among them was at least one species of the primate order, known as *Purgatorius*. The remains of this creature were discovered in the United States at a place called Purgatory Hill in northeastern Montana. Dated from around 65 million years ago, these remains consisted only of tiny pieces of bone and a few teeth,[2] so we cannot be at all sure what this ancient ancestor really looked like. Clearly, though, it lived in the shadow of the dinosaur. One might therefore wonder at the exquisite calibration of the cataclysmic event that wiped out the mighty dinosaurs and yet preserved this unprepossessing little creature that would eventually evolve into modern humans. Was it an act of God? More likely, I think, it was another example of the fateful hand played by chance. "Our current existence," writes Stephen Jay Gould, "is an extended function of enormous improbabilities."[3]

Following the demise of the dinosaurs, the mammals came into their own. The earliest known mammals were in fact primates,[4] including *Purgatorius*. They emerged from the forest floor and gradually took to the trees, where they proliferated as small nocturnal animals that lived on insects. In the course of evolution, various bodily features adapted to arboreal life; for instance, fingers with nails replaced paws with claws, and the eyes migrated to the front rather than the side of the head. The forelimbs were freed from their role as weight supporters in standing and moving around, and became adept at grasping and handling objects. The opposable thumb, which allows the hand to grasp, is the main characteristic that distinguishes the primates from other mammalian species.[5] Many primate species became diurnal rather than nocturnal and evolved color vision. By some 40 million years ago, then, at least three important characteristics of modern humans had emerged, namely, grasping hands, stereoscopic vision, and the ability to see in color.[6]

From these tree-dwelling creatures, known as *prosimians,* the monkeys evolved. The monkeys were larger than the prosimians, and their diet changed from insects to plants, including leaves and fruit.

During this era the population of monkeys divided, with one group, the so-called New World monkeys, confined to the South American continent as it drifted westward to finally join up with North America. It was from the remaining colony of monkeys in the Old World, some 30 million years ago, that the apes evolved. These were larger, in turn, than the monkeys and were of several different types.

Remnants of prehistoric apes, dating from the Miocene period of some 20 million to 8 million years ago, have been found in far-ranging sites in the Old World. These are known as *hominoids* and are the ancestors of the pongids (great apes) and the hominids, who include modern humans. The earliest fossil in the hominoid record was a specimen known as *Proconsul,* found in Africa and dating from about 20 million years ago. This creature was at first thought to be an African subspecies of another creature known as *Dryopithecus,* but the dryopithecines are now regarded as a later family of apes that ranged across Europe and Asia from Spain to China. Another group, including *Sivapithecus* and *Ramapithecus,* were first discovered in the Siwalik Hills north of New Delhi in the 1930s. In 1961, Louis Leakey found another *Ramapithecus* skull in southern Kenya, and several skull remnants from India and Pakistan were then also classified as belonging to the same species. Since then, traces of *Ramapithecus* have been found in a wide area encompassing Africa, Asia, and southern Europe. Until recently, it was thought that *Ramapithecus* represented the line from which humans are descended, but recent evidence suggests that it was probably the ancestor of the modern orangutan.[7]

Unfortunately, there is a break in the fossil records from the late Miocene, some 8 million years ago, until the early Pliocene of something under 4 million years ago. It has therefore not been possible to reconstruct the events that led to the emergence of the first true hominids. As we shall see, however, it now seems likely that the split between the apes and the hominids did take place within this mysterious window of time. It also seems clear that the emergence of hominids did not take place in India, Asia, or Europe, but was confined to Africa. Indeed, as we shall see, that remarkable continent has more than once been the source of evolutionary change along the path to modern humans. Africa is truly the cradle of humanity.

From Apes to Hominids

The Split with the Apes

In *Descent of Man,* Darwin had argued that humans originated in Africa and that our closest living relatives are the chimpanzees and gorillas. These astute speculations were supported by subsequent evidence. Chimpanzees and gorillas live naturally only in Africa, and discoveries of the fossils of the earliest hominids have also been confined to Africa. Since gorillas look much like enlarged chimpanzees, it has generally been assumed that they split as a pair from the human lineage, and later split from one another to become separate species. However, this assumption has been challenged by evidence from molecular biology.

In 1967, V. M. Sarich and A. C. Wilson published a now classic paper on the molecular similarity between apes and humans. They based their study on a protein called *albumin,* on the grounds that the structure of proteins closely reflects the structure of the genes themselves. They reached the controversial conclusion that, at the molecular level, humans and chimpanzees were only about 1 percent different. Assuming a constant rate of mutational change, they calculated that the common ancestor of the human and the chimpanzee existed as recently as 5 million years ago. This too was a controversial claim, since at that time it was generally assumed that the split occurred much earlier, perhaps as long as 20 million years ago.[8]

In 1984, Charles Sibley and Jon Ahlquist, then at Yale University, published further evidence on the evolutionary relationships among humans, chimpanzees, and gorillas based on a technique known as *DNA hybridization.*[9] Their results suggested that humans and chimpanzees formed a pair that diverged from gorillas before they diverged from each other. Sibley and Ahlquist also estimated the approximate dates at which divergences took place (see Figure 2.1). They suggested that the gibbons split off from the hominoid family tree some 16.4 to 23 million years ago, followed by the orangutans some 13.0 to 16 million years ago and then by the gorillas 8.0 to 9.9 million years ago. Humans and chimpanzees, they inferred, split some time between 6.3 and 7.7 million years ago, which is somewhat earlier than the estimate that Sarich and Wilson had made.[10] However, it also lies within the period of the fossil void between the late Miocene and early Pliocene.

Another, more recent technique has been to compare the actual sequences of bases on DNA molecules from humans, chimpanzees, and gorillas. Michael Miyamoto, Jerry Slightom, and Morris Goodman identified 7100 base pair sequences in equivalent parts of a beta-globin gene in each species and found the smallest divergence, only 1.6 percent, between humans and common chimpanzees (*Pan troglodytes*), with chimpanzees showing a divergence of 2.1 percent from gorillas.[11] That is, chimpanzees appear to resemble humans more than they resemble gorillas.

These conclusions based on molecular evidence remain some-

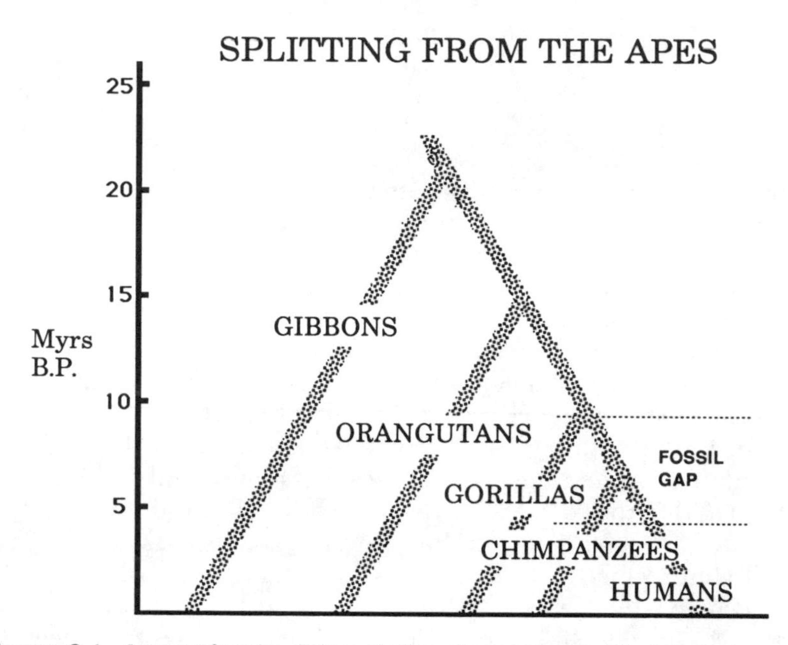

Figure 2.1. Approximate dates of the divergences of gibbons, orangutans, gorillas, chimpanzees, and hominids, estimated from DNA hybridization. Time is measured in millions of years (Myrs) before the present (B.P.)

what controversial, since they appear to conflict with anatomical evidence. In particular, chimpanzees and gorillas share two characteristics that are not possessed by humans. One is a distinctive mode of locomotion known as *knuckle-walking,* and the other has to do with a distinctive formation of enamel on the molar teeth. By this form of analysis, known as *cladistic analysis,* chimpanzees and gorillas form a separate grouping distinct from the hominid line. It is possible that these characteristics were also present in a common ancestor of humans and African apes, but if so, there appear to be no signs of them now. Indeed, there is no evidence for knuckle-walking in the very earliest hominid fossils, those of *Australopithecus afarensis*.[12]

The differences between humans and chimpanzees in bodily size and shape also belie the close genetic similarity between ourselves and our primate cousins.[13] This can only be explained, it seems, if the differences at the genetic level involve genes that can have a disproportionate influence on structure. It has been suggested, for example, that the main differences have to do with so-called regulatory genes rather than with genes that code for specific structural proteins.[14] Regulatory genes control the pace and pattern of growth by turning developmental processes on and off, and so can profoundly influence overall size and shape.[15] We shall see in the next chapter that humans are distinguished by an unusually long period of infancy and childhood; this itself is no doubt a result of a genetic change, and may have allowed for other regulatory genes to exert their effect on the pattern of development.

Whatever the explanation for the discrepancy between the anatomical and genetic evidence, there can be little doubt of our closeness to these other living apes, however disturbing that may be to our sensibilities or to our conceptions of human uniqueness. Analyses of the chromosomes of the three species suggest that 18 of the 23 chromosomes of modern humans are virtually identical to those of the common precursor of the orangutan, gorilla, and chimpanzee.[16] The close resemblance in DNA structure between humans and chimpanzees even suggests that a hybrid species would be viable[17]—a chastening thought. And the common ancestor of humans and chimpanzees may have existed as recently as 5 million years ago, a mere eyeblink on the scale of evolutionary time.

Australopithecus

In 1924, a young Australian anatomist, Raymond Dart, came into possession of a skull like that of an immature ape in a cave at Taung, South Africa. Although his colleagues were at first skeptical, Dart hailed his find as the "missing link" between chimpanzee and human, and christened it *Australopithecus africanus*. (One wonders if the choice of name was dictated in part by the fact that Dart was Australian.) The brain of the Taung child was not particularly large, but its teeth and forehead resembled those of a human rather than an ape. Moreover, the position of the foramen magnum, which is the aperture through which the spinal cord descends from the braincase, suggested an upright, two-legged stance. The specimen was dated at about 2 million years from the present time.[18]

Other remnants of *Australopithecus* were discovered in Africa, lending further support to Dart's claim. However, matters were complicated when different species of *Australopithecus* were discovered, including the "robust" form *A. robustus* in the 1930s and the even more robust *A. boisei* in the 1950s. It is clear that different species of australopithecine coexisted, some 1.5 to 2.5 million years ago, with perhaps the main distinction being that between the *gracile* and *robust* forms.

The robust forms were distinguished from the "gracile" *A. africa-*

nus by large posterior teeth and facial muscles associated with heavy chewing. They probably lived mainly on coarse plant food of low nutritional value. It has been suggested that the dispersal of plant food on the savanna would have favored organization into large groups in order to combat predators.[19] The gracile forms, from which the species *Homo* is presumed to have descended, may have adopted a different survival strategy involving scavenging for meat. It was this strategy that led to the manufacture of tools and an increase in brain size. I shall discuss these matters in more detail in the next chapter.

The earliest form of australopithecine was not discovered until 1976, when Donald Johanson and his colleagues found the skeleton of a primitive female in the Afar region of Ethiopia. They named her Lucy, after the Beatles' song "Lucy in the Sky with Diamonds."[20] Later, however, this form was to be renamed *A. afarensis*. It is the most primitive of the known australopithecines, and one remnant, dated at about 3.9 million years, is thought to lie close to the point of divergence between the hominids and the African apes.[21] There is, however, some evidence that this early australopithecine may also have existed in both robust and gracile forms.[22] Whether or not this is so, it is a reasonable guess that this is the earliest hominid, and the one from which subsequent forms evolved.

Compared to humans, the australopithecines had fairly small brains, about the size of an ape's brain, although their faces somewhat resembled human faces. But their most distinctive attribute was that they walked upright on their two hindlegs, and in this respect they clearly resembled humans rather than apes. Evidence for upright walking, or bipedalism, comes in part from the structure of the bones. C. Owen Lovejoy has recently described a detailed analysis of the pelvic anatomy of Lucy, showing that bipedalism must have been established well over 3 million years ago. "Lucy's ancestors," he writes, "must have left the trees and risen from four limbs onto two well before her time, probably at the very beginning of human evolution."[23]

Even more remarkable evidence that *A. afarensis* walked upright was discovered by Mary Leakey. While digging in northern Tanzania, in a barren area of badlands known as Laetolil, she and her colleagues came across a set of footprints that were over 3.5 million years old. These were partially obscured, but when the overlying deposits were cleared away, a trail of more than 70 footprints, made by three individuals, was laid bare.[24] These footprints look remarkably human, with a well-developed arch and a big toe pointing forward rather than sideways, as in the apes. But for their antiquity, they could have been made by modern children. They confirm beyond reasonable doubt that the hominids of 3 to 4 million years ago strolled about very much as we do today.

From Hominids to Humans

Homo habilis

Until recently, it was thought that the australopithecines were indeed the missing link, and that the line of descent ran from *afarensis* to *africanus* to *robustus* to *boisei*, and thence ultimately to humans. However, this simple theory was dashed by yet further discoveries from East Africa. In 1962 Louis Leakey claimed to have found a skull fragment at Olduvai in Tanzania that belonged not to *Australopithecus* but to a true (if primitive) human. He named this species *Homo habilis*, or "practical man," since its remains were associated with the remnants of primitive tools. It was dated at about 1.75 million years and therefore coexisted with *Australopithecus*.

Still, the classification of this skull remained controversial until, some 10 years later, Louis Leakey's son Richard found a nearly complete skull, dated at about 1.8 million years, on the shores of Lake Turkana in Kenya. It had a cranial capacity of some 752 cc, almost double that of the australopithecines, although later discoveries suggest that this was an unusually big-headed specimen. This skull is generally referred to simply by its neutral code name, ER 1470, but was identified as an example of *H. habilis* by Leakey. It seems clearly to belong to our own genus. Leakey has argued that the species *Homo* arose much earlier, and that some earlier fossils are in fact primitive members of this species, although not originally labeled as such.

In any event, further examples confirmed the existence of *H. habilis*, who was also an upright walker and was probably descended from a gracile australopithecine.[25] The heyday of this creature was about 1.8 million years ago, which means that it almost certainly coexisted with the later australopithecines, suggesting that they belonged to different lineages.[26] But there were yet further complications. Another skull found on the shores of Lake Turkana has been identified as an example of *A. boisei*, yet dated at 2.5 million years before the present. This suggests that *boisei* coexisted with *africanus* and was not the end of an evolutionary chain of australopithecines.[27] Subsequent to *A. afarensis*, therefore, there were evidently several different species of hominid that lived at the same time. Figure 2.2 shows the different species, along with rough estimates of their time spans.

I have not attempted to show the lines of descent among the differ-

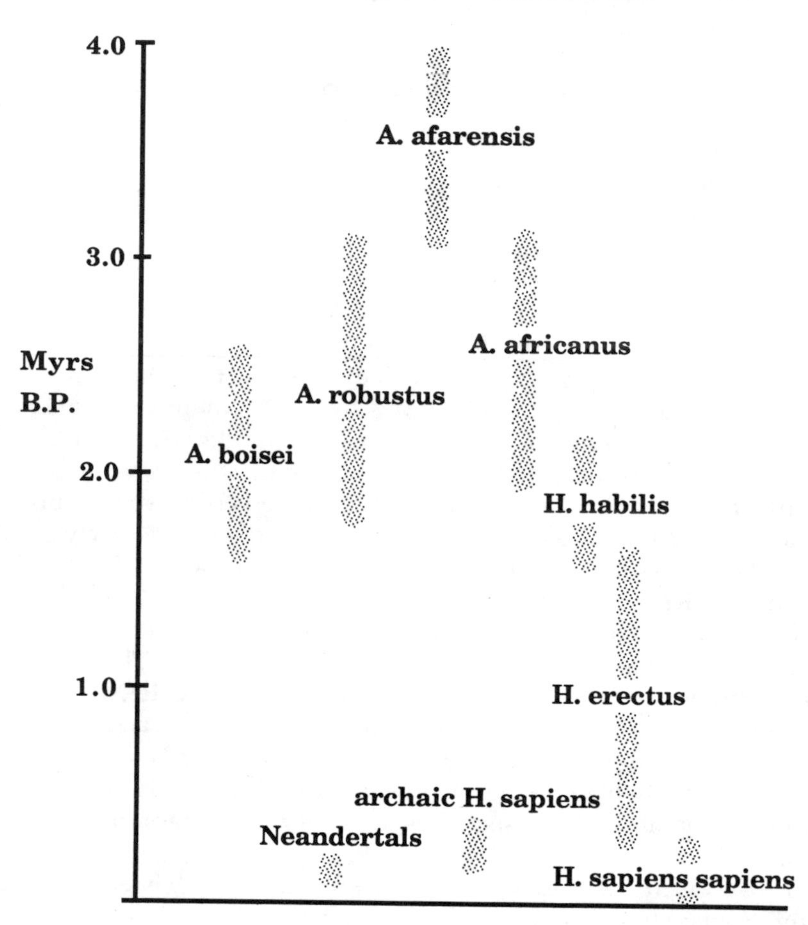

Figure 2.2. Different species of hominid and their approximate time spans. Time is measured in millions of years (Myrs) before the present (B.P.).

ent australopithecines and *H. habilis*, since this is a matter of considerable controversy, although there is probably general agreement that more than one hominid species coexisted around 2 million years ago.[28] Moreover, given the rapid pace of discovery and the development of classification and dating techniques in recent years, interpretations are bound to change. If anything, the classification presented above is likely to prove too simple. For example, there is a school of thought that suggests that there may have been two species of *H. habilis*, one gracile and the other robust.[29] Stephen Jay Gould suggests that there may have been even more than three hominid species at one time, despite the fact that only a single one, ourselves, survives today.[30] In his view, evolution is not a question of gradual transformation so much as one of *speciation*, in which a new line splits from a parent stock. This tends to happen in *punctuate* rather than gradual fashion, and is likely to occur in very small populations at the periphery of the main population, so that genetic changes are rapidly transmitted through the population. For reasons that are not entirely clear, it seems that there was a burst of speciation between some 2.5 and 2.0 million years ago.

In any event, there seems little doubt that *H. habilis* was distinct from the australopithecines. Its brain was clearly larger, averaging 659 cc, compared with the average of 451 cc for the gracile australopithecines.[31] It may also have been differently shaped. This has been inferred from the pattern of imprints on the inside of fossil skulls. These imprints can be revealed from endocasts made by filling the inside of the skull or of skull fragments with latex, a technique pioneered by Ralph L. Holloway of Columbia University. Dean Falk has shown that an endocast of the ER-1470 skull (*H. habilis*) shows a distinctively human-like appearance, with squared frontal lobes and evidence that a fold known as the *lunate sulcus,* which separates the parietal lobe from the occipital lobe, has been pushed rearward.[32]

The location of the lunate sulcus has played a prominent but controversial role in theories about hominid evolution. Its rearward migration is associated with an increase in the size of the parietal lobe of the brain, which in turn is thought to be critical in evolution because of its role in learning, memory, and language. The controversy has to do with whether the lunate sulcus had already begun its rearward march in the australopithecines. On the basis of the available evidence, which is flimsy, Holloway has maintained that it had,[33] while Dean Falk has argued that what Holloway and others have identified as the lunate sulcus is in fact a different feature known as the *lambdoid suture.*[34] According to Falk, the pattern of sulci in the australopithecines was ape-like rather than human-like. Falk and Holloway agree, however, that the rearward location of the lunate sulcus was clearly present in *H. habilis* by 1.8 million years ago.

In other physical respects, *H. habilis* differed rather little from the australopithecines. Its face, jaws, and teeth were similar. The very recent discovery of arm and leg bones that belonged to a female *H. habilis* also reveals that this creature was surprisingly small, standing just over 3 ft tall.[35] Moreover the arms are long relative to the legs, a characteristic that is more ape-like than human. This feature has suggested that *habilis* remained in part a tree-dwelling creature, despite its bipedalism.

The small size of this female specimen also suggests that the extreme sexual dimorphism of *A. afarensis*, in which males were about twice as big as females, persisted in *habilis* as well as in its other descendants, the later australopithecines. The exaggerated size of males relative to females is generally attributed to competition among males for access to females, suggesting that *habilis* had not yet achieved a social pattern of cooperation among males.

Homo erectus

The heyday of *H. habilis* is said to have been about 1.8 million years ago. Thereafter, we find the emergence of a more advanced form, known as *Homo erectus*. It is generally assumed that *erectus* evolved from *habilis*, although there is no sure proof of this. The earliest fossil that has been identified as *erectus* dates from about 2 million years ago,[36] although it has been claimed that this specimen might equally be identified as *habilis*.[37] Specimens dating from about 1.6 million years ago are more unequivocally identified as *erectus*.[38] However, even this later dating suggests that the change from *habilis* occurred within a narrow time period.

The most complete skeleton of *erectus* was discovered in 1984 by Kamoya Kimeu at western Lake Turkana in Kenya.[39] It dates from about 1.65 million years ago and is of a boy aged about 12 years. One remarkable feature is its height, estimated at about 1.64 to 1.68 m, which is tall even for a modern 12-year-old. It suggests a mature height of around 1.82 m, or 6 ft. This is rather taller than most other estimates for this species, which are usually put at around 5 ft. The cranial capacity of *erectus* ranges from some 800 to 1000 cc, with an average among early specimens of 942 cc. This is clearly larger than that of *habilis*, for which the average is 659 cc. There is evidence that the sizes of males and females became more equal, suggesting an altered social order with greater emphasis on cooperation.

The tools developed by *erectus* were more sophisticated than those of *habilis*, and included handaxes and stone chopping and scraping tools with symmetrical bifaces. The distinctive handaxes of this species were first found only in Africa but later were discovered in sites throughout the Old World.[40] *Erectus* was evidently restricted to Africa for about 0.6 million years but then spread to other parts of Europe and Asia, beginning about 1 million years ago.[41] The skull of the so-called Peking man, found in China, is an example of one that got away, and is very much like a skull found at the Koobi Fora camp on the shores of Lake Turkana in Kenya. It has been estimated that there were as many as 40,000 bands of these early hunter-gatherers roaming Europe, Asia, and Africa.[42]

There is also evidence that *erectus* developed the use of fire, which may have facilitated migration from tropical to temperate zones. Besides warming the cave, fire was useful for keeping predators at bay and for roasting meat. Evidence from China suggests that Peking man dined mainly on venison but also secured such varied game as elephants, rhinoceroses, water buffaloes, horses, camels, antelopes, sheep, tigers, leopards, bears, and hyenas! This enterprising creature probably used spears and perhaps pit traps to capture its quarry. Regrettably, it also appears that Peking men and women consumed the flesh of their fellows as well as of other animals.[43]

Archaic *Homo sapiens*

Discoveries from some 300,000 years ago indicate a further expansion of brain size to within the range of modern humans, some 1200 to 1400 cc. Again, the assumption is that this species, an early form of *Homo sapiens*, evolved from *H. erectus*. One famous example is the Swanscombe Man, whose fossilized skull was discovered in the English village of Swanscombe, near London. The first fragment was discovered in 1935 by local cement workers, who had stopped digging for the day and noticed a piece of bone sticking out of a gravel bank. A second fragment was found a year later, and a third was found in 1955, some 75 ft from the location of the other two. By a miracle, the three pieces belonged to the same individual and fitted together to form the entire back half of the skull.

This specimen was matched by another, found in 1933 in a gravel pit in Steinheim, in Germany. This nicely complemented the Swanscombe find, since it consisted of the face and upper jaw. Both skulls are thought to belong to the same species and to date from about 250,000 years ago. The cranial capacity of these skulls was estimated to be about 1300 cc, well within the modern human range. Nevertheless, the skulls are not shaped like those of modern humans, and are characterized by large brow ridges and sloping foreheads. Indeed, careful measurements and comparisons with modern human skulls suggest that this species was transitional between *H. erectus* and modern humans, the product of gradual change rather than of a sudden spurt. Similar skulls have been found in a wide variety of locations, including Africa, Greece, Hungary, and China. The species represented by these scattered discoveries has been named *archaic Homo sapiens*.[44]

With this archaic form, there was some increase in the sophistication and variety of tools—this is discussed further in the following chapter. There is also evidence of ritual burial, perhaps signifying an awareness of the significance of death. The earliest suggestion of ritual associated with death actually predates *sapiens* and comes from a site in China dating from about 500,000 years ago.[45] There, the evidence suggests, a group of our ancestors indulged in a cannibalistic feast, eating the brains of some of their fellows. They evidently gained access to the brain by widening the hole at the base of the skull, instead of smashing the skull and scooping it out. This difficult and meticulous procedure suggests an element of ritual and respect for the dead.

By the end of the so-called archaic world, prior to the emergence of anatomically modern humans some 200,000 years ago, there seem to have been at least three locally distinct populations of archaic *H. sapiens*, each having presumably derived from the groups of *H. erectus*, who in turn came originally from Africa. First, there were the archaic hominids of eastern Asia, with their characteristic pebble tool technology. Then there were the archaic hominids of Europe, the Mediterranean, and central Asia, who probably gave rise to the Neandertals. Finally, there were the hominids of sub-Saharan Africa, who were probably the ancestors of anatomically modern humans.[46]

Before I discuss this last transition, however, it is worth digressing to consider that mysterious group, the Neandertals.

The Neandertals

The Neandertals seem to have emerged some 100,000 years ago and to have disappeared about 35,000 years ago. The remains of this creature were first discovered in 1856 in the Neandertal valley, near Dusseldorf in Germany. At first it was thought that the remains were those of a sick, freakish, but modern person, and one anatomist even pronounced that the flat forehead had been caused by blows to the head. An English scholar of the time indulged an even wilder fantasy:

> It may have been one of those wild men, half-crazed, half-idiotic, cruel and strong, who are always more or less to be found living on the outskirts of barbarous tribes, and who now and then appear in civilized communities to be consigned perhaps to the penitentiary or the gallows, when their murderous propensities manifest themselves.[47]

As John E. Pfeiffer remarks, the Neandertal "came into the world of the Victorians like a naked savage into a ladies' sewing circle."[48]

Subsequent discoveries have shown the Neandertals to have been a distinct subspecies of *H. sapiens*, although there is still some controversy as to whether they were a separate species that became extinct, or whether they were the ancestors of modern humans. They lived through much of the last major period of glaciation in Europe, and were probably well adapted to intense cold. They had receding chins and very large noses, which may have served as radiators to warm and humidify the air. It has also been said, however, that if one were to clothe and shave a Neandertal man he would be indistinguishable from many of the travellers on the New York subway.[49]

The Neandertals also had more durable teeth and more muscular limbs than do modern humans. E. Trinkaus suggests that they were adapted to long hours of walking about, and that they may in fact be representative of an earlier but undocumented transition between *H. erectus* and archaic *H. sapiens*. His view is that they were more poorly equipped to subsist than the early anatomically modern humans because they lacked certain weapons for hunting.[50] They may also have been relatively poor at communicating with one another—a point taken up in Chapter 6.

Nevertheless, one of the striking aspects of Neandertal life was their use of ritual. In a site in Iraq occupied by Neandertals some 60,000 years ago, there are the remains of a corpse laid out on a bed of flowers.[51] In several Neandertal graves discovered in southern France, the bodies were laid out alongside stone tools. Some burial sites suggest a symbolism that has not been deciphered. For example, in a cave near Monte Circeo, on the coast between Rome and Naples, a Neandertal skull with a hole bored into it lies at the center of a circle of stones.[52] It seems that the Neandertals understood the significance of death.

For all that, the artifacts left by the Neandertals and archaic *H. sapiens* are meager in comparison with those of even the earliest anatomically modern humans. Reviewing the evidence, Paul Mellars of Cambridge University has recently concluded that "the most significant observation . . . is the very sparse evidence for almost any form of complex, clearly symbolic behavior amongst Neandertals and other archaic human populations."[53] Things evidently changed quite dramatically with the appearance of our next guest, *H. sapiens sapiens*.

Homo sapiens sapiens—and the Search for Eve

The final transition was from archaic *H. sapiens* to the so-called anatomically modern variety, or *H. sapiens sapiens*. These included the Cro-Magnons, who seem to have replaced the Neandertals some 35,000 years ago. The transition appears to have involved considerable changes in morphology, if not in brain size; it has been said, for instance, that the Neandertals were much less like modern humans (with the possible exception of riders on the New York City subway) than they were like the earlier *H. erectus*.[54]

There are two conflicting theories as to how this final stage in human evolution came about. William Howells of Harvard University has called these the *candelabra model* and the *Noah's Ark model*.[55] According to the candelabra model, the evolution from *H. erectus* to archaic *H. sapiens*, and thence to *H. sapiens sapiens*, occurred independently at the different locations in the Old World to which *erectus* had migrated. Different geographic populations would therefore have been independent of one another for perhaps a million years.

According to the Noah's Ark model, by contrast, the evolution from archaic to modern *H. sapiens* occurred at a particular geographic location. These anatomically modern humans then migrated to other territories, replacing the incumbent archaic variety. This would explain the demise of the Neandertals. While it conjures up images of ravaging hordes of invaders wiping out the indigenous populations, Ezra Zubrow has pointed out that replacement of one group by the other need not have involved force. A subtle competitive edge in survival tactics may have been sufficient to make the difference. For instance, Neandertals and anatomically modern humans foraged in overlapping territories, and the modern humans were probably simply better adapted to efficient foraging, with better planning and more efficient hunting techniques.[56] There is also evidence that the Neandertals did not possess a vocal tract capable of producing the rapid and flexible speech of modern humans,[57] although as we shall see in Chapter 6, this has been disputed.

Although the truth may lie somewhere between these two models, recent evidence from molecular biology does tend to support the Noah's Ark model—sometimes also called the *Out of Africa* model, for reasons that will become clear. This evidence is based on mutations in mitochondrial DNA, which mutates at the rate of 2 to 4 percent every million years. By geological standards, this constitutes a fast-ticking molecular clock. Mitochondrial DNA is inherited only through the mother, and is therefore immune to change by the recombination of genes from each parent. That is, it is simply passed on through the female line, and the only changes that occur are due to mutations.

Rebecca L. Cann of the University of Hawaii, with her colleagues Allan C. Wilson and Mark Stoneking, collected samples of mitochondrial DNA from the placentas of newborn infants whose ancestors lived in five different regions of the world: Africa, Asia, Europe, Australia, and New Guinea. By comparing the samples from different regions, they were able to work out how many mutations had taken place since the samples had evolved from a common ancestor. This enabled them to construct an evolutionary tree linking the different samples. At the base of the tree, dating from some 200,000 years ago, lies the common female ancestor—no less than Eve herself.[58]

The evidence also indicated that Eve lived in Africa. The evolutionary tree showed an early split into two branches, one of which consisted only of African samples. There were more diverse forms of DNA on this branch than on the other, which included samples from all five geographic regions, indicating that the mitochondrial DNA on the Africa-only branch was older. This also ties in with the fact that the oldest anatomically modern fossils have been found in Africa, dating from some 150,000 years ago. The demise of the Neandertals also appears to have occurred at different times in different parts of Europe, suggesting gradual migration of competing populations, perhaps of *H. sapiens sapiens*.[59]

The much publicized discovery of Eve, mitochondrial mother of us all, does not actually prove that *H. sapiens sapiens* evolved at this time. Unlike nuclear DNA, mitochondrial DNA is essentially a passenger in the genetic process and tells virtually nothing of the species that carry it. Allan C. Wilson and his colleagues have argued in fact that Eve probably belonged to the archaic species *H. sapiens*, with the transformation to the anatomically modern variety occurring in Africa at a later time, some 100,000 to 140,000 years ago. All present-day humans, they suggest, are descended from that African population. Alternatively, the modern variety might have predated Eve, with some bottleneck occurring 200,000 or so years ago that reduced our ancestry to the fortunate Eve and her lover.[60]

Regardless of precisely when *H. sapiens sapiens* emerged, both the biochemical and archeological evidence suggests that present-day humans originated in Africa and migrated to different parts of the globe. Migration patterns may have been quite complex; for instance, New Guinea samples of mitochondrial DNA are scattered on several limbs of the evolutionary tree, mixed up with samples from all of the other locations. Early *H. sapiens sapiens* evidently moved around quite extensively before settling down in sufficient isolation to evolve distinctive racial features. Even so, the fossil evidence suggests that different parts of the world were fully colonized by these modern humans at different times, with Asia, Europe, and Australia being settled by some 40,000 years ago, New Guinea by 30,000 years ago, and the New World by 12,000 years ago. The earliest fossils of anatomically modern humans outside of Africa were found at Qafzeh, in Israel, and date from some 92,000 years ago.[61]

An "Evolutionary Explosion"

The period known as the Upper Paleolithic, lasting from about 35,000 to about 10,000 years ago, saw what has been described as an "evolutionary explosion,"[62] recorded chiefly from artifacts found in Europe and the Near East. Up to this time, tool culture had remained essentially Acheulean and remarkably stable, although there was some development among the Neandertals. But with the Cro-Magnons of the Upper Paleolithic, tool making took on new dimensions, as we shall see in the next chapter.

The earliest known cave drawings are also associated with the Cro-Magnons and likewise date from something over 30,000 years ago. The great majority of caves in which drawings are found are located in France and Spain. Most of the drawings are of animals, especially bisons, horses, deer, bears, and lions. They are often depicted with darts or spears sticking into their sides, reflecting the Cro-Magnon concern with hunting. There is also evidence that the Cro-Magnons used cosmetics and wore jewelry.[63]

Randall White of New York University has examined the ornaments and images of this period in some detail and has argued that they mark a new-found ability to transfer qualities from one context to another.[64] For example, a spear point is shaped to resemble the snout of a seal, with holes resembling the eyes. White points out that this transfer of qualities is the essence of metaphor, which in turn may be the basis for technological innovation. The concepts of pointedness or barbedness may be taken from their natural contexts and applied to the construction of new objects. Visual thinking of this sort may still underlie the development of new technology. It has also been argued that this phase in human evolution saw the emergence of language as a distinctively human attribute, although I shall argue in Chapter 6 that language emerged earlier.

Although there was an apparent increment in the variety and sophistication of artifacts dating from this period of some 30,000 to 35,000 ago, it is unlikely that this period heralded any significant biological change. Rather, *H. sapiens sapiens* arrived in the region only shortly before that time and probably brought with them at least the potential for new technology. These anatomically modern humans probably came originally from Africa, as we have seen. Similar developments, associated with *H. sapiens sapiens*, may well have occurred elsewhere and at earlier times but may have left no traces. In any event, there is some evidence that there was a sophisticated culture in Africa that predated the Cro-Magnons. In his recent review, Mellars writes that "it is possible to point to at least certain features of the archeological record of the Middle Stone Age (roughly between 100,000 and 40,000 years ago) in Southern Africa which suggest a significantly more 'complex' (and perhaps more 'advanced') pattern of behavior than that reflected in the parallel records of the Middle Paleolithic in northern Eurasia over the same time range."[65]

It should also be noted that even today, humans have a *potential* for skills and symbolic thought that is probably never, in any one of us, fully exploited. We *could* learn languages, aspects of mathematics, crafts, or sporting skills that we do not in fact learn. We are endowed with considerably more potential than we can use in a mere lifetime. The variety of skills that humans can learn is exploited culturally rather than individually, so that different people are specialized for different trades—the plumber, the baker, the potter, the mathematician, the professional tennis player, and so on. Again, this might be taken as evidence that the final step in human biological evolution was an adventitious change rather than an incremental adaptation. Exploitation of that change may have taken tens of thousands of years to develop and is still developing.

Overview

The aim of this chapter was to provide a broad-brush picture of human evolution through time. Many of the conclusions have been speculative, based as they are mainly on a few old teeth, bones, and stones. Although the common ancestors of humans and apes existed but a few million years ago, a minuscule gap of time on an evolutionary scale, that gap nonetheless produced the difference between humans and apes. If humans do indeed possess some unique quality that separates them in some fundamental way from the apes, then that gap becomes something of a chasm. If only some other hominids had survived, to give us a better sense of the evolutionary steps that have made us what we are!

Nevertheless, it now seems possible to establish at least the rough sequence of events in the emergence of humans from apes. The decisive split seems to have occurred somewhere in that transition between the Miocene and the Pliocene, 8 million to 4 million years ago, when a gap in the fossil record occurred. Prior to that gap, there is evidence only of ape-like creatures, but after the gap we find the first signs of an upright-walking hominid, *A. afarensis*. By 2 million years ago there were several species of hominids, including the australopithecines and *H. habilis*, who eventually evolved, with one or two diversions, to become modern humans.

Despite the extinction of the australopithecines, the step from ape to hominid was in the long run probably worth taking. The apes that did not take this step have not fared particularly well. Even prior to the appearance of the hominids, the ancestors of the modern pongids (gorillas, chimpanzees, and orangutans) had spread throughout the Old World; we have seen, for example, that *Ramapithecus*, thought to be an ancestor of the orangutan, lived at one time in Africa, India, and southern Europe. However, present-day pongids are found naturally only in restricted areas of West Africa (chimpanzees, gorillas) and in parts of Indonesia (orangutans); unlike the hominids, they responded to geological events by retreating to safe but ever-decreasing habitats. Over the past 12 million years, the Old World monkeys became dominant over the pongids and today inhabit a wide area, including most of Africa, India, Southeast Asia, Japan, and Indonesia. But the hominids—or at least that resourceful branch of them that managed to survive—were even more successful, to the point where a few of them have launched themselves off the earth altogether.

By 1.6 million years ago, the persevering *habilis* had given way to *H. erectus*, some of whom migrated from Africa to Europe and Asia about 1 million years ago. *H. erectus* eventually evolved into archaic forms of *H. sapiens*, including the Neandertals. However it appears that out of those who remained in Africa there emerged the first anatomically modern human, *H. sapiens sapiens*, better known in some circles as Eve. This creature embarked on yet another series of migrations, eventually replacing the archaic varieties and populating the globe.

In the Upper Paleolithic, beginning some 35,000 years ago, there was evidence of further development, an evolutionary explosion, that was largely manifest in Europe and Asia. In many ways, this seems to have set the stage for the subsequent developments that culminated in modern human society. Whether it represented a further genetic advance, or was simply a cultural consequence of genetic changes that were already present in *H. sapiens sapiens*, is one of the critical questions of human evolution.

Notes

1. See Alvarez, Alvarez, Asaro, and Michel (1980). The asteroid theory is also discussed by Gould (1984) in his essay "The belt of an asteroid."
2. Van Valen and Sloan (1965).
3. In "The belt of an asteroid" from Gould (1984).
4. Simons (1963).
5. Buettner-Janusch (1963). There are other distinguishing characteristics, such as large anterior teeth. Thus *Purgatorius* was classified as a pri-

mate in terms of resemblances between its teeth and those of present-day primates.

6. See Chapter 2 of Leakey and Lewin (1979).

7. In their 1979 book, Leakey and Lewin picked *Ramapithecus* as the human ancestor, but very soon the picture was to change; see, for instance, Ciochon and Corruzzini (1983) and Cartmill, Pilbeam, and Isaac (1986).

8. Sarich and Wilson (1967).

9. See Sibley and Ahlquist (1984) or the more popular account in the article "Molecular clocks turn a quarter century," published under *Research News* in the 5 February 1988 issue of *Science* (Vol. 239, pp. 561–562). Basically, the technique involves taking DNA from one of the species to be tested, cutting it into lengths of about 500 base sequences, and separating the double-stranded molecule into single strands. Repeated sequences are removed, leaving behind unique-sequence DNA, which is then made radioactive. This *tracer* DNA is then added to a quantity of single-stranded DNA, known as *driver* DNA, from the second species to be tested. The mixture is allowed to anneal when tracer sequences that are sufficiently similar to driver sequences form double strands, or hybridize. The more similar the sequences, the more closely they bond. The mixture is then subjected to step-wise increases in temperature, which progressively cause the strands to break apart. This process is monitored by the number of radioactive counts that are lost at each step. The higher the temperature needed to release all the radioactive tracer, the closer the bond and the closer the similarity between the two species.

10. The hybridization technique pioneered by Sibley and Ahlquist has recently been the focus of intense debate. Much of the controversy centers on the actual measure of distance between species that is derived from the plot of the percentage of single-stranded DNA remaining at each temperature following hybridization (see note 5). For a popular account of the debate, see two articles published under *Research News* in the issues of *Science* of 23 and 30 September 1988 entitled "Conflict over DNA clock results" (*Science*, Vol. 241, pp. 1598–1600) and "DNA clock conflict continues" (*Science*, Vol. 241, pp. 1756–1759), respectively.

11. Miyamoto, Slightom, and Goodman (1987).

12. This issue is discussed in popular fashion in an article entitled "My close cousin the chimpanzee," published under *Research News* in the 16 October 1987 issue of *Science* (Vol. 238, pp. 273–275).

13. Passingham (1982).

14. Kohne (1975).

15. Gould (1977).

16. See Yunis and Prakash (1982).

17. Lovejoy (1981).

18. Originally reported by Dart (1925), but see also Dart's 1967 book, *Adventures with the missing link*.

19. Foley and Lee (1989).

20. See Johanson and Edey (1981).

21. Asfaw (1987).

22. Falk (1987b).
23. Lovejoy (1988, p. 89).
24. Leakey (1979).
25. Conroy, Vannier, and Walker (1990) have shown that the pattern of veins around the basioccipital region of the brain in *A. africanus* is similar to that in the *Homo* lineage but different from that in the robust australopithecines.
26. See Leakey and Lewin (1979).
27. See Walker, Leakey, Harris, and Brown (1986) or the more popular account "New fossil upsets human family," published in the August 1986 issue of *Science* (Vol. 233, pp. 720–721). For another recent review, see Wood and Chamberlain (1987).
28. In a careful phylogenetic analysis (also known as *cladistic analysis*) of traits observed in hominid fossils, Skelton, McHenry, and Drawhorn (1986) find evidence for no more than two lineages. They favor a scheme in which *A. africanus* is the common ancestor of *H. habilis*, in one line, and *A. robustus* and *A. boisei*, in another. *Africanus* is, in turn, descended from *A. afarensis*. They note five other two-lineage models suggested by other contemporary authors. It seems that there is considerable room for disagreement.
29. Simons (1989).
30. In the essay "Bushes and ladders" from Gould (1980).
31. Blumenberg (1983). In a recent article, in which they use computerized tomography to estimate the size of a skull of *A. africanus*, Conroy, Vannier, and Tobias (1990) give the mean brain size of six samples of *A. africanus* as 440.3 cc.
32. Falk (1983a).
33. Holloway (1981b, 1985).
34. Falk (1982, 1983b).
35. See the discussion in the article "The earliest hominids were more like apes," published under *Research News* in the 29 May 1987 issue of *Science* (Vol. 236, pp. 1061–1063).
36. Day, Leakey, Walker, and Wood (1975).
37. Trinkaus (1984).
38. Foley (1987).
39. Brown, Harris, Leakey, and Walker (1985).
40. Foley (1987).
41. Simons (1989).
42. Pfeiffer (1973).
43. Clark (1969).
44. The Swanscombe and Steinheim skulls are described in Pfeiffer (1973).
45. Clark (1969).
46. Foley (1987).
47. Cited anonymously by Pfeiffer (1973, p. 160).
48. Op. cit., p. 162.
49. See the article "A new look at an old fossil face," published under *Research News* in the 12 December 1986 issue of *Science* (Vol. 234, pp. 1326).

50. Trinkaus (1983).

51. These examples are taken from Leakey and Lewin (1979).

52. Pfeiffer (1973).

53. Mellars (1989, p. 364).

54. Cartmill, Pilbeam, and Isaac (1986).

55. The following discussion of the evolution of anatomically modern humans is based on the report of a conference on "The origin and dispersal of modern humans: Behavioral and biological perspectives," held in Cambridge, England, in March 1987 and summarized in an article entitled "Africa: Cradle of modern humans," published under *Research News* in the 11 September 1987 issue of *Science* (Vol. 237, pp. 1292–1295).

56. See the article "A sharp competitive edge," published under *Research News* in the 11 September, 1987 issue of *Science* (Vol. 237, p. 1293).

57. Lieberman (1984).

58. Cann (1987).

59. A useful review of both genetic and paleontological evidence in support of the theory that *H. sapiens sapiens* originated in Africa is provided by Stringer and Andrews (1988).

60. For a discussion of some of the controversy over this theory, see the article "The unmasking of mitochondrial Eve," published under *Research News* in the 2 October 1987 issue of *Science* (Vol. 238, pp. 24–26).

61. See "Species questions in modern human origins," published under *Research News* in the 31 March 1989 issue of *Science* (Vol 243, pp. 1666–1667).

62. Pfeiffer (1985).

63. Pfeiffer (1973).

64. White (1989).

65. Mellars (1989, p. 367).

— 3 —

The Human Condition

We see things not as they are,
We see things as we are.
CHINESE FORTUNE COOKIE[1]

IN THE PREVIOUS CHAPTER, I TRACED the evolution of hominids through time. In the 5 million or so years since we split from the apes, we have diverged from them in a number of ways. The most obvious of these are cultural and environmental; we live in a world very different from that of the apes, partly as a consequence of our ability to construct our own environments (and to destroy the natural one). Despite the remarkably close genetic similarity between ourselves and our nearest relatives, the chimpanzees, we have also diverged from them physically. This chapter reviews the main changes that took place in hominid evolution and that have made us what we are. Again, the focus is broad; more detailed questions about language, and how humans perceive and think, will be dealt with in later chapters.

Bipedalism

The most decisive event that set the early hominids on the path toward humanity was their assumption of the upright stance. As we saw in the previous chapter, bipedalism seems to have been a characteristic of the very earliest hominids, going back more than 3 million years. It was not present in the prehistoric apes whose remains date from before the fossil gap that occurred about 8 to about 4 million years ago.

Since the australopithecines had brains no larger than those of apes, bipedalism evolved before the increased brain size that characterizes modern humans. Bipedalism freed the hands for other activities, such as carrying things, using and manufacturing tools, and gesturing. This freeing of the hands was in a sense only the last step in a process that had already begun; as pointed out in Chapter 2 the hands had already been emancipated from their function as weight supporters earlier in primate evolution and had assumed multipurpose roles in climbing, grasping, and handling. Bipedalism seems, nevertheless, to have been the decisive step toward the evolution of modern humans, leading to the manufacture of tools, language, and perhaps a restructuring of thought.

So what was it, then, that led to bipedalism in the first place? There is a general assumption that it had to do with changes in the geological conditions during the late Miocene and early Pliocene—that curious period in which there is a gap in the fossil record. Recall that the apes had spread quite far afield in the Old World to India, Asia, and southern Europe. Remnants of early hominids, by contrast, have been found only in Africa. Perhaps this is simply because conditions in this region were uniquely suited to the preservation and subsequent discovery of fossils, and because hominid remains from other regions were either destroyed or have remained inaccessible. By contrast, remains of other prehistoric animals have been found outside of Africa, as we have seen. Moreover, it is generally supposed that geological events in East Africa were so extraordinary that they created precisely the conditions necessary for evolutionary change, although we shall see that there is room for imaginative license in specifying precisely what those conditions were.

I begin with the conventional account. Early in the Miocene, which spans the period from about 20 million to about 8 million years ago, both East and West Africa consisted primarily of forests and woodlands. But then the continent of Africa collided with Eurasia. In Kenya and Ethiopia, the earth was heaved upward by more than 3000 feet to form the highlands of these two nations. With the upward pressure of molten lava the earth's crust eventually cracked, creating what was to become the Great Rift Valley. Huge volcanoes rose so high that they created a rain barrier, and much of the forest and woodland to the east gave way to savanna, or open terrain. Beginning some 12 million years ago, East Africa has been characterized by an unusual diversity, ranging from the original dense rain forest to semiarid desert, with varying shades in between. This diversity, rather than any particular geographical stratum, may have been the key to hominid evolution.

Along the Rift Valley as it sweeps north from Kenya into Ethiopia lies the Lake Turkana basin. The River Omo drains the Ethiopian highlands and deposits silt in the lake, a process that has been going on for some 4 million years. It is this gradual accumulation of silt that has preserved the fossils found in the region and that allows them to be dated with reasonable accuracy.[2]

There has been much speculation as to why the changed environment of East Africa might have favored bipedalism. It has been suggested, for instance, that an upright stance enabled the early hominids to see farther across the open savanna, and so to receive earlier warning of possible predators.[3] Standing upright also freed the hands, allowing them to wield weapons for staving off attackers. To achieve these ends, however, it would have been enough to stand upright only occasionally, as other primates do. These factors therefore do not seem sufficient to explain our habitually upright stance.

C. Owen Lovejoy has suggested that the evolution of bipedalism and other hominid traits may have depended on the variety of terrains created by geological events in East Africa, rather than on any particular kind of terrain (such as savanna). Greater variation in the seasons may have also encouraged the early hominids to move to different habitats at different times of the year. They therefore probably lived a nomadic life, moving from one habitat to another in the search for food. However, they also had to keep their numbers up by increasing the birth rate and reducing mortality as far as possible. This required intensive nurturing of infants, which conflicted with the nomadic lifestyle. Lovejoy argues that these pressures led to the creation of the nuclear family, with the female caring for the offspring and the male roaming for food to bring back to the family base.[4]

Bipedalism may have evolved, Lovejoy suggests, because it freed the hands and arms for carrying food. This would not have been a big step (so to speak) in evolution, since other primates are capable of limited bipedalism. Chimpanzees, for example, can walk for short distances on their hind legs, although they rapidly become fatigued. But the persistent pressure on the early hominids to gather food may have eventually selected for the alterations in bone structure that allowed upright walking to become the natural means of getting about.

Lovejoy compares the hominid lifestyle with that of other species, such as birds, that also maintain nuclear families and gather food for their dependents. Among hornbills, for example, the male collects food for his female mate and their offspring. In some 90 percent of bird species there is a monogamous relation between the parents. However, birds are able to carry food in their mouths or to regurgitate it for their young, whereas these options were impractical for the ancestral hominids by virtue of their anatomy. Consequently a special adaptation, bipedalism, was needed for hominids to solve the problem of transporting food.

Lovejoy's scenario appears to give the major role to the adventurous but monogamous males, who brought back food to the sedentary females and their offspring. A more extreme male-centered view was suggested earlier by Robert Ardrey in his book *The Hunting Hypothesis* (1976). He argued that bipedalism would have freed the hands for carrying primitive weapons, and later for manufacturing more sophisticated ones, in order to slaughter other animals for food. Ardrey assumed that it was the males who went hunting, and since hunting required cooperation, social bonding, and invention, human destiny was placed firmly in the hands of the males of the species.

Needless to say, this idea has not gone unchallenged. Two American anthropologists, Nancy Tanner and Adrienne Zihlman, have argued that the females were at the hub of evolutionary adaptation; it was they who invented tools for collecting, who carried food to share with offspring, and who feminized males by choosing the more cooperative, social ones as mates. Moreover, the critical activity for which tools were invented was not hunting, but rather the gathering of foodstuffs such as plants, eggs, termites, and small burrowing animals.[5]

A different scenario again has been proposed by Mary Leakey and her colleagues. They suggest that the early hominids were migrating scavengers who followed huge herds of ungulates, such as wildebeests and zebras, over fairly long distances in order to gain access to their carcasses. In order to do this, they had to carry their offspring, and it was this, rather than the carrying of foodstuffs, that was critical in the evolution of bipedalism. An argument against this view is that the teeth of the early hominids seem better adapted to a vegetarian diet than to a diet of meat, and would certainly have been inadequate for breaking through the tough hide of a zebra carcass, for example. Perhaps, like vultures, the hominids relied on better equipped scavengers, such as hyenas, to bite through the hide, but this could have been a dangerous strategy given that the hyena was also a potential attacker. Leakey and her colleagues suggest, however, that it was precisely this danger that selected for the ability to make stone tools, so that carcasses could be cut and scraped without having to rely on other animals to do it.[6]

These rather different scenarios illustrate the speculative character of the explanations for bipedalism. Nevertheless, they share the common theme that habitual bipedalism was selected because it facilitated carrying, whether of offspring or foodstuffs, while the more constructive use of the hands for the development of tools and for communication probably came later. But while this is probably the prevailing opinion, there is another theory that is altogether more exotic, although it has not yet found general acceptance.

An Aquatic Phase?

In 1960, Sir Alister Hardy proposed that the early hominids might have passed through an aquatic phase in which they lived in the sea. The idea was taken up and developed by Elaine Morgan in her books *The Descent of Woman* (1972, revised in 1985) and *The Aquatic Ape* (1982). Morgan is a freelance writer rather than a professional physical anthropologist, which may explain why the theory has not been taken very seriously by academics. One influential critic dismisses it entirely, asking why we should have abandoned an aquatic life if we evolved successfully in it and why we do not have flippers, like other aquatic animals.[7] Even Stephen Jay Gould was scornful: "Elaine Morgan's *The Descent of Woman* is a speculative reconstruction of human prehistory from the woman's point of view—and as farcical as more famous tales by and for men."[8]

Still, the theory has some plausibility.[9] It may explain our relative lack of bodily hair, unique among the primates. Our hominid ancestor evolved as the naked ape,[10] just as "the porpoise turned into a naked cetacean, the hippopotamus into a naked ungulate, the walrus into a naked pinniped, and the manatee into a naked sirenian."[11] To retain warmth, bodily hair was replaced by a layer of subcutaneous fat. This feature is also unique among the primates, but common to all aquatic animals. The shape of the nose is another human peculiarity. The nostrils are covered with an elaborate cartilage so that they open downward, toward the feet. This adaptation would have allowed the aquatic hominids to swim and dive without having water forced up their noses.

Humans are the only primates that weep, and we weep salt tears. Again, however, this is a characteristic that we share with marine animals. Morgan quotes from R. M. Lockley's book *Grey Seal, Common Seal* that the seal's tears "flow copiously, as in man, when the seal is alarmed, frightened, or otherwise emotionally agitated."[12] Again, human sweat is salty, and our sweating is unique among the primates as a cooling system. Humans are also uncommonly attracted to water, as a visit to any popular beach will testify.

Morgan suggests that aquatic life would have favored bipedalism. Initially, an upright posture would have been helpful in wading, and it is also natural in treading water. More important, though, swimming is most efficiently performed with the lower limbs in line with the body. Penguins and beavers also adopt a bipedal walk when on land. Swimming might also explain our highly developed sense of balance, since gravity no longer provides a cue and orientation must depend on internal mechanisms. Only the sea lion can match, and indeed surpass, the human sense of balance.

Why, then, do we not have flippers like these other aquatic mammals? Morgan's answer is that they were not needed, since swimming was probably accomplished by a frog-like kick rather than the use of flippers. Indeed, human infants perform a natural breaststroke action, and it has often been shown that children can be readily taught to swim even before they can walk. Even so, Morgan quotes Sir Alister Hardy as stating that 9 percent of boys and 6.6 percent of girls have some webbing between the second and third toes. The human hand is also unlike that of an ape in that there is a web of skin between the finger and thumb that prevents us from stretching the angle between them more than about 90 degrees.

These various physical differences between ourselves and the apes do suggest an aquatic past, neatly recapitulated in the practice of baptising infants by immersing them in water. But Morgan is less convincing in her speculations as to when and why our ancestors may have taken to the sea. Her original notion was that this occurred during the Pliocene when there was a prolonged drought[13], causing the trees to wither and the forests to shrink. Since our ancestors were smaller and less aggressive than some of the other hominoids, they were driven out onto the open savannah, where they were poorly adapted to cope. The only solution was to take to the sea, where there was an abundance of accessible food.

In a recent about-turn, however, Morgan has argued that the aquatic phase was probably a response, not to drought, but to flooding![14] She refers to research by Leon P. La Lumiere that suggests there was extensive flooding in northeast Africa during the fossil gap of 4 to 8 million years ago, and that large areas were inundated and remained below sea level for long periods. An area of high ground now known as Danakil Horst would at that time have been an island, and it is a possible site where a group of hominoid apes might have found themselves cut off. Dwindling resources would have forced them to seek sustenace from the sea, and to gradually adapt to an aquatic existence.

Such a scenario might explain the divergence off the early australopithecines from the African apes, and may even account for the fossil gap. According to Stephen Jay Gould, the evolution of new species typically occurs in very small populations at the periphery of the main population, and the idea that our ancestors were trapped on islands meets those conditions rather precisely—more so, in fact, than traditional accounts that emphasize terrestrial changes. The earliest hominid fossils probably date from a return to a more terrestrial habitat, although both the australopithecines and *H. habilis* are often depicted as having lived beside rivers or lakes. An aquatic phase might also help explain why we humans seem to be the odd ones out among the hominoid apes in terms of external morphology, while at the molecular level we seem to be closer to the chimpanzee than the chimpanzee is to the gorilla.

Even if there is some truth to the aquatic theory, Morgan has undoubtedly overstated the case and exaggerated its importance. She suggests, for example, that speech and the manufacture of tools might be attributed to the aquatic phase, but these characteristics probably did not emerge until much later. But I suspect that the issue is not a cut-and-dried (or should I say wet-and-dried?) one. I think it is a fair observation that, even today, we are conspicuously more aquatic than our ape cousins, and the question may not be so much whether our ancestors passed through an aquatic phase, but rather how complete was their immersion in an aquatic environment, for how long, and to what extent did it alter our morphology?

Sexual Practices

There is another important set of characteristics that mark humans off from the other primates and that might have appeared early in hominid evolution. Compared with the apes, humans indulge in decidedly unusual sexual practices. The estrous cycle in women is virtually imperceptible from the outside, as it were, and humans copulate more or less independently of ovulation. In nonhuman primates, by contrast, sexual receptivity in females is very largely confined to the period when the female comes visibly into estrus; at other times, males and females show little interest in one another. The total time that a female primate may be sexually receptive in the course of a 20-year life span is probably less than 20 weeks.[15] This is a far cry from the sort of human behavior reported by gossip columnists.

These changes are usually interpreted in terms of the importance of the pair bond in maintaining the nuclear family. The progressive decline in estrus and the more or less continuous sexual receptiveness of human females increase the demand for *copulatory vigilance,* to borrow Lovejoy's delightful phrase,[16] in order to ensure fertilization. That is, it favors monogamy and regular copulation, and so keeps the family together. In Lovejoy's view, this was critical to the nomadic life that the early hominids pursued, and was also essential to the nurturance and protection of the infants.

The physical characteristics that distinguish men from women are also more marked than those that distinguish the sexes in other primates, and they show a different pattern. Among other primates, the degree of physical difference between males and females seems to be correlated with the extent to which males compete with one another for mates.[17] For instance, males may be larger than females and may have especially large canine teeth, which are important in threatening and bluffing. Among humans, by contrast, the characteristics that distinguish the sexes seem to apply as distinctively to women as to men, and appear to relate to the pair bond rather than to any distinctively male behavior. The enlarged breasts of the human female are at least as conspicuous as the male penis, although both features are usually modestly covered in contemporary societies. Men and women are also distinguished by marked differences in bodily shape and in the pattern of growth of hair.

It might be thought that such *pair-bond enhancers,* to borrow another phrase from Lovejoy,[18] are less important than the role of culture in determining mating practices and differences in appearance between men and women, and it is true that culture can greatly modify such things as the clothes people wear, their hair styles, and the perfumes and makeup they use. According to Lovejoy, however, the causal relation goes quite the other way; the very fact that culture plays such a role is evidence of the primacy of the pair bond itself.[19]

As might be anticipated, Elaine Morgan presents a rather different view of the origins of human sexual behavior based on the aquatic theory.[20] Life in the sea, she suggests, was responsible for the decline in estrus. The upright stance also led to a change in the positioning of the vagina, which came to be located more ventrally (i.e., toward the front of the body). This, in turn, forced a change in the positions adopted during copulation. The normal procedure among primates is for the male to mount the female from behind, but among humans, of course, the act is usually accomplished with the couple facing one another.

Morgan goes on to suggest that this altered sexual posture made it difficult for women to achieve orgasm, which is elicited most effectively following stimulation of the ventral (i.e., forward) wall of the vagina. Frontal copulation shifts the pressure of vaginal stimulation, so that it is the dorsal (rear) wall that receives maximum stimulation. The New Zealand anthropologist Peter J. Wilson has suggested that nonhuman primates seem to derive neither pleasure nor disgust from intercourse; there is no foreplay, and copulation is completed in a few seconds.[21] Morgan's view is that primate copulation is rapid because it is efficient, and females rapidly achieve orgasm.

In humans, by contrast, it is often prolonged and many women never reach orgasm.

Perhaps it is because successful sex is so difficult that we humans are so obsessed with it. In his novel *The Lyre of Orpheus*, the Canadian novelist Robertson Davies has one of his characters remark: "Unfortunately Man is the only creature to have made a hobby and a fetish of Sex, and the bed is the great playpen of the world."[22]

Morgan also suggests that the so-called pair-bond enhancers actually have little to do bonding. The breasts of the human female are larger and more conspicuous than those of other primates, not as an enticement to ensure that her mate keeps returning each evening, but to provide the feeding infant with something to hold onto, lest it slip away from the mother's largely hairless body. The enlarged buttocks of the human female enabled her to sit more comfortably, especially when feeding. The hair on women's heads also provides something for the infant to grab hold of, especially in the water; Morgan notes that women generally retain the hair on their heads, while men are apt to go bald. Compared with other primates, men have unusually large penises, but according to Morgan, this too is not expressly a pair-bond enhancer. Rather, it is an adaptation to frontal sex, which simply requires a longer penetrative organ.

The reader may take it or leave it.

The Use and Manufacture of Tools

Jane van Lawick Goodall defined the use of tools as "the use of an external object as a functional extension of the mouth or beak, hand or claw, in the attainment of an immediate goal."[23] By this definition, humans are by no means unique in using tools. In his book *Animal Tool Behavior* (1980), Benjamin B. Beck documents many examples of the use of tools by other animals, including invertebrates, fish, amphibians, birds, and mammals. Even the making of tools is not uniquely human. For example, blue jays kept in a laboratory were observed to use newspaper to make a tool with which to rake in food pellets that were out of their reach, and one bird also made pellets of wet paper to mop its food dish.[24]

Most mammals pick up food and other objects with their mouths, but in primates the hand has become specialized for this purpose. Primates vary, however, in the degree to which they can modify their grip according to the size and shape of the object. For example, prosimians and most New World monkeys grasp an object, regardless of its size or shape, by simply closing the hand around it. This is known as a *power grip*. All Old World monkeys and apes, by contrast, are able to use what is called a *precision grip*, in which small objects may be picked up and held by the finger tips. The precision grip depends on the ability to oppose the thumb and forefinger by rotating the thumb at the base.[25] This ability is most highly developed in humans; according to Richard E. Passingham, "The skill and precision of the human hand is without doubt the result of selection for greater dexterity in the use and manufacture of tools."[26]

Primates, however, are also inveterate tool users. Cebus monkeys seem to be exceptional among monkeys for the variety of ways in which they use objects to serve their own ends. They use sticks to rake in food, arrange sticks so that they can climb them to reach food, use twigs to push food from tubes or to pry open lids, stack boxes in order to reach food, throw objects at people or other animals, throw objects to knock down suspended food, and use stones or sticks to break the shells of nuts.[27] Chimpanzees also use tools in similar ways, adding some variants of their own. Modern chimpanzees in the wild use twigs to extract termites from their holes and, indeed, fashion the twigs for this purpose by stripping them of leaves and bark.[28] They also use stones for cracking open palm nuts, even transporting them to where the nut trees grow.[29] Robert Foley writes that "tool use can thus be said to be a plesiomorphic (ancestral, primitive) trait of the African hominoids."[30]

Foley suggests that the next development was the manufacture of *stone* tools, and that this was unique to the genus *Homo*. The earliest stone tools that have been found date from about 2.5 million years ago and were discovered in the Hadar region of Ethiopia.[31] Similar tools have been discovered at other African sites. Since the most extensive documentation of early tools has come from those found at Olduvai Gorge in Tanzania, this tool culture is known as the *Oldowan culture*. Oldowan tools were made by striking flakes from a stone core, giving rise to two sorts of tools. *Core tools* consisted of the core itself, sharpened along one edge by the removal of the flakes. An example is the stone chopper, which made up some 65 percent of the early Oldowan toolkit. The flakes could also be modified by trimming the edge to produce *flake tools*. An example is the stone scraper, which made up some 19 percent of the Oldowan kit. About seven different tools can be identified from the earliest Oldowan phase, rising to over ten in the so-called Developed Oldowan phase.[32]

Although these tools are usually attributed to *H. habilis*, it has recently been claimed that at least one australopithecine, *A. robustus* (sometimes known as *Paranthropus robustus*), may also have developed tools. This claim is based on recent fossil evidence concerning the shape of the hand. Unlike the earlier *A. afarensis*, *A. robustus* and *H. habilis* had hands that resembled modern human hands, with broad fingers and a relatively long thumb, specialized for a precision grip rather than a power grip. It has been inferred from this that *robustus* may also have manufactured stone tools.[33] However, the precision grip in *robustus* may have had to do with the use of bones or sticks as instruments for digging, rather than with the manufacture of stone tools. Certain bones discovered in association with *A. robustus* reveal a distinctive pattern of wear that can only be reproduced, it is claimed, by digging.[34]

The development of stone tools is therefore probably characteristic only of the line that led to *Homo*. Stone tools seem to have been primarily associated with the cutting and scraping of carcasses, and thus with the eating of meat. Whereas *robustus* adapted to life on the dry grassland and woodland environment by seeking plant foods, *Homo* probably scavenged for animal meat, and by 1.6 million years ago, *H. erectus* is though to have consumed animal meat at a higher level than any living nonhuman primate.[35] Since sources of meat would have been widely dispersed, this would have encouraged cooperation among males in the search for meat. We begin to see, then, the emergence of a social order in which specialized tools and perhaps ways of communicating became important.

There has been some debate as to whether the manufacture of stone tools represents a conceptual advance, or whether it was essentially at the level of, say, the shaping of twigs for termiting, as practiced by chimpanzees. The actual manufacturing process was scarcely more complex; the tools were made by simply striking the core stone with a hammerstone, and it has been suggested that no more than three or four blows were required,[36] a technique that could probably be learned by simple observation. Indeed, it has been successfully taught to an orangutan[37] and even to graduate students in California.[38]

However, the manufacture of Oldowan stone tools seems to have been more purposeful than tool construction by other primates. For example, the tools used by chimpanzees are typically found within reach of the food source, whereas there is evidence that *H. habilis* carried stone tools to sites up to 10 km away.[39] Ralph L. Holloway has argued that Oldowan tools were made according to a standard plan, implying a creative ability that is uniquely human.[40] In similar vein, Mary D. Leakey has suggested that the mere *modification* of tools should be distinguished from tool *manufacture*; the chimpanzee modifies a twig to make it more effective, whereas *H. habilis* actually manufactured stone tools according to a prescribed plan.[41]

Manufacture may also be distinguished from modification in that it involves operating on one object with another; as Beck notes, "Unquestionably man is the only animal that to date has been observed to use a tool to make a tool."[42] It may not be too fanciful to see in this process the first sign of recursion, which is critical to the nature of human language, as we shall see in Chapter 5. Beck goes on to warn, though, that attempts to define what is special about the manufacture of tools by hominids can be dangerously anthropomorphic—part of the often desperate campaign, perhaps, to characterize ourselves as superior to other animals.

It has also been noted that tools are often related to *extractive foraging,* in which food is removed from something else.[43] Termiting by chimpanzees is an example. Oldowan stone tools were used to extract meat from carcasses, now thought to have been scavenged rather than hunted. Beck points out that the manufacture of tools is also extractive in the sense that the tool is derived from a piece of stone. The idea that one object may be embedded in another may have been another precursor of language; as we shall see in Chapter 5, human language characteristically involves the embedding of structures (phrases, clauses) in other structures.

Whatever the importance of Oldowan tool culture, it does seem to have marked the beginning of a progression. With *H. habilis'* successor, *H. erectus,* and the emergence of the so-called Acheulean tool culture, we find tools that are more sophisticated than those of the Oldowan culture. The main development in the so-called Lower Acheulian was the emergence of the bifacial handaxe, made by removing flakes from two faces of the stone core to produce a sharper, more pointed tool that could fit comfortably into the hand.[44] Indeed, the handaxe was the hallmark of *H. erectus* and was part of the baggage when *erectus* migrated from Africa to Europe and Asia. In the late Acheulean, cores were also prepared so that flakes of specified sizes and shapes could be struck from them to produce specialized tools for such activities as cutting, scraping, boring, and skinning.[45] The technique of using prepared cores in this way is known as the *Levallois technique,* and was more fully exploited by the archaic *H. sapiens* that followed.

Unlike Oldowan tools, Acheulean tools were probably sharp enough to cut flesh. It has been suggested that the skill required to make the tools of this period could not easily be gained by mere observation but required instruction. In any event, the Californian graduate students introduced earlier, who were able to copy the manufacture of Oldowan tools from mere observation, were evidently unable to learn Acheulean technique in this way.[46]

With the spread of *erectus* to different locations throughout the Old World, and with the emergence of archaic *H. sapiens,* the tool cultures developed characteristics peculiar to individual populations. For example, in eastern Asia there was a reversion to pebble tool industries, while in India the handaxe tradition was maintained. The biface industry prevailed in Africa, Europe, and western Asia, with distinctive variations, such as the Levallois technique in sub-Saharan Africa.[47] Throughout this period, however, the development of tools was very slow, with little change for tens of thousands of years at a stretch. Tools were associated with the different populations themselves. That is, *H. habilis* had their distinctive tool technology, and the different groups of *H. erectus* and archaic *H. sapiens* had theirs, but there was very little change *within* these groups, at least up to about 70,000 years ago. Tools were as distinctive of different populations as the morphology of the fossils themselves.

Things began to change dramatically with the end of the *archaic* world and the emergence of so-called blade technologies. This occurred first about 70,000 years ago in Africa, where it emerged from the earlier flake industry that had developed there. Recall that this involved the population that had probably evolved into anatomically modern humans. The appearance of blade technology was more abrupt during the so-called evolutionary explosion[48] of the Upper Paleolithic, beginning some 30,000 to 40,000 years ago, no doubt coinciding with the arrival of *H. sapiens sapiens*. Among the Cro-Magnons of that era in Europe, tool making took on new dimensions, with over 130 different tools being identified. The Levallois technique was developed further so that a number of different tools could be made from the flakes struck from a single core—tools for such varied activities as carving, engraving, chiseling, gouging, and so on. This activity may reflect something of the hierarchical property of language, whereby sentences can be elaborated into words or words into the primitive alphabet of spoken speech (phonemes).

At this stage in evolution we also find the clearest evidence of the joining of parts, as in the hafting of a wooden handle to an axehead or a stone point to a wooden spear. It is possible, however, that hafting emerged earlier,[49] and, of course, components of tools that were made of wood or other perishable substances have not survived for our inspection. In any event, the joining of parts to form new entities is also a characteristic of human language, as we shall see in Chapter 5.

From that time on, tool technology never looked back. It became central to human survival and adaptation. Technologies change rapidly both within and between populations, adapting flexibly to local demands. According to Foley, these characteristics hold for all anatomically modern humans, whereas for all other forms of hominids, including the Neandertals and archaic *H. sapiens*, tool technologies remained stable for long periods.[50] I suggested above that some of the primitive features of language may be glimpsed in primitive stone tools, but with the appearance of blade technology and the ensuing explosion of manufacture and invention, we see a property that is more unequivocally associated with human language. That property is *generativity*—the ability to construct an unlimited number of different forms from a finite number of elementary parts. This theme will play a major role later in this book.

Brain Size

As we have seen, the brain has virtually tripled in size since the time of the earliest hominids, an increase that is much too great to be attributed to the much less dramatic increase in body size.[51] It is due almost entirely to the species *Homo*, since there was very little increase in brain size among the australopithecines during the 1.5 million years of their recorded history.[52] However, the brain size of the earliest clear form of *H. habilis*, ER-1470, was already considerably larger than that of any of the australopithecines. Dean Falk suggests, however, that this creature probably descended from a gracile australopithecine,[53] in which case there must have been a fairly rapid increase in brain size in the process of transition.

Initially, at least, the larger brain was probably a consequence rather than a cause of behavioral changes brought about as a result of bipedalism. For example, we have seen that the manufacture of stone tools was associated with *H. habilis*, but probably not with the australopithecines. This may in part explain *habilis'* larger brain. Nevertheless, it would be wrong to attribute increasing brain size wholly to a selective pressure to develop more sophisticated tools. We saw that tool technology changed relatively little from *habilis* through *erectus* to archaic *sapiens*, and actually remained static for long periods, despite the fact that brain size increased dramatically over this same interval to reach its present level. Falk suggests, moreover, that the increase in brain size was not punctuate but proceeded gradually, though in an accelerating fashion,[54] such that the greatest changes occurred during a period when tools scarcely changed at all. The real explosion in manufacturing took place only after the emergence of *H. sapiens sapiens* and was not accompanied by any increase in brain size.

Although the manufacture of stone tools may have given the initial kick to an increase in brain size, then, the subsequent development must have had other causes. One possibility is that the hominids were forced increasingly into a mode of existence that required cooperation rather than competition. We have seen that this was probably especially true of the genus *Homo*, who apparently adopted a meat-seeking strategy, compared with the robust australopithecines, who opted for a vegetarian diet. The cooperative foraging for sources of meat would have favored more sophisticated forms of communication and perhaps the development of specialized skills within the social group. It may have been these factors that were largely responsible for the increase in brain size.

The possession of a large brain is not itself a sufficient guarantee of superior intelligence. If it were, we should have to yield pride of place to the whale, whose brain weighs a massive 6800 gm. In fact the human brain, which weighs around 1400 gm, ranks no higher than fourth, following those of the elephant (about 4700 gm) and dolphin (somewhat over 1700 gm). It is also amusing to consider something of the range of variation between humans. The philosopher Immanuel Kant is said to had a brain weighing 1600 gm, while the poet Johan Christolph Friedrich Schiller carried a brain of 1785 gm. Most massive of all was that of Ivan Serveyvich Turgenev, the Russian novelist and short-story writer, whose spinal column had to support a brain weighing 2012 gm. However, the novelist Anatole France is said to have had a brain weighing only 1017 gm, a size normally associated with idiocy! In fairness, it should be said that he was 80 years old when he died, and the brain is known to grow smaller from about the age of 30 on. Men have an average brain weight of about 1440 gm, compared with about 1230 gm for women—a difference that is presumed to be of no importance, except that boys may be slightly more susceptible than girls to minor brain injury at birth.[55]

Anxious no doubt to restore humans to the top of the evolutionary tree, a number of scientists have tried to devise measures involving brain size that compensate for differences in body size. Karl S. Lashley argues that the ratio of brain tissue to total body weight should indicate the amount of tissue in excess of that required to serve the *integrative centers* of the body, and so should reflect more intellectual function.[56] Harry J. Jerison has developed a somewhat more sophisticated index, called the *encephalization quotient* (*EQ*), which is the ratio of the actual brain size to the size expected for an animal of equivalent body size.[57]

It is with relief that we learn that the EQ restores humans to the top of the pile, with a value of 7.4416. The dolphin, interestingly enough, comes second, with an EQ of 5.3055, followed by the chimpanzee with 2.4865. The elephant, sometimes praised (probably erroneously) for its powers of memory, has a quotient of 1.8717, and the rat, once the darling of experimental psychology, has a quotient of 0.4029.

Richard E. Passingham and George Ettlinger have also examined the relation between brain size and body size among the primates.[58] If one leaves out humans, there is an orderly increase in brain size as body size increases. However, there is a striking discontinuity between humans and the other primates, described by Passingham as "perhaps the single most important fact about mankind. *Our brain is three times as large as we would expect for a primate of our build.*"[59]

Another quotient of some interest is the ratio of the size of the cerebral cortex, the outside layer of the brain and the one most recently evolved, to that of the brain as a whole. In humans the cerebral cortex constitutes about 80 percent of the brain, in apes about 74 percent, in monkeys about 68 percent, and in prosimians about 50 percent. In this case, however, there is no suggestion of any *discontinuity*; for example, humans differ from apes by as much as apes differ from monkeys.

Changes in Stages of the Life Span

Another development that was of undoubted significance in human evolution was a change in the durations of stages in the life span. In the well-known passage from Shakespeare's *As You Like It*, Jaques identifies seven "ages" of man, but for present purposes five will do. First is the *gestational* phase, between conception and birth. Second is *infancy*, with much mewling and puking, betraying a total dependence on the parents. Third is the *subadult* phase, when the whining school-boy unwillingly follows his more agreeable sister to school. Fourth is *adulthood*, the sighing lover and the soldier, the stage when reproduction is possible, at least when the sighing and the fighting are over. Fifth is the *postreproduction* phase, the lean and slippered pantaloon, and the oblivion of second childishness.

Figure 3.1 shows the duration of each phase in humans and in four other primate species ordered according to their similarity to humans. All five phases are longest in humans and increase in duration in the primates as they approach humans in similarity. Thus each phase in the chimpanzee, our closest relative, is shorter than the corresponding phase in humans, and the postreproductive phase appears to be unique to humans.[60] Perhaps the most striking aspect of human development, though, is the prolonged period of infancy and childhood. Although the gestational period is also longer than in other primates, it is comparatively *short* relative to the other stages of development.

It can be argued in fact that human infants are born prematurely.[61] The newborns of other species of primate are relatively well developed and capable at birth, while human newborns are as help-

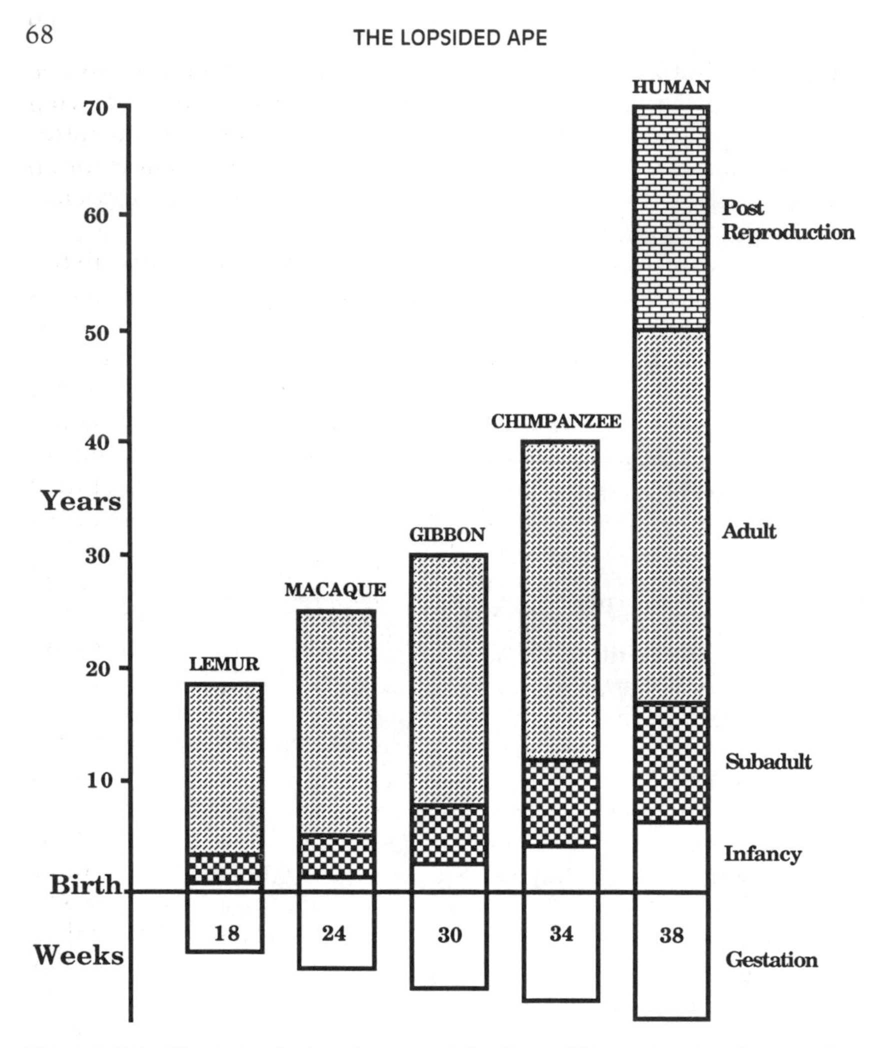

Figure 3.1. Phases of development in five different primate species, including humans (after Lovejoy, 1981).

less and undeveloped as those of more primitive mammals. To hold to the pattern of primate development, women should give birth after a gestation of about a year and a half. This means that human babies are in effect embryos for about the first 9 months after they are born. One reason for the early birth of humans is probably mechanical. Our large heads make birth difficult enough as it is, and to delay birth beyond the nine-month gestational period, when the brain is undergoing rapid growth, would surely make it impossible. The size of the pelvic canal, through which the newborn must pass, is also limited by the structural requirements of bipedal walking.

According to David Premack, human infants are at a disadvantage relative to apes and monkeys only for the first 10 months following birth. The monkey infant is better than the ape, and the ape better than the human, on such skills as grasping an object, reaching for an object, or sitting up unassisted. After 10 or 11 months, the superiority of the human infant begins to assert itself.[62]

There is some debate as to when the prolonged infancy that distinguishes humans from other primates evolved. Patterns of dental development provide one source of evidence. In apes, for instance, the three molar teeth appear at ages 3.3, 6.6, and 10.5 years, whereas in humans the corresponding ages are 6, 12, and 18 years. Given a fossil specimen, one could therefore theoretically determine the duration of infancy if one could relate the presence of molar teeth to the age of the specimen at death. It has generally been concluded from evidence of this sort that even the earliest hominids resembled humans rather than apes with respect to the duration of infancy, but this conclusion has recently been challenged. The difficulty is to determine the age of a fossil at death. For instance, the Taung child, the fossil *A. africanus* discovered by Raymond Dart, was until recently assumed to have been around 6 years old at death, but it is now claimed that the age was closer to 3 years. This recent evidence suggests in fact that the dental patterns of *A. africanus* and *H. habilis* were more ape-like than human-like, implying that the prolonged childhood of our species evolved after the time of *H. habilis*.[63]

On the other hand, there is evidence that it was present in *H. erectus*. Judging from the skeleton of the 12-year-old boy found at western Lake Turkana and from other evidence on sex differences in morphology, the birth canal in *H. erectus* was significantly smaller than that of *H. sapiens*, yet *H. erectus* grew to a size approaching that of modern humans. Frank Brown and his coauthors, who describe the Lake Turkana skeleton, write that "by 1.6 [million years] ago, the secondary altricial condition (which leads to increased fetal dependency) must have been present."[64] The best guess, then, is that the prolonged childhood that is a distinctively human characteristic was a development that took place in that narrow period of time between *H. habilis* and *H. erectus*.

The prolonged infancy and childhood of humans has an especially marked influence on the development of the brain, and indeed, it is through prolonged growth rather than an unusual rate of growth than the human brain reaches its exceptional size.[65] The brains of many mammals are essentially fully formed at birth, and in other primates brain development extends into early postnatal growth. The brain of a newborn chimpanzee is about 60 percent of its ultimate weight, while that of a newborn human is only 24 percent of that of an adult. Moreover, during the first quarter of the period from birth to puberty, the chimpanzee brain increases only 30 percent, while the human brain increases 60 percent.[66] Most of the growth of the human brain therefore takes place after birth, when the child is exposed to the modifying influences of the environment, whereas the brains of apes are more than half developed at birth.

Neoteny

According to some, humans not only take longer to grow up than do other primates, they never actually make it. In 1926, the Dutch anatomist Louis Bolk listed some 20 ways in which the adult human retains child-like characteristics, a phenomenon known as *neoteny* (which means "holding youth").[67] At the embryonic stage, the skulls of apes and monkeys are very like our own, rounded and bulbous, with small jaws and teeth and weak brow ridges. However, humans retain these characteristics, whereas the skulls of other primates grow flatter and smaller relative to the rest of the body, and the jaws, teeth, and brow ridges grow much more pronounced. The foramen magnum, the hole in the skull from which the spinal cord emerges, is at the bottom of the skull in the embryos of most mammals. In humans it remains so, but in other mammals it migrates to the back of the skull so that the head becomes tilted relative to the spinal cord. There are other changes that occur in mammalian development, such as the backward rotation of the vaginal canal and the sideways rotation of the big toe, but in human development these features retain their embryonic form.

Such features suggest that human growth may be retarded, an idea explored to good effect by Aldous Huxley in his novel *After Many a Summer*. A character in the novel, the fifth Earl of Gonister, has managed to prolong his life and that of his lover well beyond 200 years by ingesting carp guts. However, this prolongation of life is achieved at a price, for the two aged lovers have grown into apes. So much for dreams of immortality.

The concept of neoteny is not without problems. That we should be apes is bad enough, but to suggest that we are retarded apes borders on the insulting. Nevertheless, the idea has been defended by Stephen Jay Gould.[68] Some of our retarded features, like the position of the foramen magnum or the non-opposable toe, are simply convenient adaptations to our upright, two-legged stance. The more interesting aspect has to do with the brain. The human brain is exceptional in that it continues to grow long after birth, which is why the human skull retains its bulbous appearance. This may provide humans with an exceptional degree of adaptability to environmental influences.

Genes versus Culture

From the beginning of hominid evolution, culture has played a prominent role, and indeed cultural and biological evolution cannot always be disentangled. Ralph L. Holloway has argued in fact that the social component was paramount from the beginning. "We are unique," he writes, "because we have evolved the capacity to alter our environments to an extent unprecedented by any other creature."[69] Activities such as tool making, hunting, and gathering took place in a social context, so that hominids effectively created their own environments. Since biological evolution consists of adapting to the environment, the early hominids began in effect to control their own evolution. This mutual interaction between biological and cultural evolution, according to Holloway, explains the rapid changes that occurred in hominid evolution.

This is also essentially the point made by Lumsden and Wilson, the sociobiologists, when they write of *gene-culture coevolution*.[70] As we saw in Chapter 1, however, these authors also make the controversial claim that significant genetic changes are still occurring and can take place within 50 generations. To me, it is more likely that, genetically speaking, the human condition was reached with the advent of *H. sapiens sapiens*, and the significant changes since then have been cultural rather than genetic. The further question, however, is whether the gene-culture spiral was sufficient even to explain the arrival of *H. sapiens sapiens*, or whether there was some sudden genetic shift, the product of exaptation or some hitch-hiking gene, that gave us our distinctive characteristics.

It is a curious fact that the increase in the size and complexity of the human brain is achieved with very little *genetic* change. Chapter 2 noted the close genetic similarity between humans and chimpanzees. Even more remarkable, perhaps, the total amount of genetic information, as coded in the DNA molecule, has remained virtually constant through *mammalian* evolution. It has been estimated, in fact, that the number of genes is about 1 million, whether in the mouse, the cow, the chimpanzee, or the human. Moreover some 40 percent of chromosomal DNA appears to be redundant, and it has been estimated that there may be only about 10,000 genes available for constructing the brain and the central nervous system.[71]

This number seems remarkably small when compared to the complexity of the human brain alone, which consists of some 10^{10} neurons, and 10^{14} synapses (or connections between neurons). It therefore seems likely that the vast complexity of the human brain is not achieved solely by genetic coding, but is determined in part by interactions with the environment. This may be related, in turn, to the long period of postnatal development experienced by humans and to the closely knit family bonds that ensure exposure to appropriate environmental influences. Of the skills that are dependent upon environmental influences, language must be among the most important, although we shall see in Chapter 5 that there is probably an important genetic component as well.

It is clear that the brain and central nervous system are most "plastic" while they are still developing. Indeed, there is remarkable evidence that canaries learn new songs by growing new nerve cells even in adulthood.[72] Every summer, new neurons appear in the male canary's brain so that it can learn a new song, which it sings the following spring. This annual recapturing of youth is, as far as I know, unparalleled in any other species. In humans, the learning of language provides the most obvious example of the importance of learning while our brains are young and still developing; as we shall see in Chapter 5, there is good evidence that if language is not learned before the age of puberty, it will never be learned properly at all. Alas, nature does not supply us humans with a new set of neurons for the learning of new languages in adulthood, but to the extent that neoteny is a valid concept, humans may be better able to learn skills in adulthood than are other primates. Even so, at one time or another, most of us have cursed not having acquired some skill, be it music, tennis, or French, when we were young.

Given the varied terrains that the wandering hominids inhabited and the increasing demands imposed by rapid cultural change, it clearly became increasingly important to adapt quickly. Despite the gene-culture spiral referred to earlier, it seems unlikely that the mechanisms of genetic change could have kept pace with the changing demands of culture and technology, especially over the past few millennia. Extending the period of infancy and childhood would therefore have capitalized on the period of greatest plasticity. Significant adaptive changes can therefore occur within a generation and are transmitted culturally rather than genetically. Those who are concerned about "future shock" may wonder whether even this mode of adaptation is rapid enough.

The close sexual bonding between parents may also play a role here. Together with the helplessness of the human infant, it ensures a strong bond between parents and their children, and so maximizes the opportunity for children to learn from their elders (until the children become teenagers, that is). The nuclear family, essential to survival in the ecological niche occupied by the early hominids, may have been defined and cemented by the very helplessness of the infant.

The idea that the human brain is distinguished from the brains of other species by its plasticity and capacity for learning has been a persistent one. The very notion of intelligence has often been equated with the ability to learn, and seemingly clever behavior from animals has often been dismissed with the suggestion that it is merely instinctive. However, the point is a contentious one and raises an issue that goes back to the early philosophers. On the one hand there is an associationist doctrine, running from Aristotle, through the British associationists such as Locke, to the modern behaviorists, that the human mind is a *tabula rasa* at birth awaiting the imprint of experience. On the other hand, there is an idealistic tradition going back to Plato's notion of pure ideas, resurfacing in Kant's idea of a priori knowledge, and represented today in Chomsky's approach to language.

Perhaps the most extreme modern idealist is Jerry A. Fodor. In his 1983 book *The Modularity of Mind*, he conceived of the human mind as consisting largely of a set of independent, innately endowed modules, each dedicated to some specific computational function, such as the analysis of spoken speech or the segmentation of objects in the perception of a visual scene.

Modern biology suggests a compromise between these extremes. James L. Gould and Peter Marler write of "innately guided learning processes" that constrain both the type of learning that may occur and the developmental stage at which it occurs.[73] A good example of this is again supplied by song learning in birds. The white-crowned sparrow learns to imitate songs, but in order to do so it must hear the song before it is 7 weeks old. (Unlike the canary, it gets only one chance; it does not learn songs in adulthood.) This period of time during which the song must be heard is known as the *critical period* for learning.[74]

There are many other examples of learning that is constrained by a critical period, including, it has been argued, human language. More generally, the question of genetic versus cultural influences is especially critical with respect to the learning of language, and will be discussed further in that context in Chapter 5.

Cultural Evolution

Whatever the role of genetic changes in human evolution, it is apparent that many of our characteristics are purely cultural. We inherit the specific languages that we speak, and most of our habits and lifestyles, from the people around us—not genetically, but by example, instruction, and sheer generosity. The *content* of our lives is bequeathed to us through culture, not through our genes.

According to E. S. Deevey, there have been three cultural "revolutions," each producing a surge in population followed by a leveling off toward a state of equilibrium.[75] The first of these had to do with the development of tools and the spread of population. From about 1 million to 25,000 years ago, the tool-making revolution was accompanied by the spread of population from Africa to most of Europe and Asia, and an estimated increase in total population from about 125,000 to somewhat over 3 million. As noted above, there may in fact have been two migrations from Africa, with the later migration of *H. sapiens sapiens* in the last 200,000 or so years replacing the earlier archaic variety. By the Mesolithic Age some 10,000 years ago, the population had spread also to the Americas and the Pacific—over most of the earth, in fact—and had increased to over 5 million.

Then came the agricultural revolution, when hunters and gatherers gave way to plowmen and herdsmen, and villages and towns developed. The population increased by an estimated 16-fold, from somewhat over 5 million to between 80 and 90 million, between 10,000 and 6000 years ago, and then at a slower rate to about 545 million about three centuries ago. The last three centuries have seen the scientific-industrial revolution, with another sharp increase in population, said to have reached 5 billion on 10 July 1987.

The population increases brought about by each of these revolutions was presumably due in part to increases in the means of subsistence. It has been estimated, for instance, that the shift from animal food to plant food allows a 10-fold increase in food per unit of area even without agricultural labor, and the scientific-industrial revolution also brought about a dramatic shift in the efficiency of food production. But the increase in population is also due in part to increasing life expectancy. Life expectancy from birth was probably little more than 25 years, give or take 5 years, from the time of the Neandertals until the beginning of this century, when it rose sharply to exceed 70 years in many countries, including the United States.

The agricultural and scientific-industrial revolutions were no doubt the products of increasingly sophisticated cultures rather than changes in the biological makeup of humans. The exploitation of natural resources, the development of written languages, and the advance of technology depend on information that is taught and learned rather than coded in the genes—although the ability to learn is no doubt genetically inherited. There is little reason to believe that *H. sapiens sapiens* of 150,000 years ago differed genetically from those of today, at least with respect to intelligence or adaptability, just as there is little reason to suppose that different racial groups of people today differ genetically in these respects. Where modern humans in industrialized society differ from their forebears is in their vast resources of technology and information. Although language itself is a human universal, language also creates the critical conditions for a high degree of specialization. Humans diversify their knowledge and skill, so that society itself attains a complexity that is well beyond the capacity of the individual.

Summary

The most decisive event in human evolution was bipedalism. Precisely why the hominids stood and walked upright is still not clear. I have discussed conventional accounts, along with the controversial theory that bipedalism may have been an adaptation to an aquatic phase during the transition between the Pliocene and the Miocene. The aquatic theory may also explain the peculiar sexual habits of humans, although our obsessive sexuality may simply have been an adaptation to the early hominids' wandering, scavenging life style, which required close family bonds. The reader is, of course, free to choose from among the various scenarios, and may rest assured that the aquatic theory will scarcely intrude from this point on.

But whatever the reason for bipedalism, it allowed the forelimbs to be exapted for other purposes, such as manipulation, carrying things, and making and using tools. However, the australopithecines do not seem to have capitalized on the extra opportunities offered by bipedalism, but remained creatures with brains no larger than those of the apes and eventually became extinct. The next decisive event may have been the choice of survival strategy. With the good sense to which we have become accustomed, the line that emerged as *Homo* may have scavenged for meat, a strategy that may have favored the development of specialized stone tools for scraping and later cutting meat. The scavenging life may also have favored a more cooperative society and more sophisticated ways of communicating, which would, in turn, favor an increase in brain size. In such a society, diversification might also have been an advantage, so that different individuals might be specialized to do different things. This may have favored a prolonged childhood, increasing the contribution of experience to the shaping of skills. In *H. habilis*, therefore, we see the beginning of the positive feedback loop between biological and cultural changes, identified by Holloway as the mechanism driving human evolution.

One thing no doubt led to another. The use of the hands for the use and manufacture of tools and for communication would have had survival value of its own, but it required cooperation and communication between different individuals, leading to selective pressure for further increases in brain size. The development of tools may also have freed the mouth from some of its more arduous duties in food consumption, permitting its involvement in the evolution of oral language. The upright stance may also have eventually brought about changes in the shape of the vocal tract that could also be exploited in vocal communication. The evolution of language will be discussed in more detail in Chapter 6.

One of the issues in theories of human evolution is whether we evolved gradually over the past 4 million or so years to become what we are, or whether there was a relatively sudden change, as suggested by the evolutionary explosion of the Upper Paleolithic.[76] Holloway is representative of the gradualist approach, although control gained over the environment would have speeded the evolutionary process throughout the period of hominid development. That is, the positive feedback between environmental and biological change would no doubt have created a positively accelerating function, with both biological and cultural changes occurring more rapidly in recent times compared with earlier times. This might explain the relative stagnation in the development of tools from about 1.8 million until about 200,000 years ago or later. Others, however, have suggested a discontinuity in evolution within the last 200,000 years.[77] The argument for a discontinuity rests in part on the evolutionary explosion that occurred some 35,000 years ago and in part on the special nature of human language, which is held to represent an abrupt change from other forms of animal communication. This issue will surface again in later chapters.

Notes

1. This was originally attributed by David Pilbeam to the Talmud, but he later discovered that it came from a Chinese fortune cookie; see Lewin (1987, p. 44).
2. Summarized from Chapter 2 of Leakey and Lewin (1979).
3. For example, Kortlandt (1972).
4. Lovejoy (1981).

5. Tanner and Zilman (1976).

6. Sinclair, Leakey, and Norton-Griffiths (1986). See also the article "Four legs bad, two legs good," published under *Research News* in the 27 February 1987 issue of *Science* (Vol. 235, pp. 969–971).

7. Gowlett (1984).

8. Quoted by Morgan (1985, p. 272).

9. Morgan's theory is enthusiastically discussed by Graham Richards in his book *Human Evolution* (1987). On p. 194, Richards writes, "It must be stressed . . . that Hardy's original mooting of the idea was, according to at least one authority who was present, meant as a joke and that the present author is taking considerable risks with his academic credibility by devoting any space to it."

10. See Desmond Morris' popular book *The Naked Ape* (1967).

11. Morgan (1985, p. 29).

12. Op. cit., p. 54.

13. Morgan suggests that the drought lasted for 12 million years. This suggestion did little to enhance her credibility, since the Pliocene began only some 5 million years ago!

14. See the postscript to the 1985 edition of her 1972 book.

15. Jolly (1972).

16. Lovejoy (1981).

17. Morris (1967).

18. Whose very name is a pair-bond enhancer.

19. Lovejoy (1981).

20. See especially Chapters 4 and 5 of Morgan (1985)

21. Wilson (1980).

22. Davies (1989, p. 230).

23. Goodall (1970, p. 195).

24. Jones and Kamil (1973). For other examples of the use of tools by birds, see the review by Boswall (1977).

25. Napier (1971).

26. Passingham (1982, p. 66).

27. See Chevalier-Skolnikoff (1989) for documentation and further review.

28. Goodall (1970).

29. Boesch and Boesch (1983, 1984).

30. Foley (1987, p. 387).

31. Johanson and Edey (1981).

32. Leakey (1976).

33. Susman (1988).

34. Brain (1986).

35. Foley and Lee (1989).

36. Holloway (1981a).

37. Wright (1972).

38. Washburn and Moore (1980).

39. McGrew, Tutin, and Baldwin (1979).

40. Holloway (1969).

41. Leakey (1976).
42. Beck (1980, p. 218); I presume he meant to include women as well.
43. Beck (1980).
44. Leakey (1976).
45. Isaac (1976).
46. Washburn and Moore (1980).
47. Foley (1987).
48. Pfeiffer (1985).
49. Isaac (1976).
50. Foley (1987).
51. Godfrey and Jacobs (1981).
52. Falk (1987b).
53. See also Conroy, Vannier, and Tobias (1990).
54. Op. cit.
55. These facts and figures are taken from Russell (1979).
56. Lashley (1949).
57. It takes a brain of considerable size to work out this index, and the reader is referred to Jerison (1973) for the details.
58. Passingham and Ettlinger (1974).
59. Passingham (1982, p. 78; his emphasis).
60. Lovejoy (1981).
61. Krogman (1972); see also the chapter entitled "Human babies as embryos," in Gould (1980).
62. Premack (1988).
63. What is confusing, however, is that a human-like rather than an ape-like pattern appears to hold for *A. robustus*. Smith (1986), who reports this, suggests that it is an instance of parallel evolution. See the article "Debate over the emergence of human tooth patterns," published under *Research News* in the 20 February 1987 issue of *Science* (Vol. 235, pp. 748–751), for a discussion.
64. Brown et al. (1985, p. 792). You may not find the word "altricial" in a dictionary; it refers to premature birth and to the resulting need for extra care.
65. Passingham (1982).
66. These figures are from Lenneberg (1967).
67. Bolk (1926).
68. This discussion of neoteny is taken from the essay "The child as man's real father" in Gould (1980).
69. Holloway (1981a, p. 289).
70. Lumsden and Wilson (1984).
71. These figures come from Changeux (1980). A *gene* is defined as the amount of DNA needed to code for a protein of molecular weight 100,000.
72. Nottebohm (1989). The growth of new neurons occurs at the cost of old ones, however. A similar process in humans would only be useful to immigrants who want to lose their old language as well as gain a new one.
73. Gould and Marler (1987).

74. Gould and Marler use the term *sensitive period* rather than the older term *critical period.*

75. These ideas, and the population estimates in the following sections, are taken from Deevey (1960).

76. Isaac (1972).

77. See, for example, Davidson and Noble (1989).

— 4 —

Human Handedness

> Mine hand also hath laid the foundation of the earth, and my right hand hath spanned the heavens: when I call unto them, they stand up together.
>
> ISAIAH 48:13

IN OUTWARD APPEARANCE, human beings are bilaterally symmetrical, as are most other animals. Except perhaps for the parting of the hair or a twisted smile, each side of the body seems to be almost exactly the mirror image of the other side. As we shall see, however, this is a deceptive impression, since there are many asymmetries that lurk beneath the surface.

We share some of these asymmetries with other species, but two are of particular relevance to the theme of this book because they seem to single out human beings as being different. The most obvious asymmetry is that the great majority of human beings are right-handed. Less obvious, but no less striking, is the functional asymmetry of the human brain. This asymmetry is manifest in a number of ways, the most compelling of which is that in most of us the left side of the brain controls speech. It is also dominant for other aspects of language, including reading and writing. Could these asymmetries hold the key to our uniqueness?

Before considering the evidence on these matters, I should note that asymmetry itself is no criterion for uniqueness or superiority. The very molecules of which all animals are constructed are asymmetrical, and internal organs such as the heart, stomach, and liver are asymmetrically placed, leading the eminent biologist Jacques Monod to remark that "our outwardly 'bilateral' appearance is something of a fake."[1] Even outwardly, many primitive organisms are manifestly more asymmetrical than humans, as Fleur Adcock eloquently expressed in her poem *Last Song*:[2]

> Goodbye, sweet symmetry. Goodbye, sweet world
> of mirror-images and matching halves,
> where animals usually have four legs
> and people nearly always have two;
> where birds and bats and butterflies
> can fly straight if they try. Goodbye
> to one-a-side for eyes and ears and arms
> and breasts and balls and shoulder blades
> and hands; goodbye to the straight line
> drawn down the central spine,
> making us double in a world
> where oddness is acceptable only
> under the sea, for the lop-sided lobster,
> the wonky oyster, the creepily rotated
> flatfish with both eyes over one gill;
> goodbye to the sweet certitudes of our
> mammalian order, where to be
> born with one eye or three thumbs
> points to not being human. It will come.
>
> In the next world, when this one's gone skew-whiff,
> we shall be algae or lichen, things
> we've hardly even needed to pronounce.
> If the flounder still exists it will be king.

Far from being a natural biological state, bilateral symmetry is itself an evolutionary adaptation. To freely moving animals, the world is essentially indifferent with respect to left and right; there are no pressures or contingencies that should make us more responsive to one side or the other. Limbs are symmetrically placed so that movement, whether achieved by running, swimming, or flying, may proceed in a straight line. Any asymmetry in legs or wings would be likely to cause an animal or bird to proceed in fruitless (and perhaps meatless) circles. Sense organs, such as the eyes, ears, and nostrils, are symmetrically placed so that we may be equally alert to either side. Those parts of the brain and central nervous system that are concerned with these functions mirror the symmetry of the functions themselves.

Symmetry, however, is readily abandoned if asymmetry proves more adaptive. To a flounder at the bottom of the sea, lying flat on its side, survival is more likely if the two eyes gaze upward from the same side of the body than if one of them is forever doomed to stare at the blackness of the ocean bed. Similarly, the asymmetrical placement of the internal organs of the body no doubt allows for more efficient packaging, and there is no survival value to be gained from bilateral symmetry in such matters.

What is potentially interesting about human handedness and cerebral asymmetry is that they represent asymmetries in systems that had presumably been largely symmetrical in our ancestors. Again, this is not unique—the eyes of the flounder provide another example—but it does suggest a possible basis for a discontinuity between humans and other primates. Moreover, the functions to which those asymmetries apply, namely manipulation and speech, seem themselves to bear something of the stuff of human uniqueness.

Let me begin, then, with handedness.

The Nature of Human Handedness

Nearly 90 percent of humans are right-handed. That is, they prefer the right hand for most tasks requiring a single hand, and the right hand dominates in tasks involving both hands, such as unscrewing the lid of a jar. Those who prefer to use a particular hand are nearly always more skillful with that hand, suggesting that relative preference is determined by relative proficiency.[3] The difference in skill is quite striking in activities such as writing, drawing, throwing, or playing games such as tennis or squash.

There are, of course, individual variations. Some people prefer different hands for different activities, and a few are ambidextrous rather than right- or left-handed. It is possible to construct what is called a *laterality quotient* from a comparison of right- with left-hand preferences on various tasks thought to be representative of human manual activities. One widely used inventory is shown in Table 4.1, along with instructions for scoring. Although instruments such as this reveal more or less continuous gradations of handedness from extreme right to extreme left, the quotients are heavily skewed toward extreme right-handedness, and individuals tend to fall toward one or the other extreme rather than in the middle of the range.

It is sometimes claimed that handedness is not unitary, but rather that hand preferences on different clusters of activities are relatively independent of one another. For example, it has been found that hand preference on skilled activities, such as the use of tools, is distinct from that on unskilled activities, such as picking up either

Table 4.1
Edinburgh Handedness Inventory (Short Form)

Instructions
Please indicate your preferences in the use of the hands in the following activities by *putting + in the appropriate column*. Where the preference is so strong that you would never try to use the other hand unless absolutely forced to, *put ++*. If in any case you are really indifferent, *put + in both columns*.

Some of the activities require both hands. In these cases the part of the task, or object, for which hand preference is wanted is indicated in brackets.

Please try to answer all the questions, and only leave a blank if you have no experience at all of the object or task.

	LEFT	RIGHT
1. Writing		
2. Drawing		
3. Throwing		
4. Scissors		
5. Toothbrush		
6. Knife (without fork)		
7. Spoon		
8. Broom (upper hand)		
9. Striking match (match)		
10. Opening box (lid)		

Scoring
To find your *laterality quotient*, add up the number of +s in each column. Subtract the number under LEFT from the number under RIGHT, divide by the total number, and multiply by 100.

very small objects or relatively heavy ones. Preference on two-handed activities, such as using an axe or a baseball bat, may constitute another dimension.[4] But regardless of whether handedness is viewed as a single entity or as a cluster of dimensions, the overriding fact is that the right hand is the dominant or preferred one in the great majority of people.

At least superficially, a striking aspect of handedness is the discrepancy between structure and function. As the French anthropologist Robert Hertz, in an essay first published in 1909, exclaimed: "What resemblance is more perfect than that between the two hands! And yet what a striking difference there is!"[5] That is, the two hands are *physically* almost perfect mirror images of one another, but in *performance* they are very different. It may be impossible to tell whether a person is right- or left-handed by inspecting the hands, but if you ask the person to throw a ball or write a sentence with each hand in turn, then handedness is at once apparent.

This apparent discrepancy between structural symmetry and functional asymmetry may explain the symbolic potency of left and right. Almost universally, the right side has been seen as sacred and the left side as profane.[6] We need look no further than the Bible to find examples, as in Matthew 5:25:

> And He shall set the sheep upon His right hand and the goats upon His left. Then shall the King say to those upon His right, "Come ye blessed of my Father, and inherit the kingdom prepared for you from the beginning of the world." . . . Then shall He also say to those on the left, "Depart from me, ye accursed, into everlasting fire prepared for the Devil and his angels."

According to Michael Barsley, in his book *Left-handed Man in a Right-handed World* (1970), there are over 100 favorable mentions of the right hand in the Bible and about 25 unfavorable references to the left.

The remarkable differences in skill between the hands, overriding their structural similarity, may also remind us of Descartes' notion of a nonmaterial soul that transcends the physical body. Moreover, the fact that right-handedness seems to be restricted to humans implicitly supports the Cartesian idea that the nonmaterial soul is itself uniquely human, and so sets us above the other animals.[7] In modern times, with the discovery of the functional asymmetry of the human brain, the fascination with left and right has gone to our heads, and it is commonly claimed that the specialization of the left side of the human brain holds the key to human uniqueness. This is a major theme of this book, and it will unfold in later chapters.

I must hasten to say, however, that it is not my purpose to argue for a nonmaterial basis for functional asymmetries. There can be little doubt that handedness does have a structural basis, not in the hands themselves but in the brain structures that control them. Detailed analysis of brain structure has revealed asymmetries, which will be discussed in later chapters, and I shall speculate in Chapter 8 about the way in which handedness might relate to specialization of the left side of the brain. Even so, human functional asymmetries are of a degree that still commands respect and even awe; they would not be guessed from a mere inspection of a human brain relative to that of, say, a chimpanzee.

Table 4.2
Questions to Assess Hand, Foot, Eye, and Ear Preference

Hand Preference
1. With which hand would you throw a ball to hit a target?
2. With which hand do you draw?
3. With which hand do you use an eraser on paper?
4. With which hand do you give out the top card when dealing?

Foot Preference
1. With which foot do you kick a ball?
2. If you wanted to pick up a pebble with your toes, which foot would you use?
3. If you had to step up onto a chair, which foot would you place on the chair first?

Eye Preference
1. Which eye would you use to peep through a keyhole?
2. If you had to look into a dark bottle to see how full it was, which eye would you use?
3. Which eye would you use to sight down a rifle?

Ear Preference
1. If you wanted to listen in on a conversation going on behind a closed door, which ear would you place against the door?
2. If you wanted to hear someone's heart beat, which ear would you place against the chest?
3. Into which ear would you place the earphone of a transistor radio?

Some Related and Unrelated Asymmetries

Besides handedness, there are other lateral biases that are easily observable in most people. For example, most of us like to kick a ball with the right foot or look through a telescope with the right eye. These asymmetries are rather less pronounced than is handedness. To assess the proportions showing a right-sided preference for the hand, foot, eye, and ear, Clare Porac and Stanley Coren asked more than 5000 individuals the questions listed in Table 4.2, and found the percentages favoring the right to be 88.2 for handedness, 81.0 for footedness, 71.1 for eyedness, and 59.1 for earedness.[8]

They also found that these biases tended to be *congruent*, that is, that the preferred side on one measure tended to be the same as the preferred side on another. For example, 84 percent showed the same preference (left or right) for the hand and foot. The percentage of congruence was lower for the other pairings and lowest of all for the ear and eye, where only 61.8 percent preferred the same side. In a sense it is not surprising to find congruence, since in each case there is an overall bias in favor of the right side, and even if two measures were unrelated, chance alone would ensure congruence of over 50 percent.[9] Porac and Coren show, however, that the actual congruence is in each case greater than that expected by chance.

It seems likely, then, that these various asymmetries reflect some common underlying bias, but that the expression of the bias is strongest in handedness. Thus the British psychologist Marian Annett notes from several surveys that some 95 percent of right-handers are right-footed, compared with about one-third of left-handers.[10] Eyedness is more weakly related to handedness; about two-thirds of right-handers are right-eyed, compared with about one-half of left-handers. In general, left-handers seem to be more mixed than right-handers with respect to other asymmetries. This includes cerebral dominance for language, which will be discussed in Chapters 7 and 8.

Some asymmetries do not seem to be related to handedness at all, although it is often supposed that they are. For example, most people clasp their hands with the left thumb on top, but this seems to be equally true of right- and left-handers. In a survey of students in large classes in undergraduate psychology, Marian Annett found that slightly more left-handers (62 percent) than right-handers (54 percent) clasped their hands this way, but the difference was not statistically reliable. Similarly, about 59 percent folded their arms with the left arm on top, about 75 percent parted their hair on the left, and about 65 percent had a clockwise hair whorl, but right- and left-handers did not differ reliably in these respects.

One asymmetry that is related to handedness is cerebral dominance for language. This relation will be discussed further in Chapter 8, where I shall try to develop a unified view of human laterality.

The Universality of Human Handedness

Andrew Buchanan, in the *Proceedings of the Philosophical Society of Glasgow* of 1860–1864, was bold enough to write as follows:

> The use of the right hand in preference to the left must be regarded as a general characteristic of the family of man. There is no nation, race, or tribe of men on the earth at the present day, among whom the preference does not obtain; while, in former times, it is shown to have existed, both by historical documents and by the still more and authentic testimony of certain words, phrases, and modes of speaking, which are, I believe, to be found in every spoken language.[11]

This assertion holds as true today as it did in Buchanan's time.

There are occasional suggestions that some nation or race of people may have been predominantly left-handed, but the evidence is typically indirect and fails to withstand closer scrutiny. For example, in a popular article written in 1956, Trevor Holloway asserts that "the Antanalas of Madagascar are unique among the races of the world for almost every member of this tribe of 100,000 is left-handed."[12] I have been able to find no basis for this extraordinary claim.

It has also been argued that the ancient Hebrews must have been mostly left-handed because Hebrew is written from right to left.[13] But so are a lot of other scripts. Up to about 1500 A.D. there were in fact about as many right-to-left scripts as left-to-right ones, and the gradual predominance of left-to-right scripts is probably due to historical factors unconnected with handedness.[14] Although it may seem more "natural" for the right-hander to write from left to right, this applies primarily to cursive script, and is in any event due in part to the fact that most of us have been taught that way. To the ancients who formed symbols by engraving them on stone tablets, it may have been more natural to start on the right, simply because that is the side of the preferred hand.

It was once commonly believed that the ancient Egyptians were predominantly left-handed because they usually depicted humans and animals in right profile, whereas it is more natural for right-handers to draw left profiles (that is, with the left side facing the viewer). Figure 4.1 shows a typical example of a person depicted in right profile from an Egyptian tomb relief dated at about 3,500 B.C. But notice that the person himself is evidently right-handed, for he is holding a stylus in his right hand. A survey has in fact shown that, despite the predominance of right profiles, the percentage of instances of right-hand use depicted in ancient Egyptian art is comparable to that observed in modern societies.[15] The preference for right profiles may simply reflect a widespread cultural belief that the right side is sacred and the left side profane, a dichotomy that is itself a manifestation of the universality of right-handedness.

Against these isolated claims for lost tribes of left-handers, the evidence for universal right-handedness is overwhelming. It comes not just from the measurement or even the observation of handedness but is woven into virtually all human cultures in their myths, their superstitions, and even their languages. The preference for the right hand is manifest in the positive values associated with the right and the negative values associated with the left. In the Pythagorean Table of Opposites, recorded by Aristotle, the right was associated with the limited, the odd, the one, the male, the straight, the light, and the good, while the left was linked with the unlimited, the even, the many, the female, the curved, the dark, and the evil. Remarkably similar tables can be drawn up from the symbols of quite unrelated cultures.[16]

The Maoris are said to have paid special attention to tremors

Figure 4.1. Egyptian tomb relief, dating from about 3500 B.C., showing the characteristic right profile and right-handedness.

during sleep as warnings of future events. A tremor on the right side of the body foretold life and good fortune, whereas a tremor on the left warned of ill fortune and possibly death.[17] To the native people of Morocco, the involuntary twitching of an eye carried a very similar message; twitching of the right eye foretold the return of the member of the family, or some other good news, while twitching of the left eye warned of impending death in the family.[18]

Our customs and language provide further testimony of the symbolic potency of left and right, and so of the universality of right-handedness. The right hand is used for shaking hands, making the sign of the cross, saluting, laying on of hands, and swearing on the Bible, while the left hand is the "hand of the privy" and the hand that traditionally delivers the coup de grace to a dying opponent. We speak of a "right-hand man" but a "left-handed compliment." The terms "adroit" and "dexterous" have positive connotations, while "gauche" and "sinister" have decidedly negative overtones.

Given the negative associations with the left, it is not surprising that the left-handed have often been victims of abuse and persecution. In many cultures, left-handers have been forced to change, especially for activities like eating and writing. According to Barsley, the left-hander has been largely "emancipated" in North America and parts of Western Europe, but enforced use of the right hand has remained the rule in Spain, Italy, and the Iron Curtain countries with the exception of Czechoslovakia.[19] Pressures to use the right hand seem also to persist in the East. A recent survey of schoolchildren in Taipei in China revealed that only 3.5 percent were "frankly left-handed" on a test involving a variety of activities,[20] and another study in Taiwan revealed that only 0.7 percent used the left hand for writing.[21] This last figure contrasts with the 6.5 percent reported for Oriental school children living in the United States,[22] where the cultural pressure has been relaxed. In a study of schoolchildren in Japan, 7.2 percent were non-right-handed, but this figure rose to 11 percent if those who had changed their handedness from left to right were included.[23] The proportions using the left hand for writing and eating were only 0.7 and 1.7 percent, respectively, again showing the strong cultural pressure.

In Australia and New Zealand, the percentage of individuals writing with the left hand rose from about 2 percent at the turn of the century to about 13 percent in the 1960s,[24] and there was a similar change in the United States from the 1920s to the 1960s.[25] These changes reflect an increased tolerance of left-handers. The emergence of shops selling special utensils for the left-handed, such as left-handed scissors and corkscrews, indicates a new awareness that left-handers do have special problems in a world largely constructed by and for right-handers.

It is sometimes argued that human right-handedness is due entirely to cultural pressure and that the natural condition is ambidexterity. Plato, for instance, attributed right-handedness to "the folly of nurses and mothers," and in the eighteenth century Jean Jacques Rousseau wrote in *Emile* that "The only habit the child should be allowed to acquire is to contract none. He should not be carried on one arm more than the other or allowed to make use of one hand more than the other."[26] The English novelist and propagandist Charles Reade was even more forthright:

> Six thousand years of lop-armed, lop-legged savages, some barbarous, some civilized, have not created a single lop-legged, lop-armed child, and never will. Every child is even and either handed till somegrown fool interferes and mutilates it.[27]

Around the turn of the century, especially in England, there was a movement to encourage ambidexterity. John Jackson founded a group known as the Ambidextral Culture Society, which included Lord Baden-Powell, founder of the Boy Scout Movement, among its supporters. It was partly to foster ambidexterity that the Boy Scouts shake hands with the left hand. Robert Hertz, the French anthropologist, was another who believed that the two hands possessed equal potential for skilled action and urged educators to develop this potential. "One of the signs which distinguish the well-brought-up child," he observed sardonically, "is that its left hand has become incapable of independent action."[28]

Yet ambidexterity has never come about, save in a few individuals who were probably predisposed that way anyway. Instead, right-handedness has stubbornly remained part of the human condition, with an equally stubborn minority of left-handers. The very universality of right-handedness strongly suggests that it is a biological rather than a cultural condition. It seems likely that if cultural pressures were totally eliminated, right-handers would still constitute a substantial majority, with the proportion of left-handers reaching an asymptotic value of about 12–13 percent. A survey of handedness as depicted in works of art going back over 5000 years reveals a virtually constant proportion of right-handedness at a little over 90 percent. This figure probably reflects a small degree of cultural bias, and a breakdown by geographical location reveals values ranging from 96 percent in the Middle East to 88 percent in the Americas.[29]

The Left-Hander

There nevertheless remains the awkward question of why some 12 percent or so of the population are left-handed. History has not always been kind to the sinistral minority, but despite this, many of them have retaliated by leaving their marks on history. According to Michael Barsley, there is evidence that Alexander the Great, Julius Caesar, Cicero, Charlemagne, and King David were left-handers. So was Leonardo da Vinci, one of the most talented artists and inventors of all time. Contemporary examples include three Presidents of the United States (Harry Truman, Gerald Ford, and Ronald Reagan), two of the Beatles (Paul McCartney and Ringo Starr), and many actors, such as Charlie Chaplin, Harpo Marx, and Rock Hudson.[30] Left-handed athletes include Babe Ruth, and numerous Wimbledon tennis champions, including Rod Laver, Jimmy Connors, John McEnroe, and Martina Navratilova. The New Zealand cricketer Richard Hadlee, who has taken more test wickets than anyone else, is a right-armed bowler but a left-handed batsman.[31]

Despite this array of talent, left-handers are often depicted as clumsy or inferior. Given that the artificial world of humans has been constructed largely by and for right-handers, a certain degree of clumsiness might be forgiven. The location of door handles, the arrangement of pages in books and magazines, and the access to the zipper on men's trousers are all designed for the convenience of the right-hander and the frustration of the left-hander. Many common objects, such as scissors, corkscrews, nuts and bolts, and golf clubs, have been designed almost exclusively for right-handers, at least until recently, when special shops began to cater exclusively to the left-handed. In common modes of greeting, like shaking hands or saluting, the right-hand rule must be obeyed—except, of course, for the left-handed hand shake of the Boy Scouts. In short, we have constructed a world that "fits" the right-hander. The right-hander might gain some feel for the plight of the left-hander from the following verse from *The White Knight's Song*, by Lewis Carroll:

> And now, if e'er by chance I put
> My fingers into glue,
> Or madly squeeze a right-hand foot
> Into a left-hand shoe . . .

Among those who have written in less than flattering terms of left-handers is Sir Cyril Burt, the British educational psychologist. Of left-handers, he wrote:

> They squint, they stammer, they shuffle and shamble, they flounder about like seals out of water. Awkward in the house, and clumsy in their games, they are fumblers and bunglers at whatever they do.[32]

Burt also thought that left-handers were wilful or "just cussed" and noted that "Even left-handed girls . . . often possess a strong, self-willed and almost masculine disposition." In this he was echoed by the American psychiatrist Abram Blau, who argued that left-handedness was a matter of "infantile negativism."[33]

The Canadian psychologist Paul Bakan, himself weakly left-handed, has proposed that left-handedness is due to minor brain damage at birth, observing that left-handers are more frequent among those born first or later than third in a family compared with those born second or third.[34] He linked this observation with evidence that first—and late-born children are more prone to birth stress than those born in between. However, subsequent surveys failed to show this association between birth order and handedness, and although there is some evidence that deviations from right-handedness are related to other measures of birth stress, such as cesarian delivery, the associations are very small, accounting for less than 1 percent of the variance.[35]

It has also been claimed that left-handers do not live as long as right-handers. In their book *Lateral Preferences and Human Behavior*, Clare Porac and Stanley Coren reported that 13 percent of 20-year-olds were left-handed compared with only 5 percent of 50 year-olds. There were virtually no left-handers at all among those aged 80 or above. (Octogenerian sinistrals, please step forward!) While this might be due in part to relaxation of the pressure to be right-handed over the course of the century, a follow-up study by Diane F. Halpern and Coren suggests that there may be more to it than this. They examined the records of handedness and age of death among baseball players listed in *The Baseball Encyclopedia* and again found a cumulative disadvantage for left-handers; mortality rates were virtually identical until the age of 33, but for every year beyond that there was about a 2 percent advantage for right-handers. While this may suggest a pathology associated with left-handedness, it may simply mean that left-handers are more prone to accidents in a world designed for right-handers, especially as they grow older.[36]

Another common view is that left-handers are especially prone to minor difficulties of language, like reading disability or stuttering. This view will be examined in Chapter 8, since it has to do with cerebral asymmetry rather than with handedness per se.

In general, statistical surveys typically show left-handers as a whole to be no worse than right-handedness, intellectually, physically, or emotionally. A large-scale study of high school students in California, for instance, revealed essentially no differences in intelligence between left- and right-handers.[37] Another survey of 188 left-handed and 186 right-handed college students did reveal some interesting differences on individual measures.[38] For example a higher proportion of left-handers claimed special mathematical and artistic talent, but the right-handers more often claimed musical and verbal talent. To the extent that there are systematic differences between left- and right-handers, then, these seem as likely to reflect the superiority as the inferiority of the left-hander. Left-handedness may sometimes be due to pathology; that is, in a few individuals, brain injury may well bring about a switch in handedness, and since right-handers are much more numerous to begin with, such switches are more likely to produce left-handers than right-handers. However, I suspect that the majority of left-handers are simply part of the normal population and that left-handedness may arise, in part at least, from genetic variation that has, so to speak, no sinister connotations.

The Inheritance of Handedness

Casual observation suggests that left-handedness does seem to run in families, although very rarely to the extent that a whole family is left-handed. There is an old tradition that those with the family name Kerr, or its anglicized form Carr, are predominantly left-handed. The name is in fact derived from the Gaelic *caerr*, meaning awkward, and the terms *car-handed*, *ker-handed*, and *carry-handed* are still used in some places to refer to the left-handed. An article in *The Times* of London of 4 January 1972 quotes the following anonymous poem:

> But the Kerrs were aye the deadliest foes
> That e'er to Englishman were known,
> For they were all left-handed men,
> And 'fence against them there was none.

One is reminded here of the 700 left-handed men of the tribe of Benjamin, referred to in the Old Testament (Judges 20:16), who could "throw a stone at an hair's breadth and not miss." For all their supposed awkwardness, left-handers appeared to have been feared as much on the battlefields of old as on the tennis courts of today.

The Times carried out a quick telephone survey of Kerrs and Carrs listed in *Who's Who* and reported that "rather more than half" were right-handed. Nevertheless, other evidence does show a systematic tendency for left-handedness to run in families. In one classic survey of 2187 respondents, 92.4 percent of those born to right-handed parents were themselves right-handed, but if one or the other parent was left-handed this proportion dropped to 80.5 percent, and if both parents were left-handed it went down to 45.5 percent.[39] These figures are probably a little low, as they are based on a very conservative definition of right-handedness. More recent studies have shown that the children of two left-handed parents are divided almost exactly into left- and right-handers.[40]

Annett's Theory

That handedness is in part inherited does not mean that the genes are responsible, any more than the inheritance of wealth implies that it comes entwined in a DNA molecule. Nevertheless, Marian Annett has proposed a simple genetic theory that captures a good deal of the evidence in a parsimonious manner.[41] She suggests that genetic variations in handedness could be due to the operation of a single gene, which can exist in either of two forms (or *alleles*). The dominant allele produces a "right shift" in those who carry it, such that the distribution of handedness in this population is heavily skewed toward right-handedness. We may label this the RS+ (or "right-shift positive") allele. Due to environmental influences, some tiny fraction of this population will be left-handed, but not enough to explain the incidence of left-handedness in the population as a whole. Annett suggests that the RS+ allele is expressed more strongly in women than in men, which accounts for the fact that most (but not all) surveys reveal a slightly higher proportion of left-handers among males.

The interesting feature of Annett's theory is that the recessive allele (RS-, or "right-shift negative") produces not left-handedness but the *absence* of any consistent bias toward either left- or right-handedness. This explains why the children of two left-handed parents are themselves equally divided into left- and right-handers. It also explains why left-handers show a very mixed pattern of asymmetry on *other* measures, such as eye dominance, footedness, cerebral asymmetry, and even fingerprints.[42]

This theory even offers a plausible explanation for the relative proportions of left- and right-handers. If we suppose that there is no selective breeding for one or the other of these alleles, it follows that they should have equal frequency in the population. Since our chromosomes come in pairs, one from each parent, then one-quarter of the population can be expected to have two RS+ alleles, one-half to have one allele of each type, and one-quarter to have two RS− alleles.[43] Given that the RS+ allele is dominant, those possessing either one or two RS+ alleles will be predominantly right-handed. This leaves one-quarter of the population with two recessive RS− alleles and therefore no predisposition to be either left- or right-handed. In the absence of any consistent environmental or cultural bias, then, one-half of these individuals, or 12.5 percent, should be left-handed. I have already suggested that the empirical evidence converges on a figure very close to this.

However, this calculation reckons without the few left-handers, including so-called pathological left-handers, carrying the RS+ allele. In Chapter 8, where Annett's theory is extended to cover cerebral asymmetry in the representation of language, it will be argued that the proportion of left-handers carrying the two RS− alleles may be only about 9 percent, with the remaining 3 or 4 percent made up of left-handers carrying the RS+ allele.[44]

Although this suggests that the RS+ allele, with a frequency of about 58 percent, may be slightly favored over the RS− allele, there is little evidence that the RS− allele is doomed to extinction. The data suggest a roughly constant proportion of left-handers going back some 5000 years, as we have seen, implying little or no selective pressure favoring the RS+ allele. Marian Annett has suggested that there may in fact have been some advantages associated with the RS− allele, despite the cultural pressures against left-handedness. So long as they are in a minority, left-handers may often be at an advantage by virtue of their very unexpectedness, especially in warfare or competitive games. Annett argues, though, that left-handers may also exhibit superior skill in tasks where there is no element of surprise or even of competitiveness.

She goes on to suggest that the advantage may lie not with right- or left-handedness per se but with *balanced polymorphism*. Both the RS+ allele and the RS− allele may be associated with disadvantages as well as advantages. She suggests that the RS+ allele may tend to retard the development of the right hemisphere, resulting in deficiencies in mathematical skill[45] and perhaps reducing the speed with which the hands can perform skilled tasks. The RS− allele may tend to produce impairments of language, including reading disability.[46] The ideal, then, is to possess one allele of each type, a condition known as balanced polymorphism. This ideal is achieved precisely when the frequencies of RS+ and RS− alleles in the population are equal, so that the proportion of heterozygotes (i.e., those with one allele of each type) is maximized at 50 percent.[47] If in fact the split is not quite equal, but is roughly 56:44, as suggested above, the proportion of heterozygotes works out at 49.26 percent,[48] which is still very close to the maximum of 50 percent.

It is perhaps difficult to see how mathematical and reading disabilities could be as detrimental to survival as malaria and sickle-cell anemia, so Annett's theory does place some strain on one's credulity. Moreover mathematics and reading are very recent skills in human evolution, and it is difficult to understand how they could have operated over a sufficient number of generations to have been decisive in influencing the distribution of human handedness. Nevertheless, Annett does provide reasonable although still speculative evidence for her theory, and the implication is that it is not mathematics and reading per se that shaped the distribution of handedness. Rather, these abilities may be manifestations of some more basic human dimensions.

As I have already suggested, bilateral symmetry itself was no doubt the product of billions of years of evolution. The very indifference of the environment with respect to left and right has resulted in most animals possessing paired limbs and paired sense organs, symmetrically arranged, and because of this bilateral organization the controlling structures in the brain and nervous system are also largely symmetrical. However, the existence of the RS+ allele, and its association with right-handedness and asymmetry of related brain structures, threatens this symmetrical organization and the advantages associated with it. But it may also confer advantages in the development of manipulative and perhaps communicative skills. The ideal, then, may be balanced polymorphism.

The nature of the complementarity between RS+ and RS− alleles, or more broadly between symmetry and asymmetry, will be explored further in the Chapter 8, for it involves not merely the hands but also the brain. The lopsided ape we may be, but it behooves us to retain at least something of our symmetrical heritage.

There are some possible objections to Annett's right-shift theory. It seems to suggest that left-handedness should be less extreme than right-handedness, since left-handers effectively inherit the *lack* of handedness rather than left-handedness itself. Up to a point, they do; the distribution of laterality quotients is more even across the left-handed range than across the right-handed range, where it tends to skew more toward the extreme.[49] And, as noted, left-handers are more variable with respect to other asymmetries. Still, many left-handers are *very* left-handed in all that they do and do not give the impression of genotypic neutrality. However, even if the initial determination of handedness is essentially random, environmental pressures may act to augment that handedness and to favor consistency among different activities.

Earlier, I noted that handedness may not be a single entity but may reflect several dimensions. However, this does not seem to be a serious threat to Annett's theory. For one thing, the RS+ allele represents a genotypic bias, but experience may have a considerable modifying influence. For another, in those lacking the right shift, handedness is wholly at the mercy of environmental biases and may well differ for different kinds of activities. The fact that there are different categories of manual activities, subject in differing ways to environmental effects, may be sufficient to ensure that different aspects of hand preference will be somewhat independent of one another.

The Problem of Twins

A more awkward problem for Annett's theory, and indeed for *any* genetic theory of handedness, concerns twins. So-called monozygotic (or "identical") twins come from the same fertilized ovum and therefore possess identical genes. If handedness were determined wholly by the genes, then each twin of a monozygotic pair should always have the same handedness. Annett's theory does allow for some environmental variation, especially among those bearing two RS—alleles, so we need not expect *all* twins to be matched for handedness. The problem is, however, that there appears to be virtually *no* association in handedness between homozygotic twins. Knowing one twin's handedness tells us virtually nothing about the handedness of the other.[50]

It is popularly supposed that twins mirror each other, like Tweedledum and Tweedledee in Lewis Carroll's *Through the Looking Glass*. So-called Siamese twins, joined together at birth, are said to be mirror images even with respect to the placement of the internal organs, so that one has the heart on the left in the normal position while the other exhibits *situs inversus*, with the heart on the right. It has been suggested that this mirroring is more likely the later the split of the embryo to produce twins.[51] However there appears to be little documentation of the relation between twinning and situs inversus in humans, and situs inversus is sometimes observed in *both* twins. Moreover, situs inversus does not seem to be associated with left-handedness.[52] I suspect that there is nothing of significance in the notion of "mirror twins"; that twins sometimes display opposite handedness, or even opposite situs of the internal organs, may be merely a matter of chance.

There are other potential complications with twins. Crowding in the womb, or special difficulties associated with birth, may increase the probability that left-handedness is due to pathological influences rather than to genetic factors, and it has even been claimed that data from twins should not be used to test genetic theories about handedness.[53] If position in the womb also plays a role in the determination of handedness, then this too would be altered in the case of twins.

Annett has argued that her genetic theory can be made to fit the distribution of handedness among twins if it is supposed that the RS+ allele is expressed less strongly in twins than in the singly born.[54] One consequence of this is that the incidence of left-handedness itself should be higher among twins than in the population at large, and there is some evidence that this is so, although it is a matter of some dispute.[55]

Given the uncertain status of twins in testing genetic theories of handedness, and even the uncertainties of fact, my inclination is to go along with Annett's theory while recognizing that twins may yet prove to be its Achilles' heel. It is certainly the best theory of the inheritance of handedness that I have been able to discover.[56]

The Evolution of Handedness

Handedness in Other Animals

If handedness is universally human, it also appears to be uniquely human, at least relative to other primates and even other mammals. To find any asymmetry comparable to human right-handedness, we have to go to another talkative creature, the parrot. Most species of parrot seem to be *left*-footed in that they prefer to pick up bits of food with the left foot while perching on the right.[57] This has led to some discussion as to whether the parrot is left- or right-footed, but since the left foot has the more manipulative role, it seems reasonable to allow it the status of dominant foot. A few species of parrots seem to be right-footed rather than left-footed, and it is of interest to observe that the proportion who show reverse footedness relative to their conspecifics is again about 12–13 percent.

Although it has been suggested that cats show a slight preference overall for the left forepaw,[58] there is little evidence for any overall handedness or pawedness among mammals other than humans. Individuals may show quite strong preferences for one or another paw, but over the population there are as many left-pawed as right-pawed individuals. For instance, individual mice display strong paw preferences in reaching for food in a glass tube, and the paw that is preferred for reaching also usually has the stronger grip. However, left and right preferences are equally divided across individuals.[59] Even if one selects only left-pawed mice for mating, the offspring remain equally divided between left and right pawers, even after three generations of selective breeding, suggesting that pawedness is not under genetic control.[60] Similar findings have been reported for rats.[61]

There is little evidence that monkeys or apes show any overall bias either. Laboratory studies of chimpanzees and monkeys, as well as observations of gorillas and orangutans in a zoo, suggest that up to 50 percent show mixed preferences, while the rest are divided about equally between left and right preferences.[62] Jane Goodall has observed chimpanzees using sticks for digging termites out of holes, a primitive example of the use of a tool, but as many use the left hand as use the right hand.[63] In other activities, too, chimpanzees seem as likely to be left-handed as right-handed.[64] There is, however, a report that mountain gorillas tend to favor the right hand in chest-beating displays,[65] but there is little other evidence for consistent handedness in the gorilla.

Peter F. MacNeilage, Michael G. Studdert-Kennedy, and Bjorn Lindblom have reviewed the published evidence on handedness in nonhuman primates and claim that there is a slight but consistent bias in favor of the *left* hand, especially in visually guided movement.[66] This bias is clearly apparent only if the evidence is pooled over studies. MacNeilage and his colleagues suggest that it probably evolved when the hand became prehensile (i.e., able to grasp objects wholly in the hand), and is probably most evident today in animals such as galagos and tarsiers that stand upright and live on insects or small animals that are caught with the hand. Before bipedalism was fully established, then, there may have been a tendency to use the right arm and hand for postural support and the left for catching insects. But as bipedalism evolved, it was no longer so necessary to use the right hand for support, so the right hand developed a specialization for manipulation and a dominant role in bimanual coordination. MacNeilage and his colleagues note one study on monkeys that does seem to show a right-hand bias in bimanual skills.[67]

These claims are very speculative, and the biases are very weak compared with human handedness. Nevertheless MacNeilage and his colleagues may well have identified the sources of human handedness. Natural selection may well have sharpened the asymmetries, and produced the directional consistency. It is also possible that hand preferences in monkeys are more dependent on the task than they are in humans, and may reflect differences between the sides of the brain in mental processing rather than any overriding handedness.

But even if the hand preferences in monkeys and apes summarized by MacNeilage and his colleagues do allow a glimpse of the origins of our own handedness, there is still a marked discontinuity between ourselves and our nearest relatives. Even if MacNeilage and his colleagues are correct, handedness in other primates is very weak, and if anything favors the left rather than the right hand. Our own handedness, by contrast, is clear, and strongly favors the right hand. The critical events that shaped our handedness must therefore have taken place since the time that the split between humans and chimpanzees occurred. But when?

Handedness in Early Hominids

We have seen that right-handedness was prevalent at least as far back as the historical record takes us, through ancient writings and through paintings and drawings going back some 5000 years. Until recently, it was thought that right-handedness emerged only relatively shortly before that time. Several authorities around the turn of the century maintained that Stone Age people may have been predominantly left-handed, if anything. G. de Mortillet, the pioneering French prehistorian, claimed that the majority of Stone Age scrapers recovered from the Somme gravels in France were more comfortably held in the left hand than in the right hand,[68] and D. G. Brinton noted an asymmetry in bifacially flaked stone artifacts from North America that he took as evidence that their manufacturers were predominantly left-handed.[69] However, it was generally agreed that right-handedness was established by the Bronze Age, which dates from about 3000 to 100 B.C. The bronze sickle and other Bronze Age artifacts seem to have been manufactured almost exclusively for right-handers.[70]

More recent evidence suggests, contrary to de Mortillet and Brinton, that right-handedness goes back much further than the Bronze Age, possibly even to the australopithecines. It has been claimed, for example, that the Neandertals were predominantly right-handed. This has been inferred from angled striations on their front teeth. These striations do not resemble those on the teeth of modern humans; it has been argued that they were caused by a Neandertal habit of holding meat (or other matter) in the front teeth while cutting it with stone tools. The act of cutting exerted an asymmetrical pressure on the teeth, producing the angled striations. In examples from seven Neandertal skulls, five showed striations consistent with right-handedness, one showed no directional bias, and the remaining one appeared to have been left-handed.[71]

Further evidence of right-handedness in the Stone Age comes from a technique developed by the Russian prehistorian S. A. Semenov, known as *microwear analysis,* in which the working edges and surfaces of tools are examined microscopically for polishing, striations, and other signs of wear. Semenov describes scrapers that were evidently used for scraping hide, dating from the Upper Paleolithic, which covers the period from about 35,000 to about 8000 B.C. Some 80 percent of these tools, from wide-ranging sites in Eastern and Western Europe and in North Africa, show more signs of wear on the right than on the left side, indicating that they were used by right-handers.[72]

Microwear analysis also suggests a predominance of right-handedness in the Lower Paleolithic, some 500,000 to 100,100 years ago. Tools recovered from Clacton in England show a pattern of wear consistent with rotary motion accompanied by downward pressure, as in boring. On the majority of such tools, the striations indicate clockwise movement, suggesting that the users were right-handed.[73]

More evidence comes from flakes formed by the manufacture and sharpening of flint tools. In an analysis of asymmetries in flakes recovered from La Cotte de St. Brelade in Jersey, dating as far back as 150,000 to 200,000 years ago, Jean Cornford estimates the incidence of right-handedness to be about 80 to 90 percent, which is remarkably close to present-day estimates.[74] Nicholas Toth of Berkeley, California, examined flakes from Lower Pleistocene sites at Koobi Fora, Kenya, dated at 1.4 to 1.9 million years ago, and found that right-oriented flakes outnumbered left-orientated ones by about 57 to 43. The same ratio is produced if present-day right-handers are given the task of sharpening stone tools, suggesting that right-handedness goes back to our presumed ancestor *H. habilis*, who frequented the Koobi Fora site.[75]

Even *Australopithecus* might have been predominantly right-handed. Raymond Dart examined the fossil remains of several specimens of *A. africanus* and over 50 specimens of *Parapapio broomi*, a baboon, obtained from several African sites. Many of these specimens seemed to have been clubbed to death by the murderous australopithecines, wielding heavy stones or antelope bones. Of nine baboon skulls evidently struck from the front, seven were damaged on the left and two on the right, suggesting that the attackers used the right hand. A further six seem to have been attacked by stealth from the rear and were damaged on the right, again implying that the attackers used the right hand. The australopithecines were apparently not above murdering their own species, since two of their own skulls were crushed on the left side, presumably from a frontal assault by right-handed attackers.[76]

However, this evidence must be considered suspect, since the australopithecines are no longer considered the aggressive, murderous creatures that Dart supposed them to be. C. K. Brain has argued that the damage to the skulls studied by Dart could just have easily been caused by falling rocks in the caves in which the skulls were found. The antelope bones thought by Dart to have been weapons may have been the result of attacks by leopards, who probably also hunted and killed the australopithecines. Brain's view is therefore that the australopithecines were the hunted, not the hunters.[77] However, this account does not readily explain the systematic asymmetries that Dart observed.

It does not seem unreasonable to suppose, then, that right-handedness was characteristic of the earliest hominids, possibly including the australopithecines who became extinct. It may have been a concomitant of bipedalism. There is at least a negative reason for supposing that this might have been so: Once the hands were no longer involved in locomotion, there was no longer the pressure for bilateral symmetry. The freeing of the hands also set the stage for the evolution of a different *kind* of role, and one in which asymmetry rather than symmetry might have proven advantageous.

At the beginning of this chapter, I argued that the bilateral symmetry of the limbs was probably an adaptation enabling organisms to move in linear paths through an environment that exerted no overall constraints favoring one or the other side. Even if the hands and arms were used for other things, like catching insects or plucking fruit from a tree, symmetry might be an advantage—unless, as MacNeilage and his colleagues suggest, one hand was used for support. These examples refer to actions whose spatial dimensions are determined by the environment. To the bipedal hominid, however, the stable upright stance and the freeing of the hands allowed a new class of actions, those that represent operations upon the environment rather than reactions to it. Such actions might be described as manipulative. The most cursory inspection of any human terrain will reveal that we are among the most ruthlessly manipulative of creatures.

Of course we are not entirely alone in this; ants, bees, birds, and beavers all operate on their environments to their own ends. So, in a modest way, does the New Zealand wrybill plover, which has a beak that is curved to the right. This asymmetry has been a source of mystery and even scorn:

> For all we know it may be a silent grief to respectable wrybill to see their little ones grow up with this horrid distortion of the proboscis, to reflect that in the councils of the great plover family their breed has been sent to Coventry—relegated for all time to South Canterbury.[78]

Another deviant sent for punishment to the colonies? In fact, the curved bill is a useful adaptation that makes it easier for the bird to get its beak under stones to turn them over in the search for food. It also allows the bird to use its beak as a sieve in extracting tiny crustaceans from surface water.[79] This again illustrates how readily bilateral symmetry is abandoned, especially in the context of a manipulation, if asymmetry proves more adaptive.

In the case of the hominids, the advantages of asymmetry may have arisen from the fact that two hands were freed from involvement in locomotion, so that specialization was possible. Many manipulations, such as hammering a nail or peeling a banana, involve the hands in complementary roles, with one hand holding some object while the other operates upon it. Again, however, humans are not unique in this. The lobster, for instance, has one heavy claw adapted for crushing, while the other light claw is adapted for cutting.[80] The gribble, a small marine isopod that has the mildly annoying habit of destroying submerged timber by boring into it, has one mandible that is like a rasp, while the other is like a file.[81] Note again that these asymmetries have evolved in the context of actions that are essentially manipulative.

Human handedness differs from the examples of the lobster's claws or the gribble's mandibles in that the hands remain structural mirror images.[82] The primary differences between the human hands are functional, not structural. This implies that the asymmetry resides, not in the hands or arms themselves, but in the brain structures that control them. This point will be pursued in Chapter 8.

In his book *The Throwing Madonna*, William Calvin has argued that right-handedness may have evolved from throwing. More particularly, female hominids may have used their right hands to throw stones at would-be predators while holding their infants in their left arms, close to their hearts. The precise timing necessary for accurate throwing may have selected for neural circuits in the controlling left hemisphere that would later mediate speech, which also involves intricate timing. It might be objected that, in the present day at least, throwing is very largely a male pastime, although this is probably a matter of socialization rather than biology. An unpublished study carried out at the University of Auckland by two undergraduate students, Leigh Stevens and Kirsten van Kessell, shows that, at age 5, boys are only slightly better than girls at throwing a tennis ball, but that the difference increases dramatically over the next 2 years, probably as a result of involvement in ball games. A difference between the hands in throwing skill also increases sharply over this age range, especially in boys.

Throwing is indeed one of the most asymmetrical skills we possess. Although some people can write equally well with each hand, very few people seem to be able to throw equally well with each hand, even though this ability could be quite advantageous in sports like baseball or cricket. It has been suggested, however, that while it might have been more effective in primitive warfare to be able to throw stones rapidly with each hand in turn, there might have been some advantage in using one hand as a magazine.[83] However if throwing had been truly decisive in the evolution of handedness, one would have expected structural differences between the arms, as in the claws of the lobster.

It is perhaps more likely that the evolution of handedness had to do with tool making rather than with throwing. As we saw in chapter 3, chimpanzees use tools in a primitive way, but unlike hominids they do not systematically manufacture tools. The flaking of stones for use as scrapers or instruments of digging implies a deliberate, planned act, and one that might benefit from specialization of the hands. The construction of flaked tools is not subject to the whims of the environment but is under the control of the tool maker, who can effectively choose which hand to hold the flint with and which to hold the "hammer" with.

But the advantages of handedness may have been general rather than specific to tool making, or indeed to any other kind of manual activity. So long as the early hominids were engaged in manual activities that were not constrained by the spatial environment, but were deliberate, planned actions, there was a choice of which hand to use and perhaps some advantage to be gained from having one hand more specialized than the other for operations involving precision or intricate control—or, in a word, from having one hand more "dexterous" than the other. I shall suggest in Chapter 8, however, that the rationale for asymmetry had more to do with the asymmetrical representation of skill in the brain than with the functions of the hands themselves.

Conclusions

At the level of the hands themselves, handedness is functional rather than structural. This suggests that it is general rather than specific, and it is indeed manifest in a wide range of activities. This is in keeping also with the idea that humans have evolved as generalists rather than as specialists. In this respect, then, handedness may indeed offer a window on human evolution. We are general-purpose manipulators; while other species may use primitive tools in specific ways, no other species shows such a range of manipulative activities or such inventiveness in devising new methods of construction—and destruction, for that matter.

Right-handedness thus serves as a marker for humanity; it is both universally and uniquely human. And yet it is not a sine qua non of humanity, for there have been many exceptional individuals who have been left-handed or who have not shown consistent handedness. At best, then, right-handedness represents one aspect of what Annett has called "balanced polymorphism"—not an end in itself, but an ingredient that must be balanced against bilateral symmetry.

Moreover, the functional nature of handedness suggests that we must look deeper than the hands themselves, and into the controlling brain structures, for handedness is fundamentally a manifestation of cerebral rather than bodily asymmetry. However, the most conspicuous and well-documented manifestation of cerebral symmetry has to do not with handedness but with the dominant role of the left side of the brain in speech and language. But before discussing this, we need to consider the very nature of language, for as we have seen, language itself has often been considered a mark of human uniqueness. This is the topic of the next two chapters.

Notes

1. Monod (1969, pp. 16–17).
2. Adcock (1986, p. 53).
3. Bishop (1989).
4. Steenhuis and Bryden (1989).
5. Hertz (1960, p. 92).
6. Needham (1973).
7. These themes are explored further by Corballis (1980).
8. Porac and Coren (1981).
9. Suppose, for example, that 90 percent of people are right-handed and 60 percent are right-eared. By chance alone we expect $.9 \times .6 = .54$ to be both right-handed and right-eared, and $.1 \times .4 = .04$ to be both left-handed and left-eared, making a total of 58 percent who are congruent.
10. Annett (1985).
11. Buchanan (1862), quoted by Wilson (1872, p. 198).
12. Holloway (1956, p. 27).
13. For example, Blau (1946).
14. Hewes (1949).
15. Dennis (1958).
16. Needham (1973).
17. Hertz (1960).
18. Wieschhoff (1938), reprinted in Needham (1973).
19. Barsley (1970).
20. Hung, Tu, Chen, and Chen (1985).
21. Teng, Lee, Yang, and Chen (1977).
22. Hardyck, Petrinovitch, and Goldman (1976).
23. Shimizu and Endo (1983).
24. Brackenridge (1981).
25. Levy (1974).
26. Translated in Boyd (1956, p. 22).
27. In a letter to *Harper's Weekly* of 2 March 1878.
28. Hertz (1909); English translation from Hertz (1960, p. 92).
29. Coren and Porac (1977).

30. Porac and Coren (1981).

31. It would take too long to explain cricket to North American readers. Sorry. Be content with Babe Ruth.

32. From Burt (1937). Burt is better remembered for his infamous attempts to prove that intelligence is largely genetically determined, based on evidence that seems to have been fraudulent (Kamin, 1974).

33. Blau (1946).

34. Bakan (1971); Bakan, Dibb, and Reed (1973).

35. For recent reviews, see Searleman, Porac, and Coren (1989) and Harris and Carlson 1988).

36. Halpern and Coren (1988).

37. Hardyck, Petrinovich, and Goldman (1976).

38. Smith, Meyers, and Kline (1989).

39. Rife (1940).

40. Annett (1985).

41. Op. cit.

42. See Corballis (1983) for further review.

43. This follows from the binomial theorem. If you toss a pair of coins, there is a 25 percent chance that they will both turn up heads, a 25 percent chance that they will both turn up tails, and a 50 percent chance that there will be one head and one tail.

44. It follows that the proportion of *all* individuals carrying two RS− alleles is twice this figure, or about 18 percent. The frequency of the RS− allele in the population is the square root of this, or about 42 percent.

45. In a study of children aged 9–11 years, Annett and Manning (1990) found that the incidence of non-right-handedness was highest in children best at arithmetic and fell progressively through groups of children who were average and below average.

46. In another recent study, however, Annett and Manning (1989) suggest that reading disability is associated with extremes of *both* left- and right-handedness.

47. The concept of balanced polymorphism is not without precedent in genetic theory. For instance, the structure of human hemoglobin is influenced by a gene that exists in two allelic forms. The A allele is normal and dominant, while the recessive S allele produces abnormal sickle-shaped red blood cells. Those individuals who possess at least one A allele have normal red blood cells, while those with two S alleles have so-called sickle-cell anemia, a condition that results in poor chances of survival without modern medical treatment. However, those possessing one allele of each type have a better chance of surviving than those with two A alleles in regions where malaria is endemic, because the presence of one S allele provides a poorer breeding ground for malaria-transmitting parasites. In places like central Africa, where the risk of malaria is high, the population tends toward the state of balanced polymorphism, with the risk of malaria being offset by that of sickle-cell anemia. This example is given by Annett (1985).

48. This is the middle term in the binomial expansion, or $2 \times .44 \times .56$, which equals .4928.

49. Oldfield (1971).
50. Collins (1970).
51. Newman (1940); Hay and Howie (1980).
52. Torgerson (1950).
53. Nagylaki and Levy (1973)
54. Annett (1985).
55. In a review of evidence, McManus (1980) concluded that there was no evidence for a higher incidence of left-handedness among twins than among the singly born. In another review, however, Springer and Searleman (1980) concluded that the incidence of left-handedness in twins might be as much as twice that in singletons. A large-scale study by Ellis, Ellis, and Marshall (1988) found the incidence of left-handedness in twins to be just over 2 percent higher than that in the singly born, but this difference was not statistically significant.
56. For further discussion, see Corballis (1983) and Annett (1985).
57. Friedman and Davis (1938); Rogers (1980).
58. Cole (1955).
59. Collins (1970). In a book recently translated from Russian, Bianki (1988) has claimed evidence for a right paw preference in both domestic mice and cats. This work is described in Russian sources unfamiliar to Western readers and contradicts the evidence known to me.
60. Collins (1970). It is conceivable that the results of selective breeding would have been different had the mice been selected for *right*-pawedness rather than left-pawedness. As noted earlier, the evidence suggests that the offspring of left-handed *human* couples are also equally divided between left- and right-handers, whereas the offspring of right-handed couples are predominantly right-handed.
61. Peterson (1934).
62. See Annett (1985) for a review.
63. Cited in Lancaster (1973).
64. Steklis and Marchant (1987) report no significant differences in hand preference for seven different categories of hand use in 27 chimpanzees. However Bresard and Bresson (1987) report a slight bias in favor of left-handedness in 57 chimpanzees; 27 are described as left-handed, 16 as right-handed, and 24 as showing no consistent preference.
65. Schaller (1963).
66. MacNeilage, Studdert-Kennedy, and Lindblom (1987). This article is followed by commentaries from experts in the field, many of whom are skeptical of the statistical basis of the claims made.
67. Beck and Barton (1972).
68. de Mortillet (1890).
69. Brinton (1896).
70. Blau (1946).
71. Bermudez de Castro, Bromage, and Jalvo (1988).
72. Semenov (1964).
73. Keeley (1977).
74. Cornford (1986).

75. Toth (1985)
76. Dart (1949).
77. Brain (1981).
78. Guthrie-Smith (1936).
79. Hay (1979).
80. Govind (1989).
81. Examples from Neville (1976)
82. Actually, there are slight but systematic differences in the sizes of the arms and hands, summarized by von Bonin (1962), but these seem far too trivial to account for the differences in function.
83. Woo and Pearson (1927)

— 5 —

Human Language

The English have no respect for their language, and will not teach their children to speak it.

GEORGE BERNARD SHAW[1]

AND YET THEY LEARN. One of the themes of this chapter is that children acquire language with only a minimum of assistance from their parents, teachers, or other moving objects in their environments. The reason for this is that humans are innately and possibly uniquely equipped with the disposition to speak. Robert Claiborne, in his entertaining book *Our Marvellous Native Tongue* (1983), writes:

> Language, our uniquely flexible and intricate system of communication, makes possible our equally flexible and intricate ways of coping with the world around us; in a very real sense, it is what makes us human.[2]

Aldous Huxley, in *Adonis and the Elephant*, is more ambivalent:

> Thanks to words, we have been able to rise above the brutes; and thanks to words, we have often sunk to the level of demons.[3]

The idea that only humans are capable of true language can be traced to the seventeenth-century French philosopher René Descartes, who argued that the flexibility of human language placed it beyond the capacity of any mere mechanical contrivance and so set humans apart from other animals, even the apes (see Chapter 1). Over the past years, the view that language is uniquely human has been restated by the American linguist Noam Chomsky, a self-professed Cartesian[4] who continues to have a profound influence on modern cognitive science. For all his homage to Descartes, however, Chomsky did not follow the master into dualism, as the following passage makes clear:

> It is a curiosity of our intellectual history that cognitive structures developed by the mind are generally regarded and studied very differently from physical structures developed by the body. There is no reason why a neutral scientist, unencumbered by traditional doctrine, should adopt this view. Rather he [*sic*] would, or should, approach cognitive structures such as human language more or less as he would investigate an organ such as the eye or heart. . . .[5]

As we shall see, Chomsky's views are by no means universally accepted, and there is still considerable debate both as to the nature of human language and as to whether its properties are uniquely human. For example, there are those who still maintain a behaviorist position, holding that verbal behavior is simply a complex form of behavior, differing from other forms of human and animal behavior in degree but not in kind. A similar view is implied, at least, by a new approach to cognitive science known as *connectionism,* in which different aspects of thought, including language, are explained in terms of associationistic networks.

This chapter surveys contemporary views on the nature of human language and the way in which it is acquired. This will establish a platform upon which to return, in the next chapter, to the question of whether language can be found in other species.

What Is Language?

First and foremost, of course, language is a way of communicating, enabling individuals to transmit information to each other. If this were all that language involved, the question of whether it is uniquely human could be trivially answered in the negative. Even casual observation reveals that many other animals, and even insects, communicate with each other and sometimes with humans. The question is therefore whether human language has additional properties that go beyond mere communication, and whether some or all of these properties are what distinguishes human language from other forms of communication.

The Use of Symbols

One of the properties of human language is that it uses *symbols* to refer to objects, places, actions, and so on. The essence of a symbol is that it stands for or represents something else. In human language, moreover, the nature of symbols is essentially arbitrary, which is to say that the symbols themselves need bear no similarity to the objects or actions they represent. The word "water," for instance, as either spoken or written, in no way physically resembles the substance water, and indeed, the words for water in different languages bear little or no relation to each other (except insofar as they may derive from an earlier, common language).

A further property of symbols in human language is that a given symbol can serve a variety of different purposes. For example, the name of an object can be used to request the object, to name the displayed object, to respond to a request for the object, to make judgments about the object when given only its name, and so on. Similarly, the name of an action can be used to demonstrate the action given its name, to announce some future action, to describe or make judgments about a named action, and so forth. This multipurpose use of symbols guarantees that the symbol actually *represents* the object or action in questions. There is evidence that children use words in truly representational fashion, even at the stage when they use words only one at a time.[6]

Since we take it for granted that symbols are representations, we tend to overlook the possibility that animals use sounds or other communicative acts in an altogether simpler way. In particular, animals probably use symbols most often as mere *signals*. For example, a bird may cry in order to warn others of a dangerous predator. In this case, the cry is merely a signal to other birds to take action and presumably does not represent that predator in the various ways described above. There is no evidence in such cases that the cry in any sense "stands for" the predator. As we shall see in the next chapter, nonhuman animals probably rarely use symbols in more than a signaling sense.

The use of symbols as representations also gives human language the property of *displacement*. That is, we can talk about things that are not physically present, or about events that occurred or will occur at some other point in time. We can enjoy the exploits of others by simply hearing or reading about them. Language therefore provides us with a vicariousness that is indeed a characteristic feature of human existence; as illustrations of this, one has only to observe people on a commuter train, immersed in their novels or newspapers, or (except in British trains) in conversation.

The Generative Aspect of Language

Perhaps the crowning feature of human language, and the one that has received most attention, is its *generativity*. There is effectively no limit to the number of different sentences we can produce, or indeed understand. Even at the level of words, human language is remarkably open-ended. There are said to be several million words in the English language, with the number increasing at the rate of some 200,000 to 300,000 per year. Of course most of us come nowhere near mastering all of these words; the average literate adult is said to have a reading vocabulary of about 100,000 words and a speaking vocabulary of about one-tenth that number.[7] Even so, a speaking vocabulary of some 10,000 words is several orders of magnitude greater than estimates of the vocabularies attained by other species.

Human language achieves its generativity in part from its hierarchical nature. In spoken language, the smallest units that make a difference to meaning are called *phonemes*. For example, the difference between the words "rat" and "bat," in both meaning and pronunciation, depends on the initial phoneme, *r* or *b*. There is some variability from one language to another in how many phonemes are required. For example, there are some 44 phonemes in American English but only 15 in Hawaiian; this should not be taken to indicate that the one is in any way superior to the other. Alphabetic languages such as English are written in a way that is based, albeit imperfectly, on phonemes. That is, the letters of the alphabet correspond roughly to phonemes, although the mapping is not always one-to-one; the letter *c*, for example, can stand for the *ss* sound as in the word cedar, the *k* sound as in *cat,* or even the *sh* sound as in *social.* Sometimes letter pairs stand for single phonemes, as in the case of *ch, sh, th,* and so on. Phonemes should therefore not be identified with letters, which are known technically as *graphemes*.

Phonemes are combined to form *morphemes*, which are the smallest units of meaning. Morphemes may include single words, such as *dog* or *shed,* but they also include suffixes and prefixes that add units of meaning. For example, the addition of the suffix *-s* is used to signal the plural, while *-ed* signals the past tense; the prefix *-un* may be used to negate the meaning of an adjective, and so on. I leave it to the reader to count the morphemes in "antidisestablishmentarianism." Morphemes are, of course, combined to form words, which are the units that map most directly onto the objects, actions, and concepts of the real world. The written forms of some languages, such as Chinese or Japanese ideograms, are at the level of words rather than phonemes or morphemes; that is, each symbol represents a different word, although it may incorporate some phonological elements as well.

The ultimate in generativity is the combining of words to form *sentences*, for here the number of different possibilities is effectively unlimited.[8] Sentences can be mapped onto *propositions*, which are effectively conditions of the world that can be identified as true or false. For example, the proposition that there is a flea in my ear (there isn't) can be expressed by sentences such as, "I have a flea in my ear," "My ear contains a flea," and so on. The essentially infinite variety of possible sentences is no doubt necessary if we are to represent or communicate the unlimited variety of possible states of the world about us (or, indeed, of the world within).

Not all agree that sentence construction is the pinnacle of human language. David Premack has argued that the ability to *converse* stands even higher as a characteristic of human language than "the mere creation of sentences."[9] Conversation requires a give-and-take in which both participants share many implicit assumptions. For example, a person may arrive at work and say to a colleague, "That was a great game last night." The conversation might then ensue with neither party explicitly identifying the game or stating why it was great. This requires considerable powers of empathy, or the ability to understand what another person is thinking.

Language in the real world cannot be disentangled from the considerable social and mental skills than people possess. If other animals do not converse, it may be because they do not possess skills of this sort, rather than because they are incapable of acquiring the rules of sentence construction. Premack certainly has a point when he places conversation high on the scale of human achievement, but it is a skill that is more than merely linguistic.

Grammar

We have seen that the generativity of human language stems from the hierarchical fashion in which units are combined. At the lowest level, phonemes number in the tens; at this basic level the "vocabulary" of human language is very small. Phonemes may be combined to form morphemes, morphemes to form words, words to form sentences. At each stage there is an increase in the number of possibilities, so that at the level of the sentence there is really no limit; as the German philosopher Wilhelm von Humboldt put it, language "makes infinite use of finite means."[10] It is this boundless property of human language that is said to distinguish it from other forms of animal communication. But although boundless, the combining of elements is still governed by rules; outside of some modern poetry, we cannot simply string words together haphazardly and expect to be understood. Those rules make up what is known as *grammar* or *syntax*.

The rules of grammar are themselves hierarchical, which is to say that there are rules governing the combinations of phonemes to form morphemes and words, and rules governing the combinations of words to form sentences. Although native speakers of a language obey these rules, their knowledge of them is implicit rather than explicit. That is, most people cannot tell you what most of the rules are. For example, English speakers obey quite simple phonetic rules when they pluralize nouns, sometimes adding a soft *ss* sound, as in "cats," sometimes adding a hard *z* sound, as in "dogs," and sometimes adding a double phoneme *iz*, as in "finches." When I have asked a large class in Introductory Psychology the rules that they use in order to choose among these alternatives, no one has been able to give me the answer, although by the following lecture many of them will have worked it out. I have no doubt that the reader will be able to work it out too.

The most complex rules are those underlying the combinations of words to form sentences, and again discovery of those rules is handicapped by the fact that they are implicit, or unavailable to awareness. Indeed, linguists have not yet been able to agree upon the rules that determine whether a given sentence is grammatically well formed or not, even at the level displayed by a young child. It should be noted that grammar, in the sense used here, need not correspond to the sort of formal grammar we may have been taught in school. Even slang, or street language, has its own rules, and these are known implicitly to every speaker of that language.

It is also important to note that grammar is to be distinguished from *meaning*. A sentence may be grammatically correct yet have no sensible meaning, as in Chomsky's famous example "Colorless green ideas sleep furiously." Conversely, we can often make sense of utterances that we recognize as grammatically incorrect, as in the equally famous example "This sentence no verb." A good deal of effort has gone into the enterprise of determining the rules of grammar independently of meaning, although some have argued that the two cannot be completely separated.

Language achieves some of its generativity simply by virtue of the fact that words can be sorted into different grammatical categories, such as nouns, verbs, adjectives, adverbs, and so on. Consequently, even if we were to allow only one kind of sentence, such as "Article noun verb article noun," we could still generate a great many acceptable sentences and even carry out a reasonable conversation. For example, we might say "The dog bit the cat" or "A man parked the car." As we shall see in the following chapter, dolphins and chimpanzees seem able to achieve something close to this level of grammatical competence. Clearly, however, humans can deal with greater orders of complexity, as this very sentence, I hope, illustrates.

Finite-State Grammar

A higher level of complexity is achieved by so-called *finite-state* grammar. Until quite recently it was thought that human language conformed to finite-state grammar but it is now clear that it is much more complex. However, we shall see in the next chapter that some forms of animal communication may depend on finite-state grammar. It is therefore worth considering here what this implies. One version of finite-state grammar is called left-to-right grammar in which each word type governs the selection of the word to its right. That is to say, each word acts as the *context* for the selection of the next word. For example, an article might be followed by an adjective or by a noun, but not by a verb. An adjective might be followed by another adjective or by a noun. Each *transition* between one word type and the next has a certain probability associated with it.

Given the transitional probabilities between each word type and the next, one might then go about generating a string of words. One might start by choosing a word at random and then consult the transitional probabilities for that word type in order to select an acceptable next word. Given that word, one would then consult its transitional probabilities to choose an acceptable *next* word, and so on. The result is what is known as a *first-order Markov chain.* Such a chain does not in general produce acceptable sentences, however, as the following example illustrates:

> Fortunately honesty shook the boy obtained a big fault of them all things go away for Nirvana ahead. . . .[11]

Each pair of words is acceptable, but the sequence as a whole is not.

At one time, it was thought that acceptable sentences might be generated by simply extending the context. Instead of a single word acting as the context for the next, each group of four words, say, might act as the context for the fifth. The result is a *fourth-order Markov chain,* in which every group of five words is acceptable. The result still does not produce acceptable sentences, however. Here is an example:

> All of the nomads slept on sand colored by the sun haltingly disappeared. . . .

Finite-state grammar is compatible with the view that language consists merely of learned associations. The transitional probabilities can be regarded simply as associations of varying strength that might be learned through experience. As the context is extended, however, the problem of learning the associations grows more formidable. George A. Miller has pointed out that, given one word, there are on average about four acceptable options for the grammatical category of the next word.[12] For a first-order Markov chain, a child would need to learn some 4^2 associations (4 contexts×4 next words). For a second-order Markov chain, this increases to 4^3. As the context grows, the number of learned associations increases by a multiple of 4. Now consider the following sentence:

> The accountant who met the elderly man in the supermarket and helped him decipher the jam labels was later arrested for shoplifting.

The words "accountant" and "was" are clearly linked, yet they are *16* words apart. One would need a 17th-order Markov chain to capture this association, and this would require the learning of 4^{17}, or around 16 billion, associations. This astronomical figure is surely beyond the most ardent learner of language.

In his book *Syntactic Structures*, Noam Chomsky noted that there is in principle no limit to the number of words that might intervene between two linked words. In the above example, we might have extended the clause intervening between "accountant" and "was" even further and still retained an intelligible sentence. In practice, there is a limit imposed by one's ability simply to remember what has gone before. Even so, this open-ended quality of clause insertion suggested that finite-state grammars could never adequately account for English sentences, or indeed sentences in any other natural human language. By the same token, Chomsky also inferred that an associationist approach to language was misguided.

Phrase-Structure Grammar

The last example considered above illustrates an important property of human language known as *recursion*. Structures may be embedded in other structures to make sentences that are as complex as we choose to make them, at least within the limits of short-term memory. It is essentially this property that gives human language its generativity and open-endedness.

To see how recursion works, we need to move to another level of grammar known as *phrase-structure* grammar. Phrases are intermediate between words and sentences. For example, we can identify a *noun phrase* in terms of several options, such as "article noun" or "article adjective noun," or "propernoun." Specific examples include "the man," "a large bus," and "Joe," respectively. Similarly a *verb phrase* might be identified as "verb nounphrase,' as in "boarded the large bus." A *relative clause* might be defined as "relativepronoun verb nounphrase," as in "who ate the cheese." These definitions are not, of course, exhaustive, but they serve to illustrate acceptable rules.

We can now elaborate so-called *rewrite rules* that specify the generation of acceptable sentences. In the following examples, the expression on the right of each arrow represents an acceptable interpretation of the entity on the left:

(1) Sentence → nounphrase [relativeclause] verbphrase
(2) Nounphrase → article [adjective] noun [relativeclause]
(3) Relativeclause → relativepronoun verb nounphrase
(4) Verbphrase → verb nounphrase

where the entities in brackets are optional. We can now generate a sentence such as

> The man who sold the green bicycle that had the puncture bought a car.

by applying rules (1), (3), (2), (3), (2), (4), and (2), in that order.

Recursion is a property of the rewrite rules. For example, the rule for the nounphrase (Rule 2) involves a relativeclause, while the rule for the relativeclause (Rule 3) in turn involves a nounphrase. In principle, then, we can keep on invoking rules within the same sentence for as long as we like, or at least for as long as our processing capacity permits. A familiar example of a multiply embedded sentence is the children's rhyme "This is the House that Jack Built." Recursion should not be confused with mere *iteration,* in which elements are simply repeated.[13] For example even a finite-state grammar will permit the iteration of adjectives, as in "evil bad nasty person." Recursion, by contrast, depends upon a hierarchy of operations.

Another property that distinguishes phrase-structure grammar from finite-state grammar is that it requires a short-term memory. In expanding the rule for a sentence (Rule 1), for example, the processor must proceed to the rule for a nounphrase (Rule 2) and satisfy it before returning to the first rule. Expansion of the second rule may also require the processor to proceed in midstream to the rule for a relativeclause (Rule 3). It is therefore necessary to retain a record of location in the expansion of rules while other rules are satisfied. We could neither generate nor understand sentences that contain embedded structures without short-term memory.[14]

Beyond Phrase-Structure Grammar

Despite the increased sophistication of phrase-structure grammar over finite-state grammar, it still does not capture all aspects of our intuitions about grammaticality. For example, it does not describe the relations between active and passive forms of the same sentence, nor how questions or negatives are formed from affirmative sentences. *Transformational grammar,* developed by Chomsky, operates at the level of the phrase structures themselves, converting one structure to another. For example, in transforming an active sentence like "Mary ate the cake" to the passive "The cake was eaten by Mary," the nounphrases *Mary* and *the cake* are inverted, and the form of the verb is changed from *ate* to *was eaten by.* Similar tranformational rules can be implemented to create other variants, such as "Who ate the cake?" or "Did Mary eat the cake?"

We need not dwell on the details of transformational grammar, since they are not directly relevant to my concerns in this book. Besides, there is still a good deal of disagreement as to the true nature of grammar, and even Chomsky seems to have altered his ground. Some critics, led by Chomsky's former students George Lakoff and J. R Ross, have argued that grammar cannot be distinguished from meaning quite so sharply as Chomsky implied.[15]

Despite the continued wrangling over the nature of grammar, there is one point upon which Chomsky has continued to insist. This is that there is a *universal grammar* that underlies all human language, and that languages differ only with respect to surface characteristics. Certainly, different languages do have features in common. All are based on phonemes, and all are hierarchically organized and have provision for the embedding of phrases. Moreover, so far as is known, people can learn any language, regardless of race or culture, especially if they are exposed to that language at an early age. Language itself is universally human, even though the things that people talk about may differ. This implies that universal aspect is to be found at the level of grammar rather than of meaning. Even so, there is still no satisfactory theory as to the nature of universal grammar.

Modern theories of grammar, while still incomplete, are so complex that they take years to master. This is ironic, since the *use* of grammar is in fact easily achieved by most of us in early childhood. However this raises the interesting and difficult question of whether people can be said to actually "know" the grammar they so readily demonstrate in their speech and understanding, any more than a skilled ball player "knows" the complex mathematics that enables her to compute the trajectory of a ball and catch it. Although Chomsky has insisted that theories of grammar are theories of knowledge, and that linguistics is a branch of cognitive science,[16] linguistic rules may in fact be descriptive of language itself rather than of speakers and listeners.

The Idiosyncratic Nature of Grammar

One reason that the rules of grammar have been peculiarly difficult to discover is that they appear to be idiosyncratic. Natural languages differ rather strikingly from the artificial languages that have been developed in mathematics or in the programming of computers. By the same token, these languages are difficult to read, despite their logical structure. This suggests that natural languages did not evolve logically, but may have been the result instead of some arbitrary genetic rearrangement—perhaps the result of some "hitchhiking" gene or an exaptation from some other adaptation (see Chapter 1).

To give an example of the supposedly irrational aspect of natural grammar, most people have difficulty decoding sentences like

> The dog the mouse the cat chases sees bites

but much more readily understand the longer, clumsier version

> The mouse that the cat chases sees the dog that
> bites the (very) cat that chases the mouse.

Yet the first version is more compact and is compatible with the sort of hierarchical embedding that is used in programming languages such as LISP. Massimo Piattelli-Palmarini, from whom I borrowed these examples, suggests that such restrictions on our ability to decode certain constructions are due to "some quirk in our perceptual and computational makeup."[17]

In a sense, it seems gratuitous to blame the idiosyncrasies of grammar on some random but dramatic shift in our genetic makeup. It is rather like blaming certain epidemics on infections from outer space. Perhaps the restrictions on our ability to use grammar will eventually be seen to depend on quite natural cognitive limitations, such as limitations of short-term memory. But for the present, we can but recall Stephen Jay Gould's remark that we are "the product of enormous improbabilities."

Acquiring Language

Much of the debate about the nature of language has centered on the manner in which we acquire it. The behaviorist point of view, as represented by B. F. Skinner in his 1957 book *Verbal Behavior,* is simply that language is learned in essentially the same way that any complex behavior is learned. Children are reinforced for acceptable sequences of verbal output, which are then cemented into the repertoire. This fundamentally associationistic approach to the acquisition of language has received something of a boost in recent years from a new development in psychological theory known as *connectionism,* which will be discussed in the next section.

One objection to this approach is that the rules of language are, to use a phrase much beloved by Chomsky and others of a rationalist persuasion, "vastly under-determined" by the available evidence.[18] In order to illustrate this style of argument, it is useful to consider a specific example, that of constructing a question from the declarative form of a sentence.[19] For a sentence such as "The man is tall," the most obvious rule to follow would be to scan the sentence from the beginning until the word "is" is discovered and move it to the front. This yields the correct form "Is the man tall?" According to Chomsky, this is the rule that an intelligent scientist would adopt if simply exposed to examples of spoken sentences.

However, the rule does not work in the case of a sentence like "The man who is here is tall," for the rule would in this case produce the incorrect form "Is the man who here is tall?" Since children never make mistakes like this, Chomsky argues that they are equipped with prior knowledge that enables them to infer the correct rule. In order to work, the rule must specify that one must skip relative clauses in searching for the word "is" to be moved to the front; in the example, this yields the correct form of the question: "Is the man who is here tall?" To infer such a rule requires prior knowledge of phrase structure.

This is but one of many similar examples discussed by Chomsky.[20] The message is always the same: The rules that children acquire are never the simplest, most obvious ones that an intelligent but "neutral" person would naturally infer. Only someone with a priori knowledge of what language is like could possibly extract the rules from the tangle of words available. Chomsky argued that every child is innately equipped with a *language acquisition device (LAD)* that guides the learning of linguistic rules. The LAD incorporates universal grammar and is uniquely as well as universally human.

Not everyone accepts this style of argument. For example, Chomsky's colleague and friend, the philosopher Hilary Putnam, argues that Chomsky has "pulled a fast one."[21] No child would produce a sentence like "Is the man who here is tall?" simply because it makes no sense. Only "an insanely scientistic linguist" would adopt a rule that would produce such a sentence. In Putnam's view, children are less concerned with grammatical rules than with making sense. Seymour Papert, another colleague from the Massachusetts Institute of Technology, has also argued that Chomsky has underestimated the power of learning devices to learn complex rules.[22] He gives the example of the perceptron, a device constructed to learn about aspects of the visual world, which has proved capable of learning visual discriminations of surprising subtlety.

Connectionism

Papert's remarks anticipated a new movement known as *connectionism* (or *parallel distributed processing*),[23] which in fact grew out of earlier work on the perceptron. The connectionist approach is simply to take a network of interconnected units, with input and output lines, and rules that enable connections within the network to be altered as a result of experience. One then tries to "teach" the network, by altering the strengths of the connections within the network, so as to reduce systematically the discrepancy between actual outputs and "desired" outputs. Such a network might be regarded as a "pure" learning device, the modern equivalent of a *tabula rasa,* without any a priori structure beyond the units themselves.

An intimation that connectionism might prove useful in understanding the acquisition of language came from a study in which a network was taught to convert English verbs from the present to the past tense.[24] The network learned both regular (e.g., *walk→walked*) and irregular (e.g., *run→ran*) forms. It also learned the different morphemes *-t*, *-d*, and *-id*, as in the regular forms *walked, lagged,* and *batted.* It learned these forms in the same sequence that children do and made the same kinds of mistakes. For example, it went through an early stage of over-regularization, producing such incorrect forms as *bringed* and *hitted.* It also made errors like *ated* and *wented* at much the same stage that children make these errors.

This performance seemed to surprise even the connectionists, and was taken as evidence that associationistic theories were capable of greater sophistication than a priori arguments, such as Chomsky's, might suggest. The connectionists implied that all of language might be learned in this fashion. They pointed out, moreover, that the brain itself essentially *is* a network, with properties mimicked by the artificial networks they study. They suggest finally that the "rules" that Chomsky and others have sought are merely descriptive of *language*, and not of language *users*: "The child need not figure out what the rules are, nor even that there are rules."[25]

Needless to say, the connectionist approach has been attacked in turn. According to Steven Pinker and Alan Prince, the connectionist approach to the formation of the past tense of verbs, when examined closely, falls far short of mimicking a child's actual knowledge of the past tense.[26] A more general critique has been offered by Jerry A. Fodor and Zenon W. Pylyshyn, who argue in true Chomskian fashion that connectionism could never account for the generativity of human language, or indeed of human thought.[27] Indeed, to my knowledge, there is no evidence to date that connectionist networks can master anything as complex as phrase-structure grammar. Even so, the connectionist approach does raise the possibility that "dumb" networks may be taught sophisticated skills, including language skills, that obey complex rules.

Biological Constraints

The foregoing debate has been largely a matter of rhetoric. A more empirical tone was introduced in 1967 with the publication of Erich H. Lenneberg's book *Biological Foundations of Language*. Lenneberg was an admirer of Chomsky's theories, but used biological arguments to support the claim that the acquisition of language was governed by the maturation of the nervous system rather than by learning. This tied in with the Chomskian view that language was essentially innate.

Lenneberg noted that the various stages in learning to speak are closely tied to physical development rather than to the specific training a child receives. Table 5.1 shows the milestones in language development that correspond to milestones in physical development. The relationship also holds in retarded children, in whom the milestones occur later and in whom some may not be reached at all. Lenneberg noted further that the onset of language can be predicted fairly accurately from physical measures of brain development, such as gross weight, or the density of neurons, or the changing proportions of given brain substances. "On almost all counts," he wrote, "language begins when such maturational indices have attained at least 65 percent of their mature value."[28]

A "Critical Period" in Language?

As further evidence that language depends on the physical development of the brain, Lenneberg argued that the acquisition of language is governed by a critical period in development. The notion of a *critical period* was introduced in Chapter 3; it specifies an interval of time during development in which a skill must be learned if it is

Table 5.1
Relation Between Motor and Language Milestones (After Lenneberg)

Age in years	*Motor milestones*	*Language milestones*
0.5	Sits using hands for support; unilateral reaching	Cooing changes to babbling with introduction of consonants
1	Stands; walks when held by one hand	Syllabic reduplication; understands some words; applies some sounds to persons or objects in word-like fashion
1.5	Prehension and release fully developed; propulsive gait; creeps downstairs backward	Repertoire of 3 to 50 words not joined in phrases; trains of sounds and intonations resembling discourse; good progress in understanding
2	Runs (with falls); walks stairs with one foot forward	More than 50 words; two-word phrases; more interest in verbal communication; no more babbling
2.5	Jumps with both feet; stands on one foot for 1 sec; builds tower of 6 cubes	New words every day; utterances of three or more words; understands almost everything but still shows many grammatical deviations
3	Tiptoes; walks stairs with alternating feet; jumps 0.9 m	Vocabulary of some 1000 words; 80 percent intelligibility; grammar close to that of colloquial adult; fewer grammatical errors
4	Jumps over rope; hops on one foot; walks on a line	Language well established

to be learned at all. The critical period for language, according to Lenneberg, lasts from about age 2 until puberty.

Part of the evidence for this rests on the recovery of language skills following brain injury. Lenneberg summarized data from 17 cases of *aphasia*, or language disorder, showing that recovery depended very much on the age of the patient at the time the brain was injured.[29] In general, the younger the victim the better the recovery. Children who lost language between the ages of 20 and 36 months effectively started again, passing through the language milestones at an accelerated rate. In older children, language was almost invariably recovered if the disease occurred before the age of 9, provided that the injury was restricted to one side of the brain. At puberty, a turning point was reached, and by the mid-teens the prognosis for recovery was the same as that for an adult. Loss of language was not necessarily permanent in the case of injury in adulthood. Rather, cases were of two sorts: The symptoms either cleared rapidly (within 3 months), suggesting that the damage was not to the language-mediating areas of the brain themselves, or they were permanent, suggesting that when the damage was to the language areas there was no way in which language could be relearned.

It is also a common experience that second languages are more difficult to acquire in adolescence or adulthood than in childhood. A recent study of English language skills achieved by Korean and Chinese immigrants to the United States showed that proficiency increased linearly with the age of arrival up to the age of puberty, but thereafter there was no relation to age; for example, a person who had immigrated at 17 years performed just as poorly as one who had immigrated at 37.[30] This study focused primarily on grammatical skills. It is also commonly observed that if a second language is learned after the critical period, it is very difficult, if not impossible, to get rid of an accent that sounds foreign to native speakers of the language. This implies that phonological rules are also strongly dependent on the critical period.

Of course, people *can* learn additional languages in adulthood, but not without considerable effort. Such languages are essentially grafted onto a first language, rather than acquired in the easy, natural way of a child. It is ironic that the teaching of foreign languages in many countries, including my own, is delayed until precisely the age at which it is too late to be effective.

Cases of "wild children," raised in the wild without human contact, are also cited as evidence for the critical period. One of the most celebrated cases was that of Victor, the Wild Boy of Aveyron, discovered toward the end of the eighteenth century. Victor had been living alone in the woods but was captured at the age of about 12. A Professor Jean-Marc-Gaspard Itard tried for 5 years to teach him language, but although the boy could understand a number of words and phrases, he learned only to say "Oh Dieu!" and "lait," and very imperfectly at that.[31]

However, cases such as this, and the various reports of children supposedly raised by wolves, are typically anecdotal and lacking in the detail necessary to assess fully the notion of a critical period. Roger W. Brown, for example, suggests that Victor's problems may have been compounded by difficulty in hearing, and there is always the suspicion that children found abandoned may have been retarded to begin with. Still, there is one recently documented case that does provide reasonably unambiguous information.

This is the case of "Genie" (not her real name), a girl who had been locked in a room and deprived of normal human contact from the age of 20 months to the age of 13 years.[32] This was done at the insistence of her deranged father, who treated her literally as an animal, beating her and barking like a dog at her. By the time she was rescued from her ordeal she had passed the age of puberty, but was unsocialized and without language. Subsequent attempts to teach her to speak did achieve some success, but again, her comprehension was better than her production of speech, and she failed to master even simple grammar. Her mother reported that she had begun to speak single words before her incarceration, suggesting that she was probably not retarded. This early language learning may explain why she was able to attain some measure of competence. Even so, her poor achievement provides general support for the idea of a critical period, at least with respect to those aspects of grammatical competence that Chomsky and others have claimed as uniquely and universally human.

Creolization

Further evidence that children bring an innately endowed universal grammar to the acquisition of language comes from what has been called *creolization.*[33] A creole is a language that has evolved from pidgin, which is a pared-down language used for purposes of trade between people who do not speak the same language. The earliest known pidgin was the original "Lingua Franca," which was derived from southern French and the Ligurian dialect of Italian. It was used in the Middle Ages by western Europeans in the eastern Mediterranean.[34] Pidgins proliferated as a result of European colonialism in the period from 1500 to 1900, which gave rise to pidgin forms of English, Spanish, French, and Portuguese. However, there are also pidgins based on non-European languages, and the word *pidgin* itself is said to be derived from a Chinese approximation to the word "business."[35]

Pidgins are usually based on no more than five vowels and a restricted range of consonants. The basic vocabulary is less than 1000 words, although new meanings can be created by combining words. Pidgins lack inflections, that is, there is no distinction between present- and past-tense forms of verbs, for example, or between singular and plural forms of nouns. Grammar is primitive and based wholly on word order. Indeed, pidgins are like the artificial languages taught to apes and dolphins, which will be discussed in the next chapter.

In the process of creolization, inflections and syntax are added, so that the language recovers the properties of a natural language. The only place where this process can still be studied is Hawaii, which was colonized in the period from 1876 to 1920, so there are still speakers of both Hawaiian pidgin and Hawaiian creole. Derek

Bickerton of the University of Hawaii showed that the conversion took place in a single generation, and argued that it was largely invented by children and derived from their innately given universal grammar.[36] If this is so, one would expect creolization to occur only if children are exposed to pidgin during the critical period. Although there appear to be no data on this point, a critical period does operate in a related phenomenon to do with the learning of sign language.

Some deaf children are exposed to what are called *frozen* signs. Like the words in pidgin, frozen signs lack inflections, which are movements equivalent to the sounds that convert spoken words to different forms (e.g., adding *-s* to form the plural). Again, children bring their own presumably innate grammatical knowledge to bear, adding inflectional and syntactic elements. However they do this only if they are exposed to frozen signs up to the age of 6. Some deaf children are exposed later, often because their parents are not deaf and did not immediately introduce sign language. These children retain frozen signs into adulthood.[37] This critical age of 6 years is, of course, considerably earlier than the upper bound (puberty) proposed by Lenneberg and suggests that different aspects of language may be dependent on different critical periods.

The Nature and Role of Learning

Of course, neither Chomsky nor Lenneberg denies that experience determines the *particular* languages that people speak. To that extent, languages depend on learning. However, the kind of learning that determines which language (or languages) we acquire in childhood is said to be quite different from that normally studied in the psychological laboratory and does not depend on simple associations or stimulus-response bonds. Rather, it is a matter of *parameter fixing,* such that certain parameters (or switches) are set within the bounds established by the innately determined universal grammar.

Massimo Piattelli-Palmarini of the Massachusetts Institute of Technology has distinguished usefully between *instructive* and *selective* theories of learning.[38] To illustrate the difference, he notes that there are two ways in which a person might acquire a new suit.[39] One way is to have the suit made to measure by a tailor, which is an instructive process. The other is to select a suit that fits from the available repertoire of sizes and cuts. That is a selective process. With respect to the acquiring of skills, the connectionist approach described above is instructive rather than selective, since it consists of molding the network to give the desired outputs.

Piattelli-Palmarini puts forward the radical view that *all* learning might be selective rather than instructive. He notes that the general trend in biology has been away from instructive theories of adaptation and toward selective ones. This is illustrated historically in the shift from the Lamarckian to the Darwinian theory of evolution, and in more recent times by changing views in immunology as to how the body creates antibodies to fight infection. Since there are huge numbers of actual and potential antigens, it was once considered axiomatic that the antibodies must be developed to "fit" each individual case; that is, antibodies are formed by instruction, not by selection. In recent years, however, evidence has accumulated that the process is selective rather than instructive. That is, the body produces a repertoire of antibodies so huge that any specific antigen, real *or possible,* will sooner or later encounter one that recognizes it and binds to it.[40]

Such findings suggest that nature is profligate rather than parsimonious. The immune system is, to coin a phrase, *vastly overequipped* to deal with infection. As in evolution, so in immunology: Adaptation is a matter of selection, not of instruction. Piattelli-Palmarini suggests, in effect, that the Lamarckian heresy is continued by psychologists, who have failed to see that the selective principle is universal in biology. That is, the brain adapts to the environment by selecting from a preexisting repertoire of skills rather than by building new ones. In the rather anal language of one commentator, "to learn is to eliminate."[41]

Now, few psychologists are ready to believe that every human comes equipped with every word of every language, and that learning simply consists of selecting those relevant to the particular language environment. Selection is presumably not to be found at that level of specificity. Rather, choices available are at a higher level of abstraction and refer to the *rules* of language rather than to specific vocabulary. The fixing of parameters, then, has to do with the choice of which particular form a rule might take.

Again, we lack a description of universal grammar that is sufficiently complete to enable us to specify exactly what those rules are and what the available alternatives are. Nevertheless, Piattelli-Palmarini does point to some examples of parameter fixing that appear to be tightly constrained. Jacques Mehler and his colleagues have shown that by 4 days after birth, a infant can discriminate between the mother's voice and that of another woman of the same age, between a natural flow of speech and a sequence of isolated words, and even between the language of the mother and another language.[42] These remarkable instances of very early acquisition surely suggest the setting of critical parameters rather than associative learning. There is further evidence that, by 7 months, the infant can demarcate clauses as relevant units of natural speech, a development that may set the stage for phrase-structure grammar.[43] In the first year of life, then, the infant is already firmly set (or *parameterized,* for those who enjoy such words) along the path toward his or her particular language. No wonder Professor Higgins had such difficulty in retraining Eliza Doolittle toward a more refined manner of speech.

An even more intricate account of parameter fixing comes from the work of Morris Halle on the phonology of different languages.[44] Each language imposes a different set of restrictions on the possible sound sequence, on the patterns of stress, on intonation, on morphology, and on other aspects as well. Halle has shown that the various decisions that must be made can be reduced to binary choices, each of which is made on the basis of minimal exposure and, indeed, minimal evidence. The learning of a language is like a game of 20 questions, with the genetic program asking the questions and the environment supplying the answers.

There is evidence that infants are actually better at discriminating between phonemes than adults are. For example, in Hindi there are two different *t* sounds that are indistinguishable to native speakers of English. At the age of 6 to 8 months, however, infants from English-speaking families can discriminate these sounds as well as native speakers of Hindi can, but their ability to do so declines over the remaining few months, so that by 1 year of age they perform as poorly as adults.[45] Only with intensive training can adults capture their infantile ability to discriminate the phonemes of a foreign language. For example, native speakers of Japanese cannot distinguish *r* from *l,* since they use a single phoneme that is intermediate between these two; a group of Japanese adults learning English took a year of intensive training to master this discrimination.[46]

Another critical stage in the fixing of parameters is the *babbling* stage, when infants repeatedly utter syllables in an apparently meaningless fashion. This occurs at around 8 to 10 months of age. Babbling is little influenced by the sounds of the child's native language, even though the ability to discriminate heard sounds has already begun to narrow to those of the child's language environment. It is not until the child begins to utter words that the spoken phonemes begin to conform to those of the language environment.[47] There is, however, one study that shows just how broad the environmental influence can be and how flexible the concept of babbling is. Recent unpublished work by Laura Petitto and Paula Marentette at McGill University has shown that deaf children exposed to sign language babble with signs rather than with vocal sounds. This manual babbling has all the formal properties of vocal babbling and occurs at about the same age. This remarkable work suggests that the developing nervous system's quest for parameter setting need not be satisfied acoustically. That is, children may be as prepared for a language that consists of manual gestures as for one that is based on vocal sounds.

Similar examples of parameter fixing are familiar from other fields of study. Chapter 3 discussed imprinting in birds and, more relevantly perhaps, the acquisition of birdsong; both depend on a critical period that is no doubt under genetic control, but the actual behavior itself depends on environmental input during the critical period. Another example that is often likened to the acquisition of language is the development of the visual system. Through the pioneering work of David H. Hubel and Torsten N. Wiesel, which won them the Nobel Prize in 1981, it is known that there are individual neurons in the brains of cats and monkeys (and presumably humans) that respond selectively to such features of the visual world as color, lines or edges in different orientations, or binocular disparity. The responses of these neurons are also influenced by the environment.[48] For example, if a kitten is raised in an environment containing only vertical stripes, then the neurons responding to orientation will be heavily biased toward vertical rather than horizontal lines. Modern work suggests that this bias depends on a critical period of growth, and that it is selective rather than instructive. That is, the system is innately disposed to respond to lines of varying orientations, but experience determines which orientations are important. As in the development of language, experience simply fixes the parameters.

David Premack has argued that experience may actually play a larger role in the acquisition of language than in the development of the visual system, for example.[49] He suggests that the acquisition of language involves both raw, associationistic learning and "hard-wired," innately driven processes—or, in Piattelli-Palmarini's terminology, both instructive and selective aspects. At first, the child simply acquires a data base, learning the associations between words and their referents. At this stage, there are no rules, and related words such as *foot* and *feet,* or *dog* and *dogs,* are simply learned as independent units. But then, when the associations are in place, a "window" opens in the child's mind, such that regularities and rules are discovered, by means of what Premack calls a *distributional analysis.* Words are put into categories, and rules involving such things as pluralization or alteration of tense are induced. This stage is quite unconscious and hard-wired.

This is followed by another round of raw learning in which associations between the newly discovered morphemic classes (such as the suffixes *-s* and *-ed*) and environmental conditions are established.

Although Premack's analysis does not proceed beyond this point, we may envisage a further stage of hard-wired analysis in which additional rules are discovered, a further associationistic phase in which these are linked to environmental conditions, and so on.

My guess is that the acquisition of language does not proceed in quite this flip-flop manner, and that even the learning of word meanings is not devoid of the hard-wired aspect. We have seen that certain regularities are learned even as early as the first days of postnatal life, and such regularities no doubt constrain subsequent acquisition. Nevertheless, Premack is probably right in pointing out that there are both associationistic and hard-wired aspects in learning language, and that they change in relative importance at different stages. Piattelli-Palmarini may have been premature in heralding the death of instructive learning.

Hard-Wiring vs. Cognition

Premack also draws an interesting distinction between the hard-wired, parameter-fixing aspect of language acquisition and *cognition,* or the everyday processes of thought. It may seem that the discovery of the rules underlying language should require simply the application of thought, like the learning and understanding of the rules of chess or algebra. One obvious difference, however, is that we are unaware of the rules of language, and indeed unaware of the processes involved in discovering them. But since there is so much dispute over the nature of awareness, Premack suggests four other distinguishing marks.

The first is that the hard-wired aspect of learning language is dependent on critical periods, whereas most of us like to think that our normal thought processes remain with us for life. The second is that the hard-wiring component is triggered whether we like it or not, whereas we can choose when and whether to use our regular thought processes; as Premack puts it, "We can tackle one physics problem or a series of them."[50] The third difference is that the hard-wired process seems also to be triggered to *stop*. Our regular thought processes, by contrast, are tied to motivation, and we may continue intermittently to work on a difficult crossword puzzle as the mood takes us. Premack suggests that if the hard-wired component were to continue to work on the contents of language, it might well discover further regularities that could lead to more efficient representation. Why does it not do so? One possibility is that its operation is limited in order to guarantee success, and perhaps also to guarantee that all individuals achieve common knowledge. Language is, after all, a means of communication, and its rules must therefore be shared.

The fourth difference is that the sorts of inferences required in the hard-wired acts of language acquisition may well be beyond the capacities of normal thought. For example, all normal children are evidently able to discern the analogy between *bird/birds* and *dog/dogs,* to the point of generalizing the rule to the incorrect *foot/foots,* a common type of error that occurs during distributional analysis. Yet children of the same age are typically unable to discover comparable analogies when presented as overt problems to be solved.

These differences are worth noting, because they help explain why it has proven so difficult to discover the rules of language and the processes by which we acquire them. Language acquisition occurs by stealth, and we know surprisingly few of its dark secrets. In the words of Zenon W. Pylyshyn, it is "cognitively impenetrable."[51]

The Role of Growth

As we saw in Chapter 3, one respect in which we humans differ from the other primates, and indeed from any other species, is in our exceptionally long period of postnatal development. We are in effect born prematurely, and remain helpless yet exposed to the environment for the first few years of life. Puberty comes late, and there is even the suggestion that we retain child-like characteristics (neoteny) for life.

Since the brain and nervous system are especially plastic during growth, the prolonged period of postnatal growth in humans is often taken as testimony to the importance of learning in the human makeup. This argument seems especially plausible in light of the fact that the number of genes has remained roughly constant throughout mammalian evolution. The difference between mouse and human, it is suggested, lies not in the number of genes available for coding different characteristics, but rather in the way that the environment can alter the course of growth.

It should not be thought, however, that the difference between the mouse and the human is simply that the excess of brain in the human constitutes a *tabula rasa*—or an unstructured network awaiting the imprint of experience. The evidence on the development of language, reviewed above, suggests rather that experience acts to fix parameters or to *select* from among predetermined options. Jean-Pierre Changeux suggests that there is a built-in redundancy in the growth of the human brain that allows selection to occur.[52] Moreover, the choices that are made may be sequential, one building upon another, so that considerable diversity may be achieved. For example, 20 yes/no choices can generate 2^{20}, or about 1 million, different patterns. One is reminded here of Halle's work, referred to above, of the binary parameter fixing in learning the phonology of particular languages. Changeux describes this process as *gene saving,* since it allows diversity to be increased without substantially increasing the total number of genes.[53] The choices are set up by the genes but actually made by the environment.

It is also likely that the pattern of growth is itself genetically determined by so-called regulatory genes. This means that the various choices can indeed be sequential, and the results of parameter setting can have a cumulative effect. We have already seen that the development of language is cumulative, with one stage building upon another. By the same token, there is probably not just one critical period for language, but several overlapping ones. So it is that some of the parameters for language appear to be set very early in infancy, even in the first *days* after birth, while others may not be set until mid-childhood.

Conclusion

In summary, it appears that the properties of human language are very largely determined by the biological makeup of the human child. The role of the environment is simply to fix various parameters of functions that are themselves probably under genetic control. This process can result in considerable diversity among human languages, a diversity that has sometimes masked the universal properties of language. Parameter fixing is also closely linked to growth, and the very complexity of human language may result from the programmed sequence of steps, each involving the selection of options. That programming is no doubt itself under genetic control.

Although this developmental strategy is heavily dependent on innate, presumably genetically determined processes, its use of the environment nevertheless involves considerable gene-saving. The strategy is for the genes to program the questions, probably in yes/no fashion, and for the environment to supply the answers. If the *answers* as well as the questions were genetically programmed, then not only would this require more genes, it would also mean that each child would be stuck with a particular language. The strategy of allowing the environment to provide the answers is therefore highly adaptive. We humans have constructed for ourselves a wide diversity of environments, not only in language but also in other physical and social settings, and adaptation to a particular environment can tailored to the individual.

The idea that the learning of language is achieved through parameter fixing can therefore explain both the diversity of human languages and the common core that unites them. Can it also explain the diversity of possible utterances *within* a given language? In a sense, the choice of any word or sentence in any given situation can be regarded as the fixing of parameters, although in a temporary rather than a permanent sense. The on-line *generativity* of language might well depend on the same sorts of principles that underlie the learning of particular languages.

One question that was raised earlier was whether the "rules" that are so diligently sought by linguists are those of language itself or of the language *user*. In discussing this issue, Chomsky suggests that the question may not be coherent, since languages do not really exist in the absence of those who speak them.[54] But there is a way in which the question might be meaningfully framed. One might ask if an unstructured connectionist network might ever truly "learn" a language, obeying its rules even though those rules are never made explicit. If the evidence reviewed in this chapter is to be taken seriously, the answer surely is no. In order to acquire language, the network must require some built-in structure. I also suspect that it would have to mimic the pattern of human growth, adapting through a sequence of programmed choices rather than through the mere alteration of connections. In short, it would need to be selective, rather than instructive, in its mode of operation.

Given the evidence reviewed in this chapter, it is at least a reasonable inference that language is uniquely human. It depends heavily on the long period of postnatal growth that uniquely characterizes human development, and indeed is linked to specifically human developmental milestones. Its grammar seems to have an idiosyncratic quality that suggests an adventitious genetic rearrangement, rather than an incremental adaptation from some earlier skill—although I myself am skeptical of this conclusion. It is in any event dangerous to conclude that language is uniquely human without first taking account of the way other animals communicate. In the following chapter, therefore, I examine the nature of communication in other species and proceed from there to a consideration of how language may have evolved.

Notes

1. From the Preface to *Pygmalion*.
2. Claiborne (1983, p. 8).
3. Huxley (1956).
4. Chomsky (1966a).
5. Chomsky (1980, p. 37).
6. Lock (1980).
7. These figures are from Pfeiffer (1973). Presumably, they include differ-

ent inflections of words (e.g., *dog* and *dogs*; *run*, *ran*, *runs*, *running*, etc.) as different words. A count of dictionary entries gives lower figures. For example, Claiborne (1983) states that the largest English dictionaries have between 400,000 and 600,000 words. The biggest French dictionaries have about 150,000, and the Russian ones about 130,000.

8. Of course, language goes beyond sentences to paragraphs, chapters, articles, diatribes, speeches, books, and so on. Although these various manifestations raise structural issues that are not without interest, it is at the level of the sentence that linguists have sought the most fundamental bases of human language. The comparison of human language with other forms of animal communication has also focused primarily on the sentence.

9. Premack (1985, p. 242).

10. Cited by Chomsky (1966b).

11. This example, and the one in the following paragraph, is taken from Chapter 17 of Philip Johnson-Laird's (1988) *The Computer and the Mind*. This chapter provides an excellent introduction to the nature of grammar. Another accessible source is Keith Brown's (1984) *Linguistics Today*.

12. Miller (1968).

13. The distinction between recursion and iteration is emphasized by Premack (1985, p. 225).

14. Johnson-Laird (1988).

15. Lakoff and Ross (1976); see Chapter 7 of Howard Gardner's 1985 book *The Mind's New Science* for an account of the recent developments in theories of grammar.

16. Chomsky (1966).

17. Piattelli-Palmarini (1989, p. 26).

18. For the latest example of the use of this phrase, see Piattelli-Palmarini (1989, p. 13).

19. The example is taken from Chomsky (1980) and is presented here in slightly simplified form.

20. For additional examples, including some from languages other than English, see Lightfoot (1989).

21. Putnam (1980, p. 294).

22. Papert (1980).

23. McClelland, Rumelhart, and the PDP Research Group (1986); Rumelhart, McClelland, and the PDP Research Group (1986).

24. Rumelhart and McClelland (1986).

25. Op. cit. (p. 267).

26. Pinker and Prince (1988). When tested on verbs it had not encountered during the learning phase, the simulation produced some quite bizarre answers; for example, in venturing the past tenses of *squat*, *mail*, *tour*, and *mate*, it produced *squakt*, *membled*, *toureder*, and *maded*, respectively. But Pinker and Prince's critique extends far beyond such misbehaviors and is recommended reading for anyone who might be tempted to think that connectionism has all the answers.

27. Fodor and Pylyshyn (1988).

28. Lenneberg (1969, p. 635).

29. See Table 4.4 in Lenneberg (1967, pp. 147–149).
30. Johnson and Newport (1989).
31. See Brown (1957) for discussion of this and other cases of so-called wild children.
32. Curtiss (1977).
33. Bickerton (1984).
34. Hall (1959).
35. Claiborne (1983).
36. Bickerton (1984).
37. Newport (1983).
38. Piattelli-Palmarini (1989). The distinction goes back at least to an article by the Nobel Prize-winning immunologist N.K. Jerne written in 1967. Jerne writes that "looking back into the history of biology, it appears that whenever a phenomenon resembles learning, an instructive theory was first proposed to account for the underlying mechanisms. In every case, this was later replaced by a selective theory."
39. The analogy was suggested initially in the context of immunology by Edelman and Gall (1969).
40. For a recent review, see Hames and Glover (1988). It is thought that the unlimited variety of antibodies is due to the action of a particular gene, known as the "recombination activating gene," or "RAG-1." This gene makes an enzyme called recombinase, that randomly cuts and splices bits of DNA to make new genes which control the formation of antibodies.
41. Changeux (1980, p. 194).
42. Mehler, Jusczyk, Lambertz, Halsted, Bertoncini, and Amiel-Tison (1988).
43. Hirsch-Pasek, Kemler Nelson, Jusczyk, Wright Cassidy, Druss, and Kennedy (1987).
44. Halle (1988).
45. See Werker (1989) for this and other examples.
46. MacBain, Best, and Strange (1981).
47. Werker (1989).
48. For reviews of the Nobel Prize-winning work, see Hubel (1982) and Wiesel (1982). Wiesel's review also covers the work on environmental influences and critical periods.
49. Premack (1985).
50. Op. cit. (p. 237).
51. Pylyshyn (1980).
52. Changeux (1980).
53. Op. cit. (p. 193).
54. Chomsky, in Piattelli-Palmarini (1980, p. 313).

— 6 —

The Evolution of Language

> "I have," said a lady who was present, "been for a long time accustomed to consider animals as mere machines, actuated by the unerring hand of providence, to do those things which are necessary for the preservation of themselves and their offspring; but the sight of the Learned Pig, which has lately been shewn in London, has deranged these ideas and I know not what to think."[1]

NO SUCH DOUBTS appear to have dissuaded Noam Chomsky from his Cartesian view that human language is unique, although it must be said that recent research has rather neglected the pig. In his 1966 book *Cartesian Linguistics*, Chomsky wrote of the difference between human language and other forms of animal communication as follows:

> The unboundedness of human speech, as an expression of limitless thought, is an entirely different matter [from animal communication], because of the freedom from stimulus control and the appropriateness to new situations. . . . Modern studies of animal communication so far offer no counterevidence to the Cartesian assumption that human language is based on an entirely different principle. Each known animal communication system either consists of a fixed number of signals, each associated with a specific range of eliciting systems or internal states, or a fixed number of "linguistic dimensions," each associated with a nonlinguistic dimension.[2]

As we saw in the previous chapter, human language does seem to possess singular properties that make it at least plausible to suppose that it is peculiar to ourselves. Such a conclusion is premature, however, in the absence of information as to the nature of communication in other species. This chapter, therefore, focuses first on non-human communication, both in its "natural" forms and in the attempts to teach animals human-like languages. This will set the stage for discussion of the evolution of human language.

Natural Communication in Animals

As noted in Chapter 5, even casual observation reveals that animals communicate with one another. For example, birds sing to establish territory or emit cries to warn others of danger, fireflies exchange light flashes to signal sexual readiness, and leaf-cutter ants stridulate when accidentally buried so that other ants can find them by the vibrations transmitted through the soil.[3] Most linguists would no doubt agree with Chomsky that such examples do not go beyond mere communication to encompass the properties of true human language.

This conclusion has been challenged, however. In his book *Animal Awareness*, the ethologist Donald R. Griffin claims that the essential features of human language can be discovered in the communications of even quite primitive species, including insects. It is convenient to begin discussion with an analysis of Griffin's claims, for this will serve to remind us of some of the properties of human language described in Chapter 5, and to prepare us for analysis of more complex examples of animal communication.

The Waggle Dance of the Bees

In discussing the points made by Griffin, I shall focus, as he does, on a well-known and interesting form of communication among bees. This is the so-called waggle dance, first studied by Karl von Frisch.[4] The dance is carried out inside the hive on the vertical surface of the honeycomb and informs the observing bees of the location of food. The dance is in the form of a figure eight; in one "cycle" of the dance, the bee moves around a circle with a diameter about three times the length of the bee, then proceeds along a straight line, and moves finally around a cycle on the other side of the line. While it does this, it waggles its abdomen from side to side some 13 to 15 times per second. The relative orientation of the straight portion of the dance to the vertical codes the direction of food relative to the position of the sun, and the relative vigor of the waggling movement codes the distance of the food from the hive.

One property of human language, as noted in the Chapter 5, is its use of symbols to represent properties of the real world. Griffin argues that the components of the dance are indeed symbolic in this sense. That is, the elements of the dance coding direction and distance are symbols, standing in for properties outside of the dance itself.

On examination, however, the symbols do not have quite the arbitrary quality of words as used by humans. Although the direction of the dance does not point directly to the location of the food source, it does map onto direction in a one-to-one, or analogue, fashion. This kind of symbolic representation is said to be *iconic* rather than abstract. Similarly, there is an analogue relation between the intensity of the waggling movement and the distance of the source. That is, the two components of the dance correspond to what Chomsky, in the passage quoted at the beginning of this chapter, refers to as fixed *linguistic dimensions*.

In human language, by contrast, variations are not coded in this iconic or analogue fashion. One does not signal variations in distance, say, by varying the loudness of one's voice. Instead, the symbols are discrete, with variations signaled by different words or combinations of morphemes. We might say that something is "quite close" or "quite a distance away"; or, if we wanted to be more precise, we might say that it is "twenty feet away" or "sixty feet away." In these cases, distance is described by the actual choice of words, not in variations by the way a single word is uttered.

There is also no evidence that bees use symbols to actually *represent* the qualities they stand for. The symbols are used in only one context, that of instructing other bees where to go. That is, the elements of the waggle dance appear to be *signals* rather than true symbols. As we saw in Chapter 5, by contrast, humans use symbols in a wide variety of contexts, indicating a truly representational function.

Griffin has also argued that the waggle dance exhibits displacement, in the sense that the dance concerns the location of food outside of the hive. But again, it is the representational element that is at stake. The bees are not so much "talking about" the location of food as giving simple instructions to other bees about what to do. They do not use the elements of the dance in any other contexts, nor do they engage in anything resembling the give-and-take of human conversation. It is therefore very unlikely that the displacement shown by the waggle dance has the true vicariousness of human language.

Most debates as to whether or not human language is unique have focused on the property of *generativity*. The question is, can species other than humans use symbols to generate new meanings that are understood by their recipients? Again, Griffin argues that even the waggle dance is generative in the sense that a bee can communicate about some location that it has never communicated about before. However, this is a very restricted form of generativity and is qualitatively different from that which characterizes human language. Whereas bees simply modulate the two different components of the dance to signal different locations, humans achieve new meanings by forming new combinations of symbols. The components of the dance are not combined in a way that could remotely be called *grammatical.*

The waggle dance thus falls far short of anything resembling human language. Of course, one possible reason for this is that bees do not have anything very interesting to say. One is reminded of the limerick by Edward Lear:

> There was an old man in a tree,
> Who was horribly bored by a bee.
> When they said "Does it buzz?"
> He replied, "Yes it does!
> It's a regular brute of a bee."

Most other communications to be observed between members of species other than humans are, if anything, simpler than the waggle dance of the bees. That is, they are signaling systems designed for specific purposes. They serve to warn others of danger, to signal sexual readiness, or to mark out territory. There is no sense of conversation in transactions of this sort, and the signals do not have the structure or quality of sentences.

A Claim for Grammar in Birdsong

Among the more vocal and communicative of species are the birds. One of the delights of walking in the country, what there is left of it, is to hear the melodious and intricate songs of the birds—not to mention what P. G. Wodehouse once called "the intolerable screaming of the butterflies." Given the complex sequential nature of many birdsongs, might we not find some parallels here with human language?

Again, I focus on one species, the chickadee (known more formally as *Parus atricapillus*), on whose behalf perhaps the strongest claim has been made. Jack P. Hailman and Millicent S. Ficken of the University of Wisconsin have recently maintained that the calls emitted by these birds qualify as language according to the criteria of structural linguistics.[5] The calls are made up of four basic elements, which are strung together in sequences. These elements, which we may label A, B, C, and D, are qualitatively different sounds and seem to be analogous to words. It is not entirely clear what they stand for, but Hailman and Ficken suggest as a working hypothesis that each one "refers to a qualitatively different locomotory tendency."[6]

The important claim that Hailman and Ficken make, however, is that the sequences seem to obey grammatical rules. The elements are always in the same order, although any element may be repeated any number of times or may be omitted. Thus the sequences ABCD, C, ACDD, and AABBBCDD are all legitimate "sentences," while BACD, BBDCC, and AAACBDD are not. In a total of 3479 calls ranging from 1 to 24 elements in length, Hailman and Ficken counted 362 different sequences. Only 11 calls did not obey the "rules."

Hailman and Ficken show that the rules conform to the formal description of a language. That is, it is possible to decide whether any given sequence is syntactically correct or not, and there is no limit to the number of "sentences" that can be generated. Hailman and Ficken conclude that these properties make "chickadee calls far more like human language than any other animal system yet described."[7]

This claim is surely too strong, since Hailman and Ficken do not demonstrate that the song of the chick-a-dee requires more than a finite-state grammar. That is, the regularities are simply that the options available at each point in the sequence can be specified by the preceding element. In terms of transition probabilities, the "rules" are simply that any element may be followed by a repetition of the same element, by an element later in the intrinsic ordering of elements (i.e., A, B, C, D), or by the end of the sequence. This is not to say that the calls are actually generated in this way; the point is simply that we need not go beyond a finite-state grammar in order to capture the observed characteristics of the song. Human language, on the other hand, requires a grammar that is at least as powerful as a phrase-structure grammar, and probably one that is even more powerful than that.

The difference implies a number of critical distinctions that amount to an overall difference of kind rather than merely of degree. First, finite-state grammar involves no component of short-term memory—each element is generated simply from the preceding one. As we saw in Chapter 5, by contrast, the more complex grammar required for human language requires the operator to remember its location in one operation while subsidiary operations are carried out, so that it may then return to the original, unfinished operation. Second, a finite-state grammar may allow iteration but not genuine recursion. The chickadee may repeat individual elements indefinitely, but there is no evidence for the recursive embedding of phrases within phrases. Finally, much of the power of human language derives from the fact that elements are combined in different *orders*. This occurs at several levels; phonemes are combined in different orders to form words, words are combined in different orders to form sentences, and so on. While the chick-a-dee forms different *combinations* of elements, it does not order the elements differently to form different *permutations*.

Although the chickadee calls may meet some of the formal requirements of a grammar, it does not follow that they are in any sense equivalent to *sentences* in human language. Again, they are probably no more than simple signaling systems. Indeed, they may do no more than signal the strengths of four different "states," very much as the waggle dance of the bee signals the distance and direction of food.

We should perhaps leave the last word on birdsong to the poet W. H. Auden:

> As I listened from a beach-chair in the shade,
> To all the noises that my garden made,
> It seemed to me only proper that words,
> Should be withheld from vegetables and birds.

Nonhuman Primates

What of our closest ancestors, the primates? As we shall see later in this chapter, there has been considerable emphasis on attempts to teach chimpanzees and gorillas communication systems based on human language but relatively little interest in their natural forms of communication. Part of the reason, perhaps, is that other primates have seemed relatively uncommunicative, at least vocally, compared with that compulsive chatterbox, the human being.

However, recent research shows that primate communication is more subtle and even more prolific than it had earlier seemed. At first, the calls emitted by primates in the wild were to human ears indistinguishable from one another, but more sophisticated technology for the analysis of sounds has revealed important differences. For example, vervet monkeys in East Africa make different alarm calls to at least three different predators, namely, leopards, eagles, and snakes. These produce different reactions in the hearer even when played to the monkeys on a tape recorder in the absence of actual predators. That is, monkeys react to leopard alarms by running into the trees, to eagle alarms by looking up in the air, and to snake alarms by searching in the grass.[8]

Monkeys also use vocalizations in more complex social situations. Besides the alarm calls, vervet monkeys emit at least four different kinds of grunt; these, too, initially sounded alike until analyzed acoustically. One type of grunt appeared to be addressed to a monkey higher in a dominance hierarchy, another to a monkey lower in the hierarchy, another to an animal moving into an open area, and another to another group. Again, young rhesus monkeys often scream, as young children will, when they are playing with one another and things get rough. Different kinds of scream are used in different situations, one for interactions involving physical contact with an individual of higher rank, another for interactions with lower-ranking individuals but involving no contact, and so on.[9]

Although the use of calls such as these may reveal sensitivity to social contexts, they do not tell us of the animals' capacity for grammar, since they consist of single calls. In some situations, primates do emit sequences of calls. It is not always easy to determine whether these obey any syntactic rules. For example, chimpanzees sometimes get together as a group and produce a cacophony of sounds, and it is extremely difficult to isolate the contributions of individual animals in the general din. Again, the contributions of individual primates to vocal interchanges may be difficult to tease apart for other reasons, such as a forested habitat, or because some sounds are made with very little facial movement. Jean-Pierre Gautier and Annie Gautier-Hion remark that "the problem of vocalizer identification becomes one of extraordinary magnitude for research on many primate species in their natural habitats."[10]

Even so, some primate vocalizations do exhibit combinatorial properties that might be said to display elements of grammar. Peter Marler has distinguished between two levels of syntax (or grammar), *phonological syntax* and *lexical syntax*.[11] Phonological syntax governs the ordering of phonemes, or basic phonological units, and there are clear examples of different calls in the repertoires of monkeys that consist of different combinations of the same phonological elements.[12] Lexical syntax consists of the ordering of higher-level units, corresponding to words in human language, to produce new meanings, and it is this level of syntax that gives human language its distinctive quality. Marler suggests that lexical syntax is rare in nonhuman animals. However, Charles T. Snowdon has noted that tamarins occasional emit pairs of calls to convey new meanings; for example, the alarm call may be paired with the contact call to signal relaxation after danger.[13] But such examples are extremely primitive when compared with human sentences, and it is debatable whether such paired combinations are truly *grammatical*. Certainly, they do not resemble the generative, open-ended, propositional nature of human sentences.

There are also examples of more complex sequences of vocalizations that have been recorded and analyzed in some detail. For example, pairs of male and female white-cheeked gibbons indulge in what is known as *duetting,* in which complex sequences of calls are exchanged. Although this is generally likened to a song, comparable to duets among birds, there is at least some element of a conversation in that the song and behavior of the male are influenced to some extent by the female's song. Nevertheless, the exchange is highly stereotyped, and its main role seems to be in the development and maintenance of a pair bond. In both birds and primates, duetting seems to be especially important in habitats where it is difficult to maintain visual contact.[14] In this and other examples of sequential calls emitted by primates, however, there is no evidence that the sequences are governed by rules any more complex than those of a finite-state grammar. That is, certain sequences are emitted with a higher frequency than others, so that a given call is to some extent predictable from the previous one. Reviewing the evidence, Charles T. Snowdon writes: "The simplicity of these primate examples of syntax should be stressed. In no way do they approach the complexity of human rules for sequencing words and sentences or determining conversational interactions."[15]

Nevertheless, Snowdon has claimed that the vocalizations of New World monkeys do exhibit some properties that are sometimes taken to be uniquely human. One has to do with the categorical way in which they perceive their calls. Alvin M. Liberman has claimed that humans perceive phonemes in a uniquely categorical fashion.[16] For example, if one synthesizes sounds that lie between the voiced phoneme *b* and the unvoiced phoneme *p,* people hear these sounds either as *b* or as *p,* and not as intermediate sounds. The same phenomenon however, can be demonstrated in the pygmy marmoset. Synthesized calls that lie on a continuum between the open-mouth call and the closed-mouth call of this species are evidently heard as one or the other, and not as intermediates between the two.[17] Moreover, Dominic Massaro of the University of California at Santa Cruz has recently argued that the perception of speech sounds is not in fact categorical, as Liberman has claimed. Rather, people (and perhaps marmosets) *can* detect the gradations between different phonemes such as *b* and *d* but are forced to respond with one or the other because there are no intermediate *labels.*[18]

We saw in Chapter 5 that the development of speech in humans depends on the so-called critical period and goes through predictable stages of development. By contrast, the vocalizations of at least some nonhuman primates are relatively impervious to environmental influences. In squirrel monkeys, for example, the full range of vocalizations is present 6 days after birth and does not seem to depend on practice. Moreover, squirrel monkeys vocalize normally even if they are deafened in infancy or are raised by mothers who have been rendered mute. Reviewing this evidence, Robert M. Seyfarth and Dorothy L. Cheney remark that "the only evidence for learned, modifiable vocalization in species other than humans comes from studies of songbirds."[19]

Again, however, this claim has been disputed by Snowdon. He notes that infant cotton-top tamarins appear to go through a "babbling" stage comparable to that in human infants.[20] He also notes that these unusually vocal monkeys show great variability in their vocalizations in childhood, but that their calls become fixed and stereotyped with the onset of sexual maturity.[21] This may be crudely likened to the human condition, in that the ability to acquire language diminishes as children approach puberty.

The idea that monkeys might talk to one another is, of course, not new. Samuel Butler once offered a further note of caution:

> In his latest article Prof Garner says that the chatter of monkeys is not meaningless, but that they are conveying ideas to one another. This seems to me hazardous. The monkeys might with equal justice conclude that in our magazine articles, or literary and artistic criticisms, we are not chattering idly but are conveying ideas to one another.[22]

He might have mentioned learned books as well. Now read on.

Would We Recognize Language in a Nonhuman Species?

As we have seen, there is no evidence that birdsong or communication among nonhuman primates involves more than a finite-state grammar. Yet if any animal communication involved grammatical rules more complex than this, it is quite unlikely that we humans would be able to detect it. After all, it was thought until quite recently that human language could be described in terms of finite-state grammar, and the discovery of more complex phrase-structure rules presumably depended on human intuition as to what constitutes a phrase, and how phrases may be embedded in other phrases. Indeed, human intuition has still not succeeded in providing an explicit grammar capable of capturing the full complexity of human language. How much more difficult it would be to discover the structure underlying a *nonhuman* language, if such a language exists!

It is entirely possible, moreover, that animal communication is based on principles that are quite different from those underlying human language, and that may be just as complex. If this is so, then the chances of our detecting those principles are even more remote. In pursuing this line of argument, Sue Savage-Rumbaugh has been able to identify at least one respect in which humans may be unique. As we shall see below, strenuous efforts have been made to teach human language to other species, but no other species has tried to teach us their language![23] But can we even be sure of this? Could it be that the twittering of the birds is indeed a language lesson that we are too stupid to comprehend?

Perhaps not. Jake Page, in describing the behavior of his parrot, asks: "How intelligent is a creature that can amuse himself for fifteen minutes by uttering, over and over, the following sounds: uhr, uhr, uhr, Uhr, UHR, UHR, Wah, Wah, wah, wah, wah, wah?"[24] This does not sound like the activity of a teacher, unless, of course, the bird is patiently repeating the lesson in the hope that the ignorant human will understand.

Be this as it may, some current work with dolphins offers a tantalizing possibility for decoding their own natural language, if indeed it exists as such. Louis Herman and his colleagues at the University of Hawaii have taught dolphins to imitate each other, so that if one dolphin performs some complex action, a second dolphin must do the same.[25] Observations suggest that the second dolphin can imitate the first even if blindfolded while the first carries out the action, suggesting that the information might have been conveyed "linguistically." Recordings of the sounds made by the dolphins during this transaction might reveal something of the vocabulary and rules that dolphins use naturally—provided that we are clever enough to understand.

Teaching Language to Apes and Dolphins

If we cannot decode the natural languages of other animals, the obvious alternative is to try to teach them our own. If a nonhuman animal could learn human language, then we could no longer argue that the capacity for language is uniquely human. Indeed, such an animal might be regarded as superhuman if it also possessed a language of its own that was incomprehensible to us.

Quite apart from its theoretical significance, there has long been a fascination with the possibility that animals might be taught a language that humans can understand, if only so that we might discover what they think about. The interest in animal language is perhaps as much voyeuristic as scientific. But coupled with this fascination, there may also be a fear that we might not like what they have to say to us. Saki's short story "Tobermory," about a cat that learned to talk, chronicles the upheavals caused by the animal's sardonic revelations about the activities of the human household, and the relief that greeted the news that it had been killed in a skirmish with a neighboring tom.

Chimpanzee and Gorilla

In 1925, the primatologist Robert Yerkes observed that although chimpanzees were clever at mimicry, they never imitated human sounds.[26] This observation was borne out by the Hayeses, a husband-and-wife team who raised a chimpanzee named Viki from the age of 3 days to about 6.5 years in their own home. They treated her as a human child with respect to feeding, toilet training, discipline, and play, and, of course, they spoke to her. However, Viki was never able to emit more than three or four crude imitations of spoken words, including "mama," "papa," "cup," and possibly "up."[27]

In the 1960s, another husband-and-wife team, Allen and Beatrix Gardner, noticed from films of Viki that she was reasonably intelligible without the sound track. She would get her mouth into more or less the right positions to form words but could not make her tongue produce the appropriate sounds. They concluded that Viki's inability to talk was mechanical rather than conceptual and decided to try to teach a chimpanzee American Sign Language, or ASL. Since chimpanzees are adept with their hands, this seemed to provide a fairer test of language than vocal speech.

The Gardners' chimpanzee, Washoe, learned to use well over 100 gestures in a consistent fashion. Her first "word" was "more," made by bringing the hands together in front of the body. She used this sign to request more treats and tickles, and also learned to associate it with the signs for objects (such as a banana) to request more.[28] Later studies confirmed that nonhuman primates can acquire the use of signs numbering in the hundreds. For instance, Francine Patterson taught a gorilla named Koko at least 375 different signs. Patterson has claimed that Koko can use these signs in inventive, human-like ways, such as to lie, swear, and pour scorn, and to create metaphors, puns, and even rhymes. Indeed she even claimed to be able to measure Koko's IQ, which she placed somewhere between 84 and 95—not bad for a test biased in favor of humans.[29]

Another technique, developed by David Premack, has been to teach chimpanzees to use plastic tokens, stuck on a magnetic board, as signs.[30] Yet another, pioneered by Duane Rumbaugh, has been to use a computer console, with over 100 keys representing different words.[31] These studies confirm that the ability of chimpanzees to use a vocabulary of signs far exceeds their ability to imitate human speech sounds. The message, so to speak, depends on the medium.

This flurry of research activity in the 1970s led many to question whether language was unique to humans after all. The primates in these studies seemed to have little difficulty with at least one property of human language, that of using symbols to stand for objects or actions—although, as we shall see, this was later questioned. The symbols, moreover, were essentially arbitrary: For the most part, the gestural signs of ASL bore no relation to what was signed, and in Premack's studies the plastic symbols were deliberately designed to be unlike what they stood for.

What was more questionable, however, was whether the signing primates could be said to demonstrate *grammar*. According to Chomsky, of course, grammar is the distinguishing feature of human language that allows us to generate a potentially unlimited number of messages. Influenced by Chomsky (who seemed to remain above the debate), scholars such as Jacob Bronowski and Ursula Bellugi claimed that the signing primates showed no evidence of true grammar.[32]

Whether this is a fair claim seems to depend upon what is considered a sufficient definition of grammar. The signing primates often produce two-word utterances comparable to those of 2-year-old children, and occasionally seem to do so in novel ways. For instance, Washoe signed "water bird" on first seeing a swan. Once, when asked where you go when you die, Koko is said to have signed "comfortable hole bye"—a novel *three*-word utterance. Yet if these sorts of examples constitute grammar, it is grammar of a very primitive sort, scarcely on a par with the human ability to produce an unlimited number of messages. Herbert Terrace, who has taught ASL to a chimpanzee named Nim Chimpsky (no relation), has claimed that the word sequences uttered by these animals do not demonstrate grammar.[33] Much of the behavior is pure drill, with much repetition and simple sequencing of ideas. For instance, the much-heralded example of "water bird" need not represent innovation; Washoe may simply have seen *water* with a *bird* (a swan) floating on it and signed these two concepts consecutively. Indeed, little of true grammar can be inferred from any two-word utterance.

Dolphin

Most of the research on language in primates has focused on the animals' spontaneous production of signs. But as Louis Herman has reminded us, knowledge of grammar can also be demonstrated through the comprehension of sentences. Studying an animal's ability to understand complex sentences, moreover, offers greater experimental control and precision than does waiting around for an animal to emit a sentence spontaneously. Consequently Herman set out to teach dolphins to understand sentences in the form of commands.[34] One dolphin, called Akeakamai, was taught a language based on visual signs. Another, called Phoenix, was taught a language based on acoustic signals, which were chosen in terms of known features of the dolphin's auditory system and bore no relation to human speech.

Akeakamai has been taught to understand sentences of up to four words, with each sentence consisting of an instruction, such as WATER RIGHT BALL FETCH ("Fetch the ball on the right and take it to the stream of water"). The syntax depends on the order of the words. The indirect object (WATER, in the above example) is given first, followed by the direct object (BALL), with the action (FETCH) coming last. Either object can be preceded by a modifier (RIGHT). Comprehension was tested by the dolphin's ability to perform the correct action. Akeakamai was able to understand sentences she had not seen before, indicating that she had mastered the grammar.

At the time of reporting, Phoenix was able to go one step further and respond correctly to five-word sentences. She was taught a different grammar in which the direct object was given first, then the action, then the indirect object. She could cope with modifiers preceding *both* objects, although not perfectly; her accuracy was between 50 and 60 percent. An example of such a sentence might be SURFACE HOOP FETCH BOTTOM BASKET ("Go to the hoop at the surface and take it to the basket at the bottom"). Since there is inevitably some delay between receiving the sentence and performing the action, a five-word sentence might tax the dolphin's ability to remember rather than her ability to understand.

Do these findings demonstrate genuine grammar? At best, it is grammar of the most elementary sort, in which words are sorted into classes and simply inserted into standard sentence forms. The rules do not require even a finite-state grammar, let along a phrase structure. Even the sentence forms are extremely simple compared with their English equivalents, as is illustrated by the bracketed "translations" of the above examples. There are no variations of tense or mood, no articles, no embedded clauses, and so on.

David Premack has shown that the performance of the dolphins can be reduced to two simple "rules" involving three simple concepts: *property*, *object*, and *action*. One such rule is of the form (*property*) *object action*, as in the sentence SURFACE BALL TOUCH ("Touch the ball on the surface"), and the other is of the form (*property*) *object action* (*property*) *object*, as in BOTTOM FRISBEE FETCH SURFACE BALL ("Bring the frisbee on the bottom to the ball on the surface"). These properties, Premack says, owe nothing to linguistics but represent simple aspects of the world. Similarly the rules simply reflect everyday actions. True language, by contrast, requires fundamentally linguistic concepts that do not refer immediately to the real world—concepts such as *noun phrase, verb phrase, article,* and even *sentence*.[35]

We saw earlier that many of the communications between animals seem to be simple commands rather than true sentences, and the same criticism can be levied against the dolphin studies. In this respect, the one-way contact between trainer and dolphin, in which the trainer gives the message to the dolphin but there is no provision for the reverse exchange, is an impediment to the testing of true language. That is, there is no provision for the development of conversation, the two-way interchange, that is a hallmark of human language.

In discussing these matters with me, Louis Herman made the interesting point that the question of whether these dolphins have language is in some respects complementary to the question of whether humans can swim. Watching the dolphins gliding and cavorting effortlessly in the pool while we humans chatted, it was easy to believe that dolphins are born to swim, while humans are born to talk. The comparison may even be unfair to humans, who are actually not bad swimmers, and may even at one point in hominid evolution have enjoyed an aquatic phase, as discussed in Chapter 3. In any event, watching the dolphins glide through the water while their human trainers talked in the Hawaiian sun, I was not sure which was better off.

Pygmy Chimpanzee

In recent years, attention has shifted somewhat away from the question of grammar and toward the question of whether animals can use symbols in a truly representational way. Do chimpanzees or dolphins use symbols as *representing* objects or actions, or simply as requests? Sue Savage-Rumbaugh of the Yerkes Regional Primate Research Center in Georgia has noted that chimpanzees are often unable to use symbols correctly in the absence of the things they represent, suggesting that using a symbol was interpreted as a request rather than as a name for something. As we saw in Chapter 5, by contrast, when children use a word, they are able to do so in ways that demonstrate much more than a mere request. This is so even in 1-year-olds who are still at the stage of single-word utterances.[36] Savage-Rumbaugh therefore wondered if part of the difficulty in demonstrating the use of syntax in chimpanzees was due to the fact that they had not learned the proper symbolic use of words, so she and her colleagues set out to remedy this situation.

Another novel aspect of her project was that she used a rare and comparatively unknown species, the pygmy chimpanzee (*Pan paniscus*). There are fewer than 50 of these animals in captivity, and in their native Zaire they are severely threatened by logging, to the point where they may not survive for more than a few generations. Yet they are perhaps closer to humans in their general characteristics than is any other primate, including the common chimpanzee (*Pan troglodytes*). Strong male-female bonding is maintained by the continuous sexual receptivity of the female, as in humans (see Chapter 2). They commonly share food, and make frequent and complex gestures to each other, maintaining eye contact as they do so. The traits that separate them from the other apes are all traits possessed by humans.[37]

In her attempts to teach language to these animals, Savage-Rumbaugh established strict criteria for ensuring that a given word was indeed understood symbolically.[38] For example, the chimpanzee had to learn to use the symbol for an object even when the object was in a different place. The animal also had to learn to retrieve an object from a different place when the experimenter asked for it. Unlike their cousins *P. troglodytes*, pygmy chimpanzees seemed to learn the symbolic use of words spontaneously, without requiring specific training in the different uses to which words can be put. In Savage-Rumbaugh's studies, the "words" are geometric forms, known as *lexigrams*, that are arranged on a board attached to a human speech synthesizer, such that when a given symbol is touched, the synthesizer utters the corresponding word in English. Trainers also speak in complete sentences while touching lexigrams corresponding to key words, and sometimes simply speak without using the board at all.

The most impressive results have been obtained with a young male pygmy chimpanzee called Kanzi, who was assigned to the project at 6 months of age. By the age of 6 years, Kanzi demonstrated perfect or near-perfect understanding of well over 100 words spoken by the tester, as he demonstrated by selecting the lexigram corresponding to each word. He also recognized nearly all of these words when they were produced by the synthesizer, which removes all cues based on stress or intonation. The synthesized speech was far from a perfect rendition of natural speech, and Kanzi's performance was considerably better than that of a 4-year-old child, although he was more experienced with the synthesizer than the child was. Detailed analysis suggests that Kanzi's recognition of human speech was based on phonemes; only the *w* sound (as in *water* and *wash*) seems to have caused any difficulty.[39]

Kanzi's ability to understand sentences seems roughly comparable to that of Herman's dolphins. Superficially it seems better, since he is able to understand sentences like "Go to the refrigerator and get a tomato" or "Would you put some grapes in the swimming pool?" even though many of these sentences are novel. However, it is likely that he responds only on the basis of key words; in teaching Kanzi sentences, for example, the lexigrams corresponding to key words would be touched while the tester uttered the full sentence. An example is "Look, a *stick*, let's *grab* the *stick*," where the emphases correspond to words whose lexigrams are touched. The sentences are therefore effectively reduced to, at most, three-word utterances of the type "action-object-location" or "action-object-recipient" or "action-recipient-object." Although such sequences are shorter than those mastered by the dolphins, they are more flexible with respect to word order. Kanzi also shows some ability to understand general terms such as "it" or "this," as in "If you don't want the juice, put it back in."

Like Herman with his dolphins, Savage-Rumbaugh has focused on comprehension rather than production. As with young children, Kanzi's ability to understand exceeds his ability to produce sentences, and it seems that his *production* of language has not yet moved beyond two-word constructions of the form "action-object."

Again, it must be questioned whether these results demonstrate human-like grammar in a nonhuman species. Clearly, there is still a long way to go; Kanzi does not "converse" like even a 4-year-old human child, and his grammar is still exceedingly simple, more like the elementary, essentially nonlinguistic "rules" that seem to underlie the dolphins' performance. Even so, work of this sort chips away at the idea that language is uniquely human. Kanzi clearly demonstrates the symbolic use of words, and his language clearly shows displacement, as he can use words to refer to objects that are not physically present. But if the last retreat of the Chomsky school is human grammar, then they are probably still safe.

In summary, then, Chomsky's view that language is uniquely human has stood the test of time. In some respects, the ability of apes and dolphins to learn the communicative use of symbols is surprisingly good, and is improving as experimenters themselves gain in sophistication. But the gap between ape language and human language is still immense. Even by the age of 3 or 4 years, children have a speaking vocabulary of up to 1500 words and can probably understand a further 3000 to 4500 words that do not appear in their own sentences. The average adult has a speaking vocabulary of about 10,000 words and a reading vocabulary of perhaps ten times that figure. It is estimated that there are well over 1000 rules of grammar.[40] These figures are several orders of magnitude beyond those so far demonstrated by any ape or dolphin.

The Evolution of Language

On present evidence, then, it seems likely that language is distinctively human. Even Charles Darwin agreed that this was so: "The habitual use of articulate language," he wrote, "is peculiar to man; but he uses, in common with the lower animals, inarticulate cries to express his meaning, aided by gestures and the movements of the muscles of the face."[41] He went on to argue that language evolved through the imitation of natural sounds: "I cannot doubt that language owes its origins to the imitation and modification of various natural sounds, the voices of other animals, and man's own distinctive cries, aided by signs and gestures."[42]

However, the critical aspect of human language is not imitation, but the use of language as a propositional system, rather than as a set of warning signals or emotional cries. Even chimpanzees do not seem to approach humans in their ability to converse, or to discuss matters relating to objects that are not immediately present or to events that occurred at some other time. Indeed, even after several decades of quite intensive research on language in apes and dolphins, the gap seems almost too large to be compatible with evolution, as though humans have been somehow blessed with a Cartesian freedom of will that allows them unlimited freedom of utterance.

Before we resort to an explanation based on magic or divine intervention, let us try to envisage the changes that might have taken place in the evolution of the hominids and that might have led to the evolution of language. I focus first on the ideas developed by Arthur Sigismund Diamond, an English comparative linguist, in his 1959 book *The History and Origin of Language*. As we shall see, Diamond was essentially pre-Chomskian in his thinking, some of which is distinctly quaint by modern standards. Nevertheless, his account seems to me to have a backbone of truth.

Diamond's Theory

Diamond begins by suggesting that there are three types of sentences, which may be illustrated in their simplest forms by the following examples:

(1) Look!
(2) George runs.
(3) George is fast.

The first is the most primitive and is normally regarded as a imperative. Diamond suggests that this is not entirely apt, as the sentence may be an appeal as much as a command. He proposes that it be called a *suggestion of action*. In all languages, the verb form in this kind of sentence is uninflected—it represents the verb in its most naked form. The sentence consists of a single verb, with no nouns or other embellishments. It is addressed, moreover, to the hearer only, and involves no other party.

The second kind of sentence is a statement rather than a command or suggestion. It also involves an action verb, but now a third party (George) is introduced. The minimum requirement is a verb plus a subject noun (or pronoun). The third sentence is what Diamond calls a *description-statement,* and introduces a third element, the adjective. Hence it requires a minimum of three words.

Diamond suggests that the three different sentence types represent the order in which sentences evolved. The first simply suggests action and is the nearest thing in human language to an animal call. As we have seen, most animal communication, even in cases where something resembling a sentence is involved, essentially takes the form of a command. Diamond also suggests that such utterances do not really represent *thoughts* but are immediate reactions to events. One cannot think that "Look!" but one can think that "George runs." Suggestions of action typically reflect a social context, wherein individuals need to communicate with others in a group. In some cases, the request would be for some action that cannot be performed by one individual alone or that requires another individual with greater strength or expertise. Children, in particular, often need to request the help of adults in reaching for things or in performing actions that require greater strength than they can muster.

Diamond's argument is based in part on a comparison of different languages. He orders various groups of peoples whose languages are known in terms of their cultural sophistication. He begins with what he calls the Food Gatherers, whom he considers at the level of the Old Stone Age. These are represented still by such groups as the Sakai and Semang of Malaysia, Andaman Islanders, Australian Aborigines, and several others. He moves on through various Agricultural and Hunter grades, representing a neolithic stage of cultural development, and thence to Pre-Industrial and Industrial peoples. He records an increase in the total vocabularies of these different peoples, from an average of about 5000 words in the Food Gatherers to the 63,000 represented in a modern English dictionary.[43] More important, the proportion of verbs decreases from about 50 percent to 14 percent. Diamond takes this as evidence that language evolved from single verb forms to increasingly include nouns, and then adjectives and other parts of speech.

There is perhaps a touch of the nineteenth century to the idea of varying degrees of "civilization," with the English gracing the highest order. Nevertheless, there can be no denying that cultures do differ in complexity, and that this difference is reflected in vocabulary size. The relative decrease in the proportion of verbs and increase in the proportion of nouns may simply reflect the growing number of *objects* as culture becomes more complex and industrial. Diamond is also pre-Chomskian in his treatment of grammar. He notes that grammar would have become more important as sentences became more complex and then writes:

> In all this building-up of accidence and syntax, this conflict and variation from one language to another, is there anything to be found which is common and universal in language? There is nothing.[44]

So much for Chomsky's notion of universal grammar!

At this point, it is worth noting that the variations in vocabulary between one language and another have little to do with the complexity of the language itself. The languages of the Australian Aborigines, who belong to the most primitive level of Diamond's classification, are intricately inflected and, in many respects, more complex than English. Claiborne remarks that "Navaho grammar is so complicated it is virtually impossible to master unless you learn it in childhood."[45] Cultures may vary in complexity, and in the number of objects that they possess and have names for, but the biological differences among people are superficial. This is especially so with respect to language. It may be safely said that the potential to learn *any* language is independent of race or culture.

Diamond also argues that suggestions of action evolved largely in relation to violent actions, such as cutting, breaking, tearing, smashing, and crushing—activities that, in Diamond's observation, the other apes seem frequently to indulge in. As a corollary, he suggests that the originators of language were the males, who were the more violent and aggressive. His views on this fall uneasily on the modern ear:

> The feminine sex are by repute fluent, but the males are the innovators, the originators. The females take, in this as in other activities, the normal, mean position—their speech, though fluent, is commonplace: the males are the orators, but they are also the aphasics and the almost speechless. Language, then, originated in requests (or appeals) for action between males, and . . . not between spouses nor members of the family as such.[46]

We need not follow Diamond into such sexist absurdity; requests for help surely involved women as commonly as they involved men (perhaps more so), and they would have involved children of both sexes.

The Changing Larynx

Part of Diamond's argument for relating the origins of speech to violent action has to do with the changing functions of the larynx, which is shown, along with the other organs involved in speech, in Figure 6.1. This structure appeared first with the evolution of the lung, and served to exclude everything but air from the pulmonary tract. It was originally a sphincterish band of muscle around the glottis that closed that tract, much as one might tie a string around the opening of a bag. This prevented foreign substances, such as food and water, from invading the lungs.

Later, the larynx evolved further to prevent the entry of air into the lungs. This adaptation had to do with the gradual use of the forelimbs for purposes other than locomotion, such as grasping, climbing, hugging, and striking. The reason is that one of the muscles involved in moving the forelimbs, the pectoralis major, is attached to the ribs. If the ribs are to provide a firm base from which to contract the muscle, there must be some way of fixing them. This is done effectively by closing the glottis, so that when the muscle is contracted, the air does not rush in and cause the ribs to rise. This action of the larynx is most effective in arboreal animals, such as monkeys, lemurs, gibbons, and chimpanzees, which use their forearms to swing from branches of trees. In the higher apes and humans, there is an additional mechanism for preventing the exit of air from the lungs and raising the pressure within the lungs, which further steadies the wall of the rib cage. That is, for maximum steadiness one "holds one's breath."

The larynx can thus serve to hold the air in the lungs and to control its release. This was originally important in actions of the arms requiring strength and steadiness. Diamond notes that workmen who have lost their larynxes are often unable to continue manual work, and that difficult actions that require precise movement are usually performed with the breath held and the lips closed. After performing such actions, especially those involving considerable expenditure of effort, the air is suddenly released, producing an audible grunt. Again, Diamond suggests that this was largely confined to the males, but observation of the new-found fashion for grunting during tennis matches suggests that Jimmy Connors does not have it on his own. Many of the young women players are also formidable grunters.

The release of air from the lungs after activities involving the arms can be accomplished either through the mouth or through the

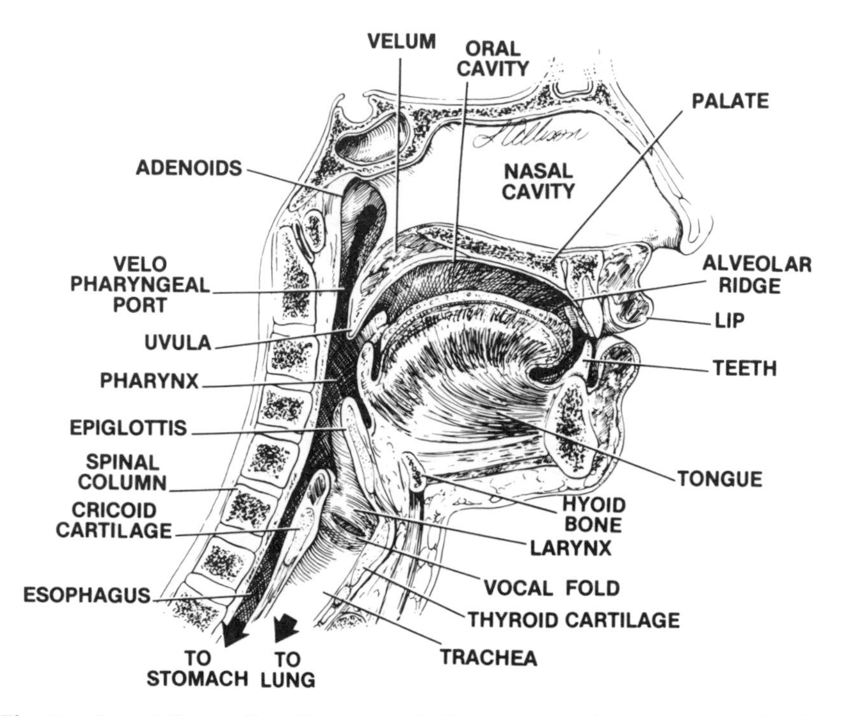

Figure 6.1. Schematic diagram of the principal organs involved in speech (from Calvert, 1980).

nostrils. Diamond notes an age-old trend toward the use of the mouth for breathing, beginning when our ancestors took to the trees, so that the sense of smell diminished and the sense of sight gained in importance. The earliest words, according to Diamond, were essentially formed by the plosive consonants, and took simple forms such as CV, CVC, and CVCV, where C is a consonant and V a vowel. The most common vowels are the nasal plosives *n* and *m*, followed by the dental (*d* and *t*) and labial (*b* and *p*) plosives. The dominant vowel was the short *a* sound. Diamond notes that in practically all known languages, past and present, the consonants *m*, *n*, or, more rarely, *ng* have signified the personal pronoun of the first person singular. This most basic of utterances may go back to the first primitive grunts emitted by our forbears.

Gestural Language

Diamond suggests that the earliest language of the hominids may have been primarily gestural rather than vocal. This idea actually goes back to the seventeenth-century French philosopher with the distinguished name of l'Abbé Etienne Bonnot de Condillac (better known simply as Condillac)[47] and has more recently been championed by Gordon W. Hewes.[48] Once the hands were freed from locomotion, they could be used for communication, as well as for carrying and tool use, and were probably more flexible and articulate than the voice as a source of communication. Different people or objects may have been indicated at first by pointing rather than vocal naming, encouraging a gestural form of reference; in his book *One Hundred Years of Solitude,* Gabriel Garcia Marquez writes: "The world was so recent that many things lacked names, and in order to indicate them it was necessary to point."[49]

Indeed, pointing with the outstretched arm and index finger at objects in visual space seems to be unique to humans.[50] Infants seem to understand the pointing of others by about 12 months of age and can themselves point at objects for the benefit of others at about 14 months.[51] The chimpanzee, by contrast, can be taught to respond to pointing by looking in the direction of a pointing finger (and not at the end of the finger), but does not itself point.[52] In the evolution of hominids, the index finger became progressively longer relative to the other fingers in the period from 4 million to 1.7 million years ago.[53] It is commonly supposed that pointing is a precursor to the use of words to refer to objects,[54] but pointing may also have served as the primary means of referring to objects in early gestural language.

Of course, gesture still plays an important role in human communication. People commonly gesture as they speak, and certain ideas are more easily expressed by gestures than by words; try, for example, explaining to someone what a spiral is. In a detailed discussion of the gestures accompanying speech, David McNeill has argued that the two "share a computational stage; they are, accordingly, parts of the same psychological structure."[55]

Even more impressive evidence for gestural language comes from recent studies of ASL. Until quite recently, it was thought that ASL was merely a primitve pidgin, at the level of "You Tarzan, me Jane."[56] However, it has become clear that ASL loses nothing in comparison with spoken language; it has essentially the same hierarchical, embedded structure, and it is acquired by children in remarkably similar fashion.[57] We saw in Chapter 5 that deaf infants exposed to sign language "babble" just as hearing children do, except that they use elements of signs rather than spoken syllables. This suggests that we are just as preprogrammed to learn sign language as to learn vocal language. We have also seen that chimpanzees and gorillas can be more readily taught sign language than vocal language.

For all that, vocal language eventually became the dominant partner, at least among the hearing community. Hewes suggests that this may have occurred by the end of the Lower Paleolithic, with the emergence of archaic *H. sapiens*, when gestural language may have been no longer able to cope with the more sophisticated culture; he suggests, though, that the talent for gestures may have later found expression in the cave drawings, often called these frozen gestures, of the Upper Paleolithic. As we have seen, however, it now appears that gestural language, at least in the form of ASL, may be as sophisticated linguistically as vocal language, so the switch to vocal language may have been for other than linguistic reasons. For example, vocal speech does not require vision, and can therefore be carried on in the dark or when obstacles get in the way. It also frees the hands and eyes for other activities, allowing people to explain what they are doing with their hands and allowing the observer to watch while listening.

These factors may have therefore created a selective pressure that led, eventually, to gestures being superseded by vocal communication. Precisely how this occurred is a matter of speculation. As Diamond suggests, grunts associated with gestures or tool making may have acquired referntial meaning by association. Hewes also suggests that there may be a natural tendency to mimic with the tongue what the hands are doing, leading eventually to vocalizations that bear at least a crude relation to the gestures they accompany. Charles Darwin had also noted that precise movements of the fingers and hands are often accompanied by thrusts and twists of the tongue, often to be observed in a person trying to thread a needle.[58] Sam Weller, in Charles Dickens' *The Pickwick Papers*, moved his tongue during the difficult and unaccustomed act of writing. The transition from gestural to vocal language probably also depended in part on further changes in the vocal tract and in the positioning of the larynx.

The Descent of the Larynx

Philip Lieberman of Brown University notes that in modern humans, unlike the other apes, the larynx is relatively deep in the neck and exits directly into the pharynx, which is essentially a vertical tube. The oral cavity (the mouth) is approximately at a right angle to the pharynx and forms a second horizontal tube. This adaptation may have been a consequence of the hominid's upright stance, which involved an angular shift in the way the head is attached to the spinal column, bringing the line of sight back to the horizontal. The tongue curves around this two-tube airway, and can alter its shape so as to produce the range of vowel sounds that the adult human is capable of uttering. The tongue can also block the passage of air at the intersection of the two tubes to produce the stop consonants *k* and *g*.[59]

Lieberman and his colleagues argue that these characteristics are unique to humans. Their computer-aided modeling of the chimpanzee's vocal tract indicates, for example, that the chimpanzee is incapable of producing the so-called point vowels *i*, *u*, and *a*. We have already seen that the Hayes' chimpanzee could not learn to talk, whereas more recent attempts to teach other communication systems to chimpanzees were relatively successful. What is more controversial, however, is Lieberman's claim, based on a computer-aided reconstruction of the Neandertal fossil discovered at La Chapelle-aux-Saints, that the Neandertals were also incapable of producing the point vowels and thus could not have spoken like modern humans. The essential feature of Lieberman's model is that the Neandertal larynx was too high in the throat to produce these critical vowels.[60] Since the Neandertals lived as recently as 35,000 years ago, the implication is that the rapid, flexible style of human speech may be uniquely associated with anatomically modern humans.

Lieberman's claims have been disputed. According to one report, chimpanzees are capable of producing at least two of the point vowels that Lieberman's simulations say they cannot produce![61] Dean Falk has also argued that if Lieberman's simulation of the Neandertal vocal tract is accurate, then the hapless creature would have been unable to swallow, and so would have perished (which it did, of course, but probably not for that reason).[62] Morphology is not always a good guide to performance; for example, mynah birds can produce passable approximations to human speech, but this would never be suspected from an inspection of the syrinx (the avian equivalent of the larynx). Humans, moreover, can recover intelligible speech after removal of the tongue or the larynx,[63] and can whisper intelligibly without using the voice at all.

Doubts as to the validity of Lieberman's claims about Neandertal speech have recently centered on the hyoid bone, a small bone that lies between the root of the tongue and the larynx and plays a critical role in the production of speech. Lieberman's reconstructions of the Neandertal vocal tract were made in the absence of any of the soft tissue or of small bones like the hyoid, and were therefore based partly on guesswork. Recently, however, a well-preserved Neandertal hyoid bone was discovered at Mount Carmel in Israel. In terms of its shape and the markings left by the attachments of muscles, it lies within the range of modern human hyoids. The inference is that the larynx itself has not altered significantly since the Middle Paleolithic, and shows no significant variation between Neandertals and modern humans.[64]

Yet this discovery may not be decisive. John C. Marshall suggests that it may not bear on the critical aspect of Lieberman's model, which has to do with the position of the larynx in the throat, rather than with the detailed anatomy of the larynx itself. He remarks that arguments about Neandertal language "will undoubtedly run and run until we discover a deep-frozen Neandertal who is susceptible to resuscitation."[65] The place to start looking might be the New York City subway.

Lieberman does recognize that vocal language undoubtedly preceded *H. sapiens sapiens*. Moreover, the selective advantage conferred by the change in the vocal tract presumably depended on at least some prior development of vocal language. It may have been the descent of the larynx that marked the switch from a language dominated by gesture to one dominated by vocal communication. This switch would have further freed the hands for the development of tools, and for the creation of ornaments and works of art that are uniquely associated with anatomically modern humans.

In summary, the foregoing account is based largely on Diamond's treatise, which includes some outdated ideas. Nevertheless, it has an overall plausibility. Simple vocal communication was no doubt present long before the split between apes and hominids, but merely served a signaling function. However, the tying of vocalization to action may well have evolved from the primitive grunts that accompanied action, and it was this development that may have led eventually to true language. The earliest "sentences" may well have been simple "suggestions of action," consisting of single verbs. In the early stages of hominid evolution, gestures may have outstripped vocalizations as a medium of communication, since the hands and arms were probably capable of more varied and flexible signals than was the vocal tract. Changes in the functioning and positioning of the larynx may have allowed increasing flexibility and control in vocalization. This development culminated in the descended larynx of *H. sapiens sapiens* and permitted the voice to reassert its dominance over gestural communication. Note that, if this account is correct, the evolution of speech depended heavily on *exaptation* rather than adaptation; that is, many of the modifications that permitted speech were selected for functions other than speech.

As the hominids developed as tool makers, objects would have assumed more importance. The roles of individuals in the group may also have diversified, so that different people may have had to be identified and addressed individually. Consequently, nouns would have gradually become more important. This process may have been accompanied by the development of new sentence types, from the most primitive suggestion of action denoted by the single action verb to descriptions and description-statements.

A Missing Link?

Yet there still seems to be something missing. Diamond's analysis deals with the changes in the extent and nature of vocabulary that may have taken place during evolution, and it touches on possible changes in the kinds of sentences that humans might have become capable of uttering. But it does not deal with the *grammatical* complexity that only humans seem capable of achieving. Diamond, essentially pre-Chomskian in his view of language, has missed the true role of grammar in the generation of sentences; his sentence types are really no more than templates, and do not accommodate the embedded nature of human language that gives rise to its infinite variety.

David Premack also wonders at the "missing links," remarking that "nature provides no intermediate language, nothing between the lowly call system and the towering human language."[66] Perhaps this is an exaggeration. Derek Bickerton suggests that there is a level higher than the call system, and that it is represented by pidgins before they have become "creolized," by children under the age of 2, by wild children such as Genie, by the sign systems of trained apes (and dolphins, it might be added), and by the early attempts of adult humans to speak a second language.[67] Premack, replying to Bickerton, adds another—the speech of an intoxicated teenager.[68] Quite so.

What these examples still lack, however, is the natural grammar of human language. They do not take us beyond the basic sentence types identified by Diamond. They might also be considered the best that *cognition* can do in the absence of those secret recipes that guided the acquisition of language during the critical period of human growth. Premack bravely suggests an intermediate form that might have preceded the evolution of language in its present form, but this is pure speculation; Bickerton suggests that no such intermediate could possibly have existed.

Premack's concern to find (or invent) missing links seems to be based on the premise that evolution should be gradual. As we have seen, however, modern evolutionary theory has largely dispensed with Darwinian gradualism. Changes can be sudden, punctuate, dramatic. Piattelli-Palmarini comments:

> We do not "have to" find language "precursors" in the apes, simply because we do not "have to" find intermediate forms in the emergence

> of all biological traits. And, in fact, there are no such language precursors in the apes. . . . Stephen J. Gould never ceases to remind us that many incomplete series in the fossil record are incomplete, not because the intermediate forms have been lost *for us*, but because they simply never existed.[69]

That is, the human capacity for generative grammar may have emerged more or less fully fledged, due perhaps to some sudden genetic rearrangement or perhaps, as Bickerton suggests, to two or more preexisting systems that tapped into one another, producing an output that was quite different from the outputs of the original systems. As I have remarked before, there is a gratuitousness in such proposals; they do resort to magic after all. Nevertheless, such dramatic changes do happen, and human language is of such distinctive quality that it may well provide another example—a flash of evolutionary insight, as it were, rather than the slow burn of gradualism.

When did true human language begin? The answer is, of course, not known, but there have been attempts to track languages back through time to their common origins. Indeed, languages may even be said to "mutate," so that it is possible to provide some estimate of when different languages split from some common source. A pioneer of this approach was Sir William Jones, a British judge in India who decided in the late eighteenth century to learn ancient Sanskrit. He noted the strong similarity between Sanskrit and Latin and Greek. The common source of these languages is now known as Indo-European and may have arisen in the region of the Danube somewhere between 5000 and 6000 B.C.[70]. It is thought that this language in turn can be traced back to the Nostratic superfamily of about 13,000 B.C., and that there were perhaps five superfamilies at that time. These have been identified as Nostratic, Dene-Caucasian, Amerind, proto-Australian, and Austro-Asiatic.[71]

Language scholars have recently begun to probe even further back to a single *super*superfamily, or "mother tongue," known as proto-World. According to Vitaly Shevoroshkin of the University of Michigan, linguists have been able to reconstruct some 150 to 200 words of this language. For example, various words for *tooth,* such as the Congo-Saharan *nigi,* the Austro-Asiatic *gini,* the Sino-Caucasian *gin,* and the Nostratic *nigi* and *gini,* leading to the English words *nag* and *gnaw,* suggest that *nigi* and/or *gini* were in the original proto-World lexicon. Again, the English word *tell* is said to be derived from the proto-World term *tal,* and later *dal,* meaning tongue.[72]

Until recently, proto-World was located in the Near East at about 35,000 years ago, on the grounds that this would have allowed enough time for the various populations to spread and formed the five superfamilies described above. However, in light of the evidence, reviewed in Chapter 2, that modern humans have a single origin in Africa, it is now thought that proto-World has an African origin perhaps dating back 100,000 years.[73] Colin Renfrew has recently remarked that "Most (but by no means all) scholars today believe that the comprehensive linguistic ability seen in human populations emerged with *H. sapiens sapiens,* the anatomically modern form of our own species."[74]

According to Shevoroshkin, though, proto-World was a rather primitive language. In its earliest form, meaning was conveyed by consonants only, and the only vowel was the short *a* sound produced naturally when consonants were produced (as foreseen by Diamond). Later, different vowels were developed in order to distinguish between different meanings. For example, it is thought that in early proto-World the word *changa* referred both to the nose and to odor, but in a later development the word *chunga* came to refer to ordor, while the older term *changa* was restricted to the nose. The language may therefore have developed in the period from the origins of *H. sapiens sapiens* some 150,000 to 200,000 years ago until the period when this species began to disperse, about 100,000 years ago.

These considerations suggest that spoken language as we know it may indeed be uniquely associated with *H. sapiens sapiens* and may have emerged in Africa in a rather narrow period of time. The question still remains, however, as to whether it was due to some dramatic genetic reshuffle, or whether it was a culmination of more gradual evolutionary changes.

Two Scenarios

To conclude, then, I suggest two possible scenarios. The first is that language did indeed first appear in *H. sapiens sapiens*, and may even explain the adaptive advantage that this species held over the archaic *H. sapiens*. As we have seen, one of the important properties of language is its generativity, and this property may have been later applied, or perhaps exapted, to the manufacture of objects and ornaments, as evidenced by the so-called evolutionary explosion of some 35,000 years ago. This scenario is in accord with the views of those, such as Piattelli-Palmarini, who believe that generative language represents a sudden but sizable jump from all other forms of communication between animals.

Although proto-World seems to have been rather primitive, at least in the early stages, this scenario still seems to imply that *H. sapiens sapiens* was suddenly possessed of a new-found generative ability, although it took perhaps tens of thousands of years to fully realize its potential. This, in turn, seems to imply some arbitrary genetic reshuffle. Although such genetic changes do occur, this account still smacks of the magical, even of the theological.

The second scenario is that generative language evolved gradually from *H. habilis* to *H. sapiens* but was gestural rather than vocal. It may have evolved first in the contexts of tool making and of an increasingly complex social organization, and may have taken the form of demonstrating tool-making techniques and of such gestures as pointing or indicating action. Such gestures would have served as commands, instructions, and propositions rather than as expressions of emotion. Vocalizations, by contrast, may have served a primarily emotional role, warning of danger or expressing anger, fear, or delight; indeed, we still use vocalizations in these essentially nonverbal ways. However, gestural actions may have been accompanied by grunts, as Diamond suggests, and these may have gradually taken on a secondary role in propositional language.

Although authors such as Piattelli-Palmarini and Chomsky have denied the possibility of the gradual evolution of generative grammar, this possibility may still be more palatable than that of resorting to a sudden, dramatic reshuffling of genes. The missing links would simply be lost in the mists of time. The quality of generativity may have evolved initially in the context of the increasingly complex events that took place in the lives of the early hominids. As social structures grew more intricate, it was necessary to differentiate the various roles of members within the group and the different combinations of interactions that could take place. These interactions might have involved pairs of group members, or individuals and their stone implements, or the roles of different individuals as tool makers, carers, hunters, and so on. Rather than code these interactions individually, it may have been more efficient to derive general principles so that various interactions could be generated.

Some more specific aspects of grammar might also be traced to this early phase of hominid evolution. For example, the act of making a tool might be described from the angle of the maker ("X makes a chopper") or of the tool ("The chopper was made by X"), giving rise to the distinction between the active and the passive. This does not explain the precise rules governing the relation between active and passive, but at least we can note that the focus of the utterance (X or the chopper) is placed first in both cases. Similarly, the embedding of phrases might have evolved with the need to elaborate on the items and to link the elaboration by proximity to the item elaborated ("The chopper that you gave me was made by X").

In a recent unpublished article, Steven Pinker and Paul Bloom have also defended the idea that language evolved through natural selection, and not as a spin-off from a selection for some other trait or as a sudden genetic reshuffle. They too stress the survival value of language in the social life forged by the earliest humans:

> In sum, primitive humans lived in a world in which language was woven into the intrigues of politics, economics, technology, family, sex, and friendship and that played key roles in individual reproductive success. They could no more live with a Me-Tarzan-you-Jane level of grammar than we could.[75]

According to this second scenario, the critical event that distinguished *H. sapiens sapiens* was the switch from gesture to vocalization as the principal medium of propositional language. This interpretation rests somewhat on Lieberman's controversial ideas about the descent of the larynx, but it need not have been as abrupt as Lieberman implies. That is, the anatomical change in the larynx might have been just enough to tip the balance in favor of vocal over gestural language. This shift would have had adaptive significance in that individuals could communicate with one another in the dark, or around large obstacles, or simply without having to look at one another. But the more important aspect of the shift may have been that it freed the hands and the eyes from any involvement in communication, and this may have had important consequences in the manufacture of tools and other artifacts. Individuals could now explain manual techniques without having to interrupt the processes themselves. More important, perhaps, the generative component, previously restricted to gestural language, could now be exapted to the processes of manufacture itself. This theme will be explored more fully in Chapter 9.

The next chapter, however, marks a return to the lopsided theme of this book.

Notes

1. From Sarah Trimmer's *Fabulous Histories Designed for the Instruction of Children* (3rd edition, 1788, p. 71), cited by Thomas (1984, p. 92).
2. Chomsky (1966a, pp. 77–78).
3. These examples are taken from Griffin (1976).
4. See von Frisch (1955); also Griffin (1976) for discussion.
5. Hailman and Ficken (1987).
6. Op. cit., p. 1901.
7. Op. cit., p. 1901.
8. Seyfarth and Cheney (1984).
9. Op. cit.

10. Gautier and Gautier-Hion (1982, p. 6).
11. Marler (1977).
12. Snowdon (1989).
13. Snowdon (1989).
14. Duputte (1982).
15. Snowdon (1982, p. 233).
16. Liberman (1982).
17. Snowdon (1989).
18. Massaro (1987). The argument involves other considerations as well.
19. Seyfarth and Cheney (1984, p. 72).
20. Snowdon (1982).
21. Snowdon (1989).
22. Jones (1919, p. 185).
23. Savage-Rumbaugh (1987).
24. Quoted by O'Grady and Dobrovolsky (1987, p. 418).
25. Herman, personal communication, February 1989.
26. Yerkes (1925).
27. Hayes (1952).
28. Gardner and Gardner (1969).
29. Patterson (1978)
30. Premack (1971).
31. Rumbaugh (1977).
32. Bronowski and Bellugi (1970).
33. Terrace (1979); Terrace, Petitto, Sanders, and Bever (1979)
34. Herman, Richards, and Wolz (1984).
35. Premack (1985). Premack argues, in fact, that the true test of language should not be merely the ability to generate sentences but to converse.
36. Savage-Rumbaugh (1984).
37. Zihlman, Cronin, Cramer, and Sarich (1978) have argued that the pygmy chimpanzee is similar to *Australopithecus*, and might be considered a prototype of the common ancestor of apes and hominids. This has been disputed, however, by Johnson (1981).
38. See Savage-Rumbaugh (1987) for a summary.
39. Again, this evidence seems contrary to Liberman's (1982) view that the manner in which humans perceive phonemes depends on special mechanisms that are uniquely human.
40. These figures are from Pfeiffer (1973).
41. Darwin (1901, p. 127).
42. Op. cit., p. 132.
43. These figures are given by Diamond (1959). According to Claiborne (1983), the largest English dictionaries contain between 400,000 and 600,000 entries. Diamond may also have underestimated the number of words in the languages of the Food Gatherers: Claiborne remarks on p. 26 that "every human language that has been studied has a vocabulary in excess of twenty thousand words—about the number Shakespeare used and far more than we find in the Bible."

44. Diamond (1959, p. 110).
45. Claiborne (1983, pp. 8–9).
46. Op. cit., p. 144.
47. Condillac (1746).
48. Hewes (1973).
49. Marquez (1971, p. 11).
50. Butterworth and Grover (1988).
51. Schaffer (1984).
52. Woodruff and Premack (1979).
53. Hilton (1986), cited by Butterworth in Weiskrantz (1988, p. 68).
54. For example, Bruner (1983).
55. McNeill (1985, p. 305).
56. See John C. Marshall's introduction to Poizner, Klima, and Bellugi (1987).
57. See, for example, Bellugi and Klima (1982) and Hoffmeister and Wilbur (1980).
58. Darwin (1872).
59. Lieberman (1984).
60. Lieberman, Crelin, and Klatt (1972).
61. Jordan (1971), cited by Marshall (1989).
62. Falk (1975).
63. Luchsinger and Arnold (1965).
64. Arensburg, Tillier, Vandermeersch, Duday, Schepartz, and Rak (1989).
65. Marshall (1989, p. 703).
66. Premack (1985, p. 276).
67. Bickerton (1986).
68. Premack (1986).
69. Piattelli-Palmarini (1989, p. 8).
70. See Claiborne (1983) for an accessible account.
71. See Roger Lewin's article "Linguists search for the mother tongue," published under *Research News*, in the 25 November 1988 issue of *Science* (Vol. 242, pp. 1128–1129).
72. Shevoroshkin (1990).
73. Op. cit.
74. Renfrew (1989, p. 90).
75. Pinker and Bloom (1990, p. 34).

— 7 —

Language and the Brain

. . . a chimpanzee is very smart and has all kinds of sensori-motor constructions (causality, representational functions, semiotic functions, and so forth), but one thing is missing: that little part of the left hemisphere that is responsible for the very specific functions of human language.[1]

As THE ABOVE QUOTATION from Noam Chomsky illustrates, it is widely believed that the uniquely human gift of language is due to some special property of the left side (or *hemisphere*) of the human brain—as though a right-handed God had reached down and touched humans with some magic wand on the left temple. The evidence for the special role of the left hemisphere came in the first instance not from scripture but from the study of patients suffering damage to the brain. And although the predominance of the right hand has long been observed, the asymmetry of the brain was not discovered until well into the nineteenth century.

Evidence from Brain-Injured Patients

Disorders of Spoken Language

In 1836, a little-known medical doctor named Marc Dax presented a paper to a meeting of a medical society in Montpelier, Prance. Many of his patients had suffered from loss of speech as a result of damage to the brain. In more than 40 such patients, Dax observed signs of damage to the left side of the brain. He could not find a single case in which there were indications that the right side alone was damaged.[2]

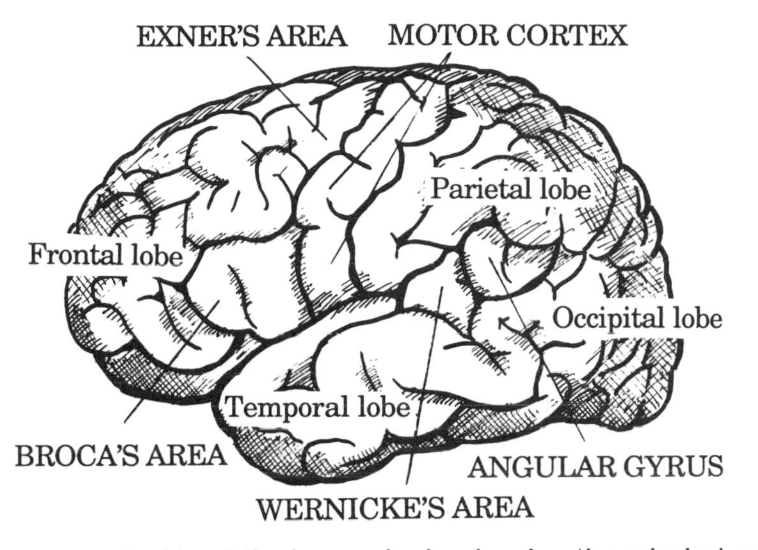

Figure 7.1. Left side of the human brain, showing the principal areas involved in language.

Dax died the following year and his contribution was neglected until the 1860s, when the young French physician Paul Broca made further observations on the representation of language in the brain. He found that patients with damage to the posterior portion of the third frontal convolution of the brain suffered loss of speech, a condition that he called *aphemia*. This area of the brain, now known as *Broca's area*, lies close to (and in front of) part of the so-called motor cortex of the brain associated with movements of the mouth. Broca's area is marked in Figure 7.1. Not until he had evidence from eight patients, however, did Broca note that in each case the damage was to the left side. With due scientific caution, he said "I do not attempt to draw a conclusion and I await new findings."[3]

By the following year Broca was convinced of the dominant role of the left side of the brain in language, and in 1865 he took the further step of suggesting that in left-handers the right side would be dominant.[4] The idea that the hemisphere controlling speech is on the side opposite the preferred hand became known as *Broca's rule*, and was widely believed until well into the twentieth century. As we shall see in Chapter 8, however, there are many exceptions to the rule.

In the meantime, Marc Dax's son Gustav had learned of Broca's work and wrote a letter claiming that Broca had deliberately overlooked his father's contribution. Broca denied the charge and claimed never to have heard of the elder Dax. Gustav located the text of his father's talk and had it published, thereby securing Marc Dax's place in history.[5]

The term *aphemia*, coined by Broca, was later replaced by *aphasia*, which refers to general disorders of spoken language. Patients with damage to Broca's area typically suffer a form of aphasia that primarily affects expressive language rather than comprehension. For example, one of Broca's famous patients was known as "Tan" because this was the only articulate sound he could make. He could understand the speech of others, however, and seemed to have no impairment to the motor organs involved in speaking; for example, he could move his tongue according to command. This kind of aphasia is known as *motor aphasia*, *expressive aphasia*, or simply *Broca's aphasia*.

Not long after Broca's discoveries, a German neurologist named Carl Wernicke discovered that damage to a posterior portion of the brain produced a deficit in the understanding of speech, leaving expression more or less intact.[6] This condition is variously known as *sensory aphasia*, *receptive aphasia*, or *Wernicke's aphasia*. Again, it nearly always occurs following damage to the left side. Wernicke's area is located in the upper posterior part of the temporal lobe, around the junction of the temporal, parietal, and occipital lobes, and is marked in Figure 7.1.

Patients with Wernicke's aphasia may speak fluently, even garrulously, but do not appear to understand even what they themselves say. Their speech is usually grammatically correct but meaningless—the sort of thing one notices occasionally in one's academic colleagues. (We might even recall here Chomsky's sentence "Colorless green ideas sleep furiously.") The late Norman Geschwind called this kind of speech "the running on of the relatively isolated Broca's area."[7] Because of its garrulous quality, it is sometimes also called *jargon aphasia*.

Broca's and Wernicke's areas are connected by a tract of fibers known as the *arcuate fasciculus*, and damage to this pathway produces what is known as *conduction aphasia* or *central aphasia*. The patient can understand speech, and can speak quite fluently except for a tendency to circumlocution and what are called *paraphasic* errors, in which phonemes are wrongly substituted into words—a typical example is "fots of fun" instead of "lots of fun."[8] But the main symptom is a poor ability to repeat back what someone has said. In standard tests of short-term memory, most people can repeat back about seven or eight randomly chosen digits, but patients with conduction aphasia can typically manage only one or two.[9] These patients evidently lack the "inner speech" that normally allows rehearsal, and so permits us to remember such things as a telephone number for a short period of time.

Yet another category of aphasias are the *transcortical aphasias*, in which the main symptom is essentially the converse of the repetition deficit of the conduction aphasic. The transcortical aphasic has a persistent and no doubt irritating habit of repeating back everything that is said.[10] This is known as *echolalia*. It is the primary symptom in two otherwise distinguishable subtypes of transcortical aphasia. One is transcortical *motor* aphasia, in which there is damage in areas marginal to Broca's area, and the patient shows a loss of spontaneous speech, although comprehension seems to be normal. The other is transcortical *sensory* aphasia, resulting from damage near Wernicke's area. This is accompanied by poor comprehension and an inability to name objects.[11]

In 1885 Lichtheim proposed a theory to explain all of these different kinds of aphasia. A simple version of this theory is shown in Figure 7.2. Basically, there are three "centers." One is Broca's area, where words are represented in their spoken (or motor) forms. Another is Wernicke's area, where words are stored in auditory forms. The third is an area where concepts are stored and elaborated. Aphasias are caused by damage to Broca's and Wernicke's areas themselves, or by damage to the connection between them (conduction aphasia), or by damage to their links with the concept area (transcortical aphasias). This theory was influential at the time but lost favor as arguments raged over whether psychological functions could be meaningfully localized in the brain. Lichtheim's basic approach has been revived, however, in modern cognitive neuropsychology.[12]

In Chapter 5, I introduced the idea of connectionism, or parallel distributed processing. One of the premises of this approach to the modeling of the mind is that representations are distributed over networks rather than localized at specific points. Taken to its extreme, this would imply that all parts of the brain represent all functions. The truth probably lies somewhere between this extreme and that of strict localization. Consequently, a theory like Lichtheim's is probably a reasonable approximation to the truth, although there is undoubtedly some diffuseness and overlap in the way the centers are represented in the brain.

This diffuseness may become apparent on a more fine-grained analysis. For example, patients with damage in the region of Broca's area sometimes suffer from what is known as *agrammatism*, or an apparent difficulty with aspects of grammar.[13] Function words such as articles, prepositions, and conjunctions tend to be omitted, as do

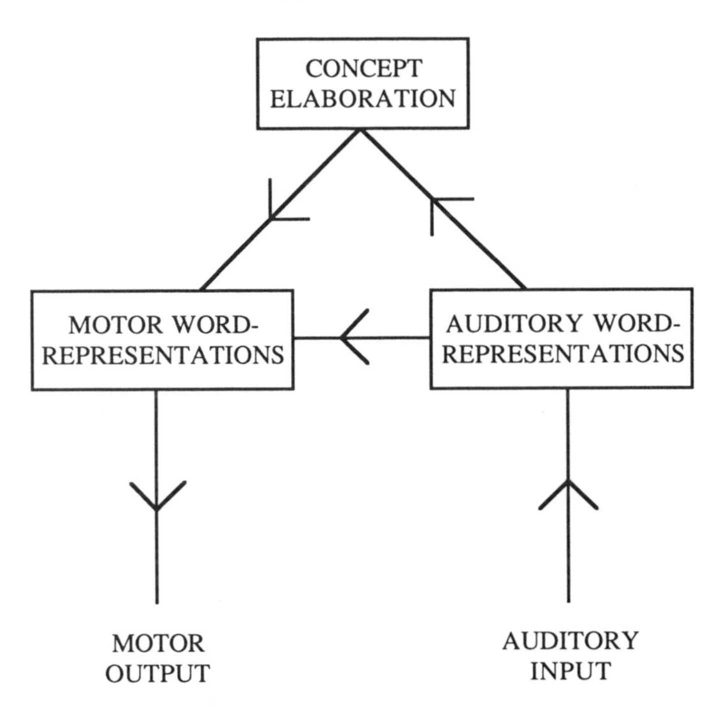

Figure 7.2. Schematic representation of Lichtheim's theory. Different forms of aphasia result from damage to the boxes or to the arrows linking them (after Shallice, 1988).

inflections such as the *-ed* ending to signify past tense or the *-s* to signify plural. Speech is reduced to a kind of "telegraphese," although individual words may be formed correctly. At one time it was thought that this was just an aspect of Broca's aphasia, but it has become clear that it is at least partly dissociable from it. For example, patients sometimes have extreme difficulty in producing phonemes (*phonetic disintegration syndrome*), but their grammatical competence remains unaffected.[14] It has also been shown that patients classified as Broca's aphasics have difficulty in *understanding* sentences when grammatical analysis is critical, suggesting that the distinction between expressive and receptive disorders may not always be absolute.[15]

The idea that agrammatism might be a form of aphasia distinct from Broca's aphasia received some impetus from the Chomskian revolution in linguistics. As we saw in Chapter 5, the Chomsky school sees grammar as the key to human language. During the 1970s a group of researchers located, not surprisingly, in Chomsky's home town of Boston documented the idea that agrammatism was the result of damage to a central input/output system having to do specifically with grammar and was dissociable from Broca's aphasia.[16]

Traditionally, agrammatism has to do with deficits of grammar at the level of morphology (deletion of morphemes, function words, etc.) rather than at the level of the ordering of words in sentences. Recently, however, Stephen E. Nadeau of the Veteran's Administration Medical Center in Gainesville, Florida, described two patients with extensive damage to the left frontal area of the brain who were fluent in speech and who were not agrammatic in the usual sense. However, they had severe difficulty with grammatical constructions at the level of sentences. For example, they could not deal with embedded phrases or with the passive voice. This suggests that there is a dissociation between the morphological and the sentential (or what Nadeau calls the *syntactical*) aspects of grammar.[17]

As these developments illustrate, increasingly fine-grained analyses add considerable complexity to the scheme proposed by Lichtheim, and it becomes increasingly difficult to locate the component processes with any precision in the brain. However, there is at least one respect in which localization of function holds true: In the great majority of people, the mechanisms of language are located predominantly, if not exclusively, in the left hemisphere of the brain.

Are there any aspects of language in which the right hemisphere plays a role? It is sometimes suggested that the right hemisphere contributes especially to *prosody*, which refers to *how* something is said rather than *what* is said. Prosody involves both emotional aspects of speech, as when one speaks angrily, fearfully, impatiently, and so on, and nonemotional aspects, as in the different intonations required in asking a question, giving a command, making an assertion, and so on. There is some evidence that the right hemisphere plays a specialized role in emotional prosody,[18] in keeping with more general evidence for a specialized right-hemispheric involvement in emotion[19]—this will be documented in Chapter 10. However, there is doubt as to whether the right hemisphere contributes in any special way to nonemotional prosody.[20] In this as in other respects, however, the role of the right hemisphere in language remains a controversial topic.

Disorders of Reading and Writing

Left-hemispheric dominance holds for reading and writing as well as for speech. At the end of the nineteenth century, the French neurologist Déjérine noted that damage to the angular gyrus, which is also marked in Figure 7.1, produced *alexia*, or an inability to read.[21] In this case, alexia was typically accompanied by *agraphia*, or an inability to write. Déjérine assumed that the angular gyrus was the center for visual images of letters, and was necessary for both reading and writing. He noted that some patients suffered alexia *without* agraphia, a rather paradoxical condition in which they could write but could not then read what they had written! Déjérine assumed that in such cases the angular gyrus was intact, but that its connections to the areas of the brain involved in visual analysis were damaged. Over the past two decades, there has been considerable interest in alexia,[22] and again, this simple picture has been complicated somewhat. In essence, however, Déjérine's basic insights have remained largely intact.[23]

Agraphia sometimes occurs without alexia, often as a result of damage to the second convolution of the left frontal lobe.[24] This area is sometimes known as *Exner's area* and is the final area shown in Figure 7.1. It lies quite close to the area of the so-called motor cortex concerned with movements of the hand. It may be considered to contain the programming for the formation of script and is in many respects analogous to Broca's area. However, pure agraphia may result from damage to other areas as well, and it has not been so extensively studied as have disorders of speech.[25]

Disorders of Sign Language

We have seen that American Sign Language (ASL) has essentially the same formal characteristics as spoken language, and that deaf children acquire it in much the same way that normal children acquire spoken language. Chapter 6, moreover, suggested that gestural language preceded vocal language in hominid evolution and remains, biologically speaking, an equally viable option. If this is so, one would expect ASL also to be mediated primarily by the left cerebral hemisphere.

Cases of one-sided brain injury in speakers of ASL are, of course, rare. Howard Poizner, Edward S. Klima, and Ursula Bellugi have analyzed six such cases, three with left-sided and three with right-sided damage, in considerable detail.[26] They conclude that ASL is indeed mediated primarily by the left cerebral hemisphere and that, as in the case of speech disorders, damage to the forward part of the left hemisphere produces expressive disorders, while damage to the rearward part produces receptive disorders. Damage to the right side of the brain does not seem to affect the ability to produce signs, even though aspects of sign language are spatial rather than sequential in form. However, damage to the right parietal lobe may produce an impairment in the ability to *understand* the spatial component of signs.

These conclusions are based on only a small number of cases, but they support the idea that the representation of ASL is essentially the same as that of spoken language. However, it is not yet known whether the actual areas involved are the same. For example, it is not yet known whether the production of signs is associated with Broca's area, which lies close to the area of motor cortex that controls the mouth, or whether it is associated with an analogous region lying closer to the area that controls the hands.

Summary

In summary, evidence from patients suffering injury to one side of the brain overwhelmingly supports the idea that the left side of the brain, at least in the great majority of people, is dominant for language. This is true of language as spoken, written, or signed. Damage to the right side of the brain seldom causes any impairment in language, save perhaps for difficulties with prosody and with the comprehension of the spatial aspect of sign language. Until fairly recently, then, it was assumed that the right hemisphere, often called the *silent* hemisphere, had essentially no role in language. This assumption was questioned, however, following studies of so-called split-brained patients.

The Split Brain

The two hemispheres of the brain are connected by several tracts of fibers, or *commissures*, the most important of which is the *corpus callosum*. There has long been interest in what would happen if the commissures were cut, separating the two halves of the brain. The nineteenth-century experimental psychologist Gustav Fechner thought that the human mind would be essentially duplicated if the brain were split, since each half has "the same moods, predispositions, knowledge, and memories."[27] Following the split, however, each half would be free to develop independently, creating two minds in the one head.

This view was challenged by the British psychologist William McDougall, a dualist who believed that the mind would transcend any physical separation of the two halves of the brain. He even volunteered to have his own corpus callosum cut if he ever contracted an incurable disease, in the hope of showing that his consciousness would remain unitary.[28] Fortunately, his offer was never taken up.

In the early part of this century, observations of the effects of disease of the corpus callosum led to a number of theories about its function. For example, a *mental callosal syndrome* was identified in 1906 as involving some loss of connectedness of ideas, problems with recent memory, a "bizarreness" of manner, and swings of mood.[29] In most cases, damage to the corpus callosum also involved damage to neighboring areas, and it was difficult to determine precisely which deficits were associated with the callosum itself. However, following an early case of surgical section of the callosum, Walter Dandy wrote as follows in 1936:

> The corpus callosum is sectioned longitudinally . . . no symptoms follow its division. This simple experiment puts an end to all of the extravagant hypotheses on the functions of the corpus callosum.[30]

Sectioning of the corpus callosum for the relief of intractable epilepsy was undertaken more systematically in the early 1940s by William Van Wagenen, a surgeon from Rochester, New York. The patients were tested by Andrew Akelaitis, who found remarkably few impairments of perceptual or motor abilities.[31] This seemed to confirm Dandy's conclusion, and it was widely accepted that "the corpus callosum is hardly connected with any psychological functions at all."[32]

This picture was soon to change. In the early 1960s two Los Angeles surgeons, Philip J. Vogel and Joseph E. Bogen, cut the forebrain commissures, including the corpus callosum and the anterior and hippocampal commissures, in a new series of epileptic patients. The idea was to prevent the spread of epileptic seizures from one side of the brain to the other, and in this the operation proved successful. Again, the patients seemed remarkably normal once they had recovered from the operation. However, subtle tests, devised for the most part by Roger W. Sperry at the California Institute of Technology, revealed that the two hemispheres were psychologically as well as physically disconnected from one another; that is, input directed to one hemisphere was largely inaccessible to the other. This provided a striking way of demonstrating and studying the different capacities of the two sides of the brain.

The most clear-cut way of studying one hemisphere independently of the other is to use visual input. The visual system is organized so that pictures or objects that fall to one side of where a person is looking are projected entirely to the opposite side of the brain. This is true regardless of whether one views with the left eye, the right eye, or both eyes (see Figure 7.3). Sperry and his colleagues had the patients fixate on a point on a screen and then flashed words or pictures of objects to one or the other side of that point. It was necessary to flash them quickly, in about a tenth of a second, so that the patients did not have time to move their eyes. (It takes about a fifth of a second for shifts in gaze to be initiated.)

It was found that the patients could not name words or objects flashed to the left side of fixation but could easily do so when they were flashed to the right side. The explanation, presumably, was that material flashed to the right side was relayed to the hemisphere responsible for speech, while material flashed to the left side was relayed to the silent right hemisphere. These findings corroborated the evidence that the left side of the brain was indeed the one that was responsible for speech.[33]

Similar findings were obtained with touch and hearing. The patients could name unseen objects felt with the right hand but not those felt with the left hand.[34] With spoken words, a similar asymmetry was observed, but only when the words were spoken in simultaneous pairs, one to each ear—a procedure known as *dichotic listening*. In this case, the patients could repeat back words spoken to the right ear but not those spoken to the left ear. These results were again taken as evidence for the dominance of the left hemisphere for speech.[35]

What was surprising at the time, however, was that the right hemispheres of at least some of the patients seemed able to *understand* words. For example, patients were able to pick out with their left hands the object from an array that corresponded to a noun flashed in the left visual field, which meant that the right hemisphere could understand the word and match it to the object. Eran Zaidel of the University of California at Los Angeles later devised an elegant contact-lens technique for restricting input to one side of the visual field during prolonged viewing, and so was able to carry out more intensive testing of the right hemisphere's language capacities. One of the tests he used was the Peabody Picture Vocabulary Test, in which the person must pick out pictures corresponding to spoken words; with Zaidel's procedure, only one hemisphere had visual access to the pictures. By this means he could test each hemisphere's comprehension of the words. The right hemispheres of two adult split-brained patients had scores equivalent to those of the average 16- and 11-year-old, respectively.[36]

Zaidel later showed that the visual vocabularies of these two patients were somewhat poorer, about 5 years behind their auditory vocabularies.[37] This discrepancy may be due to an inability of the right hemisphere to generate the sound image of a printed word.[38]

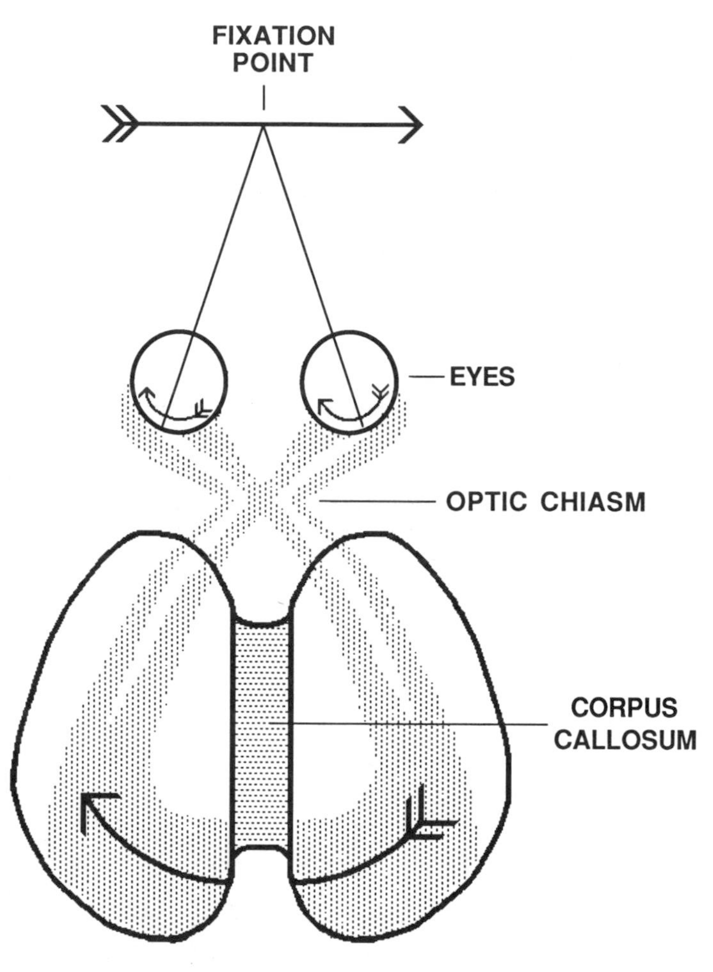

Figure 7.3. Schematic representation of the human visual system, showing the neural projections (*stippled*) from the retina of the eyes to the visual areas of the brain.

For example, the right hemispheres of the same two patients could not pick out rhyming words, such as *pea* and *key*, whose rhyming portions are spelled dissimilarly. Zaidel suggested that the right hemisphere was able to read only in a visual fashion and was unable to imagine the sounds of the words.

Until recently, these findings were interpreted to mean that the right hemisphere, although unable to produce speech, is capable of some degree of comprehension. This has been disputed by Michael S. Gazzaniga, who was initially involved in testing the California series of patients but then shifted to the East Coast, where he has been working with split-brained patients operated on by the late Donald H. Wilson at Dartmouth Medical School. Gazzaniga has become convinced that language comprehension by the right hemisphere is the exception rather than the rule, and that the two patients tested by Zaidel were unusual even among the 15 or so patients in the California series.[39] According to Gazzaniga, only three out of a total of 28 patients in the East Coast series show evidence of language processing in the right hemisphere. He suggests, moreover, that the right-hemispheric contribution to language in these exceptional patients might be attributed plausibly to early damage to the left hemisphere.

Zaidel has replied that the two patients he has tested extensively were selected precisely because they were relatively free of complications and seemed most likely to reflect the state of affairs in the normal brain.[40] At the time of writing, however, the dispute does not seem to have been resolved. All of the split-brained patients have had histories of epilepsy, and we may never know to what extent the organization of function in their cerebral hemispheres is representa tive of that in normal people. Probably, though, Gazzaniga is correct in suggesting that the extent of right-hemispheric involvement in language has been exaggerated, at least.

Tests of Normal People

The tests described above for split-brained patients have also been used to assess cerebral asymmetry in normal people. They are not so convincing as those in split-brained patients, however, since the presence of the commissures in normal people means that information projected to one side of the brain can be transferred to the other, although with some loss of accuracy and speed.

This means that most people recognize words, letters, and other verbal items slightly more accurately if they are flashed to the right of a fixation point than if they are flashed to the left. This may be termed a *right-hemifield advantage*. However the reason for this is not always to be found in cerebral asymmetry. Another possible explanation is that, in English at least, words are scanned from left to right, and this should favor presentation to the right of fixation since the beginning of the scan is nearer the fixation point. Indeed, in one of the earliest studies, it was found that Yiddish words, which are scanned from right to left, were more accurately recognized if flashed to the *left* of fixation, reversing the asymmetry found with English words.[41] However more careful studies have shown that when the influence of scanning is removed there remains a small advantage for presentation to the right side of fixation, which presumably reflects the left-hemispheric dominance.[42]

There is also evidence from dichotic listening that words and speech sounds spoken to the right ear are usually reported more accurately than words simultaneously spoken to the left ear—a so-called *right-ear advantage*.[43] Again, this effect is much less striking in normal people than it is in split-brained patients, but it can still be taken as evidence for left-hemispheric specialization.

Marcel Kinsbourne has suggested that asymmetries in both auditory and visual perception might be due, at least in part, to biases in paying *attention*.[44] There is evidence that each side of the brain controls attention to the opposite side of space. The presentation of verbal material activates the left hemisphere, and so directs one's attention to the right side of space. Kinsbourne has claimed that one can simply demonstrate switches in attention due to hemispheric activation by noting whether people initially look to the left or right when asked different kinds of questions. He reported that most people tend to look to the right when asked a verbal or computational question (such as "What is the cube root of 27?"), reflecting activation of the left hemisphere. They tend to look to the left when asked a spatial question (such as "How many windows are there in your living room?"), reflecting a specialization of the right hemisphere for spatial processing.[45] However these effects are not always reliable and are subject to a number of possible artifacts. For example it has also been suggested that people have a general preference to look one way or the other regardless of the question, with "right-lookers" classified as predominantly left-hemispheric (rational, logical, boring) and "left-lookers" and as predominantly right-hemispheric (intuitive, creative, interesting).[46] A critical review of the rather tangled evidence suggests that these trends are erratic,[47] but this need not dissuade the reader from trying it for fun. Just look into someone's eyes, ask the question, and make of the result what you will.

The voluminous literature on visual and auditory asymmetries in normal people has been reviewed elsewhere.[48] It does not clearly resolve the issue of whether the left-hemispheric specialization for language is relative or absolute. As we have seen, normal people *can* process words if they are flashed to the left hemifield, or if they are spoken under dichotic conditions to the left ear, but the evidence does not make it clear whether this is due to some degree of language processing in the right hemisphere or to transfer from the right hemisphere to the left.[49]

A Special Left-Hemispheric Speech Processor?

The right-ear advantage in dichotic listening has attracted interest for another reason. Alvin M. Liberman of the Haskins Laboratory in Connecticut has argued that it occurs because the left hemisphere of the human brain has evolved special mechanisms for the perception of speech sounds. These mechanisms apply particularly to the perception of certain phonemes. For example, perception of stop consonants such as *b*, *d*, and *g* depend on the detection of very rapid shifts in sound frequency, and these shifts are strongly dependent on the context in which they are embedded. The frequency shift in a given consonant is quite different, depending on whether the consonant is followed by the vowel *a* or *u*, for example. This means that consonants that actually sound the same to the human ear may be acoustically quite different. To cope with this, Liberman argues, the human left hemisphere has evolved mechanisms that are distinct from those involved in the perception of nonverbal acoustic events.[50]

In support of this theory, two of Liberman's associates, David Shankweiler and Michael Studdert-Kennedy, demonstrated a right-ear advantage for the discrimination of syllables that differed only in a consonant, but found no difference between ears in discriminating steady-state vowels. They concluded that the special acoustic mechanisms for the analysis of speech sounds are indeed normally located in the left hemisphere.[51] This work has been influential in the idea that the brain is organized in modular fashion, with different modules dedicated to specialized modes of processing. Indeed, the highly specific nature of speech processing seemed to provide one of the clearest examples of a module.[52]

In recent times, however, there has been an accumulation of evidence against Liberman's theory. In his recent book, *Speech perception by Ear and Eye*, Dominic W. Massaro has argued that much of the evidence for the special, modular nature of speech perception has been misinterpreted, and that the perception of speech actually follows quite general laws of perception.[53] At least some of the supposedly special characteristics of the perception of human speech sounds seem to be present in nonhuman species, including macaque monkeys, pygmy marmosets, and even the lowly chinchilla![54] We saw in Chapter 6 that a young pygmy chimpanzee named Kanzi is evidently able to understand English words, and that this understanding is based on the genuine perception of phonemes, including consonants. There is also evidence that factors other than the discrimination of consonants may contribute to the right-ear advantage in humans; indeed, it seems to be possible to demonstrate a right-ear advantage in the discrimination of vowels, provided that the discrimination is made difficult enough.[55]

These various findings need not completely rule out the existence of special mechanisms in the human left hemisphere for the perception of certain speech sounds. Liberman's elegant experiments go well beyond the evidence so far available from nonhuman animals, and there may well be aspects of the perception of speech sounds that are uniquely human. Liberman has continued to insist that this is so, and in an article published in 1989 he states that speech perception "is part of the natural human grammatical capacity that, together with syntax, distinguishes language from all other forms of communication."[56]

In summary, the predominantly left-hemispheric representation of language stands as one of the distinctive characteristics of the human brain. I next consider whether comparable asymmetries are to be found in nonhuman animals.

The Evolution of Cerebral Asymmetry for Language

Nonhuman Species

This chapter opened with a quote from Chomsky implying that the left-hemispheric specialization for language is uniquely human. If language itself is uniquely human, then this view is trivially true. However, there is growing evidence for systematic cerebral asymmetries in nonhuman animals, including in vocal communication.[57]

One of the most striking of these asymmetries occurs in certain songbirds. Birds sing by means of an organ known as the *syrinx*, which is in effect the equivalent of the human larynx. This is innervated by a nerve known as the *hypoglossal* nerve. In chaffinches, canaries, and the white-throated sparrow, singing is largely abolished if the left hypoglossal nerve is cut but is relatively unaffected if the right hypoglossal nerve is cut.[58] Since the hypoglossal nerve is controlled by a brain nucleus on the same side, it is inferred that singing in these birds is under left-hemispheric control.[59]

Since it has already been suggested that our forebears may have led an aquatic existence for a period of time, it would be a flight of fancy to claim now that we are also descended from birds. More relevant to our own evolution, perhaps, is a claim that Japanese macaque monkeys show a right-ear advantage in discriminating between the "coo" sounds of their own species. Other Old World monkeys showed no such advantage for these particular sounds, probably because the sounds were of no communicative significance to them. Although the sounds were delivered monaurally, not dichotically, it was inferred that there is a left-hemispheric specialization for the perception of communicatory sounds that are specific to the animal's own species.[60]

This was supported by a further study showing that Japanese macaques lost the ability to discriminate between "coo" sounds following removal of the left temporal lobe of the brain, but there was no such effect following removal of the right temporal lobe. The authors concluded that their results "are consistent with the notion that Japanese macaques possess an area analogous to Wernicke's area."[61] The effect of removing the temporal lobe was only temporary, however, suggesting that lateralization was not nearly as entrenched as in humans.

Even the lowly mouse may show a left-hemispheric advantage in recognizing communicative calls. Young mice emit ultrasonic cries that evoke caring from their mothers. Lactating mice showed a normal maternal response to these cries when they listened with both ears or with a plug in the left ear, but not with the right ear plugged. This suggests a left-hemispheric specialization. A second experiment showed that there was no such asymmetry when virgin females were simply trained to discriminate ultrasounds, without any association with maternal caring. This again suggests that the left-hemispheric mechanism has to do specifically with the processing of events that serve a communicative purpose.[62]

In the above examples—birdsong, monkey coos, ultrasonic squeaks—the signals themselves are specific to the species in question, and it is reasonable to suppose that the asymmetry is innately programmed. There are two further claims, both preliminary, of cerebral asymmetry in the processing of elements that are from the species' point of view entirely arbitrary, but that are given referential meaning through training.

The previous chapter described the work of Louis Herman and his colleagues at the University of Hawaii on the ability of dolphins to learn to comprehend sequences of sounds and gestures. The visual system of the dolphin is such that each eye projects only to the opposite hemisphere of the brain, so that a difference in performance between the eyes can be interpreted as a hemispheric difference. In one dolphin, there is evidence that the time taken to initiate responses to instructions based on sequences of gestures made by the trainer is shorter if the animal views with the right eye than with the left eye. Although data from one animal can scarcely establish a rule on laterality in the dolphin, this result is at least consistent with a left-hemispheric specialization for gestures comparable in fact to that in humans.[63]

To the solitary dolphin we might add three chimpanzees. William D. Hopkins of the Yerkes Regional Primate Research Center at Emory University has developed an ingenious technique for comparing the visual hemifields in the processing of shapes by chimpanzees.[64] Preliminary evidence from three adult chimpanzees indicates a right-hemifield advantage in the processing of plastic symbols, known as *lexigrams* (see Chapter 6), that stand for objects or food items. There was no such effect for novel lexigrams that had no external reference.[65]

This left-hemispheric asymmetry for meaningful lexigrams is like the human left-hemispheric advantage for printed words, just as the dolphin's left-hemispheric advantage for gestures resembles the human left-hemispheric advantage for sign language. In the case of humans, however, it is generally taken for granted that these asymmetries are tied to an innately programmed, left-hemispheric specialization for language that is manifest also, and more fundamentally, in speaking or gesturing. But dolphins do not gesture, and the chimpanzees' use of lexigrams has nothing to do with speech or with gesture. If their left-hemispheric dominance in processing these arbitrary signs is corroborated in future studies, this will raise fundamental questions about its underlying nature. Perhaps the left hemisphere is innately disposed to assume control of communicative functions, regardless of the physical form of the actual signals.

These various findings, while still fairly preliminary, suggest that Chomsky may have been premature in identifying the left-hemispheric component as unique to humans. The origins of the left-hemispheric specialization for language in humans may well precede even primate evolution, and by the same token precede true language itself. Since right-handedness, by contrast, seems to be unique to humans (see Chapter 4), this suggests that the origins of cerebral asymmetry may lie in communication rather than in manual activity.

The left-hemispheric representation of language in humans is surely more extensive than that of communicative behavior in other species, however, if only because language itself is more complex than other forms of communication. Whatever asymmetry underlies the lateralization for mouse squeaks, it is undoubtedly much simpler than the complex left-hemispheric circuits involved in the various aspects of human language. Even in the macaques, the disruption in the ability to recognize "coo" sounds following removal of the left temporal lobe was only temporary, implying that the appropriate mechanisms were at least present in the right hemisphere. Comparable damage to the language areas in adult humans typically produces lasting impairment.

Cerebral Asymmetry in Hominid Evolution

Since our forebears did not carry out the appropriate experiments, or did not publish them if they did, there is no direct information on whether their brains were lateralized for the representation of language. There is, however, indirect evidence from the imprints found on the inside of fossil skulls. As we saw in Chapter 2, these are best studied from latex endocasts, which provide models of the surface folds and fissures.

Ralph L. Holloway of Columbia University has shown from endocasts that *H. habilis* possessed a prominent Broca's area in the left frontal lobe comparable to that in modern humans.[66] Although a similar claim has been made for *A. africanus*,[67] the protrusion of this area is greater in *habilis*. As we saw in Chapter 2, Holloway has also claimed that the lunate sulcus has migrated rearward in the hominids, although Falk has disputed this in the australopithecines. At least from *H. habilis* on, though, this rearward shift can be attributed in part to an enlargement of Wernicke's area. It is not clear from the available evidence whether this enlargement is greater on the left side.[68] However, in *habilis* the Sylvian fissure, which demarcates the parietal and temporal lobes, is angled more steeply upward on the right than on the left, as it is in modern humans.

Phillip V. Tobias, a pioneer in research on the paleontology of the brain, summarizes as follows:

> The occurrence of both a strong inferior parietal lobule and a prominent motor speech area of Broca in the endocasts of *H. habilis* represents the first time in the history of the early hominids that the two most important neural bases for language appear in the paleoneurological record.[69]

Does this mean that *habilis* spoke? There is little doubt that this creature communicated vocally, as do other mammals. *Habilis* presumably also takes us beyond the lateralized representation of vocalization, perhaps already present even in the mouse, toward something of the more complex organization that we see in the left hemisphere of the human brain. However, Iain Davidson and William Noble of the University of New England in Australia have remarked on the "fallacy" of arguing from structural similarities between areas in the hominid brain and speech areas in the human brain.[70] Such similarities need not imply spoken language in the early hominids comparable to that in modern humans. Davidson and Noble suggest that these structure were associated with vocalization, not true language, but were exapted for language at a later stage.

Since left-hemispheric specialization for vocal communication may have been long rooted in the mammalian brain, perhaps going back even to the mouse, while right-handedness appears first in the hominids, it has been argued that vocal speech preceded gesture. On these grounds, indeed, Dean Falk has argued against the idea, documented in Chapter 6, that the language of the early hominids was primarily gestural rather than vocal.[71] However, since the areas of the brain controlling the hands are close to those controlling the mouth, the left-hemispheric dominance for vocalization may have spread, by what Falk calls a *field effect,* to create a left-hemispheric dominance in tool making, and perhaps in gesturing as well. Indeed, right-handed humans typically gesture much more with their right hands than with their left hands when they speak.[72] This extension of left-hemispheric dominance to include gesturing as well as vocalization may explain why left-hemispheric dominance is so much more marked in humans than in other species, and indeed may explain why humans are predominantly right-handed.

My guess is that gestural language may have dominated early in hominid evolution, with vocalization restricted largely to emotional signals, as it is in present-day primates. Gestures may have provided the basis for propositional language, perhaps first in the form of pointing to people or things or indicating actions. As suggested in Chapter 6, however, vocalizations may have gradually become associated with actions, and with changes in the vocal tract may have eventually replaced gestures as the primary form of communication. Even so, as we also saw in Chapter 6, sign language seems to be acquired just as naturally as vocal language and has essentially the same structural characteristics.

In Chapter 6, two scenarios were suggested for the evolution of language. In one, true language emerged only with *H. sapiens sapiens,* perhaps as a result of some genetic reshuffle. In the second, language evolved gradually, from *H. habilis* through *H. erectus* and archaic *H. sapiens,* to reach its true potential in *H. sapiens sapiens.* The evidence for left-hemispheric language areas as far back as *H. habilis* supports this second scenario.

Conclusions

In most humans, language is mediated predominantly, if not exclusively, by the left hemisphere of the brain. There seems little doubt that this left-hemispheric specialization is programmed biologically rather than acquired as a result of experience. Indeed, there is nothing about vocal language *per se* that is spatially lateralized, and that might have induced any asymmetry in its representation; in this respect, of course, language is unlike handedness. The representation of language in the left hemisphere involves areas that are quite widely distributed and intricately interconnected. The more it is studied, the more this system matches the complexity of language itself.

Rather surprisingly, however, evidence for left-hemispheric control of vocalization does not seem to be restricted to human speech but applies to other species as well. Although little research has been done, there is evidence for left-hemispheric control of at least some aspects of vocalization in songbirds, monkeys, and mice. We see also from hominid fossils some evidence that the brain circuits involved in human language may have been present in *H. habilis*, but probably not in the australopithecines.

This evidence must still be considered somewhat speculative, since it is based on very few cases and on crumbling evidence. Nevertheless, it does tie in with the other evidence that *habilis*, the first of the *Homo* line, had embarked on the road to humanity. This was the first hominid to show some increase in brain size, the first to develop the systematic manufacture of tools, and perhaps the first to develop a form of propositional gestural language. But it was probably not until the emergence of *H. sapiens sapiens* that we find true generative language, as well as the truly generative construction and use of tools.

The relations among handedness, gesture, and spoken language will be considered further in the following chapter.

Notes

1. Chomsky, in Piattelli-Palmarini (1980, p. 182).

2. Dax (1865). Benton (1984) has pointed out that the evidence for a left-hemispheric dominance for language was available even before Dax's time in case studies going back to the seventeenth century. Indeed, the association between speech and the left hemisphere is clearly evident in extensive treatises by Morgnani (1769), Bouillaud (1825), and Andral (1840) but was not noted by any of these writers. Dax himself had been alerted early to the possibility of left-hemispheric dominance by the case of the naturalist

Broussonet, who became aphasic after a left-hemispheric stroke, and he thereafter took special note of the side of brain damage in his aphasic patients.

3. Broca (1864), cited in Joynt (1964).
4. Broca (1865).
5. Dax (1865).
6. Wernicke (1874).
7. Geschwind (1969, p. 111).
8. See Buckingham and Kertesz (1974) for this and other examples.
9. See Chapter 3 of Shallice (1988) for a review.
10. Rubens (1976).
11. Buckingham and Kertesz (1974).
12. For more discussion, see Chapter 1 of Shallice (1988).
13. Isserlin (1922).
14. Lecours and Lhermitte (1976).
15. Goodglass, Blumstein, Gleason, Hyde, Green, and Statlender (1979).
16. For example, Lapointe (1985).
17. Nadeau (1988).
18. Ross (1981).
19. Gainotti (1972).
20. See Heilman, Bowers, and Valentine (1985) for a review. A recent paper by Behrens (1989) suggests that the right hemisphere might contribute to some nonemotional aspects of prosody.
21. Déjérine (1892).
22. In modern neuropsychological parlance, the term *acquired dyslexia* is usually preferred to *alexia.*
23. See Shallice (1988) for an up-to-date summary.
24. See, for example, Aimard, Devick, Lebel, Trouillas, and Boisson (1975).
25. See Roeltgen (1985) for a review of evidence on agraphia.
26. Poizner, Klima, and Bellugi (1987).
27. Fechner (1860), cited in Springer and Deutsch (1985, p. 26).
28. Springer and Deutsch (1985).
29. Raymond, Lejonne, and Lhermitte (1906). See Bogen (1985) for a more extensive historical survey.
30. Quoted by Bogen (1985, p. 303).
31. Akelaitis (1941, 1944).
32. Tomasch (1954), quoted by Bogen (1985, p. 304).
33. Gazzaniga, Bogen, and Sperry (1965); Gazzaniga and Sperry (1967); Sperry (1982). A general review of the early work may be found in Gazzaniga (1970).
34. Gazzaniga, Bogen, and Sperry (1967).
35. Milner, Taylor, and Sperry (1968).
36. Zaidel (1976).
37. Zaidel (1981).
38. Zaidel and Peters (1981).
39. Gazzaniga (1983).

40. Zaidel (1983).
41. Mishkin and Forgays (1952).
42. Bradshaw, Nettleton, and Taylor (1981). Barton, Goodglass, and Shai (1965) also found that if words were presented vertically rather than horizontally, there was a right-hemifield advantage for both English and Yiddish words.
43. This has been shown with spoken digits (Kimura, 1961, 1967), nonsense words (Kimura and Folb, 1968), prose passages (Treisman and Geffen, 1968), and even prosody and syntax (Zurif, 1974).
44. Kimura (1961).
45. Kinsbourne (1972).
46. Day (1964).
47. Ehrlichman and Weinberger (1978).
48. See, for example, the books by Beaton (1985), Bradshaw and Nettleton (1983), Bryden (1982), Corballis (1983), and Springer and Deutsch (1985)!
49. The issue is addressed, but not resolved, in Cohen's (1982) review.
50. The argument that "speech is special" is presented in Liberman (1982). One formulation of the theory is the *motor theory of speech perception*, according to which speech is perceived with reference to how it is produced, rather than with reference to the actual acoustic signals. See Liberman, Cooper, Shankweiler, and Studdert-Kennedy (1967) for early documentation of the evidence leading to this theory, and Liberman and Mattingly (1985, 1989) for the latest version of the theory.
51. Shankweiler and Studdert-Kennedy (1967).
52. Fodor (1983). For a critical discussion of the modularity theory, see Shallice (1988, Part IV).
53. Massaro (1987).
54. We saw in Chapter 6 that one characteristic of human speech perception is that synthesized sounds that vary in small increments from one consonant to another are perceived in "categorical" rather than continuous fashion. Liberman claims that this property is uniquely human. However, it has been demonstrated in the perception of *human* phonemes by chinchillas (Kuhl and Miller, 1975) and by macaques (Kuhl and Padden, 1982), and in the pygmy marmosets' perception of certain of their own calls (Snowdon, 1989—see also Chapter 6). Kuhl and Padden (1983) have also shown that macaques perceive the continuum from *ba* to *da* to *ga* in categorical fashion, just as humans do.
55. Godfrey (1974); Weiss and House (1973).
56. Liberman and Mattingly (1989, p. 491).
57. For general reviews, see Glick (1985) and Denenberg (1981, 1988). Bianki (1988) has also reviewed Russian work, some of it apparently in conflict with Western findings.
58. Lemon (1973); Nottebohm (1977).
59. Curiously, recordings from neurons in singing birds, as well as observations of the syrinx itself, fail to show any asymmetry (McCasland, 1987). McCasland suggests that these observations cast doubt on any analogy be-

tween lateralization of birdsong and the asymmetrical control of human speech.

60. Peterson, Beecher, Zoloth, Moody, and Stebbins (1978).
61. Heffner and Heffner (1984, p. 76).
62. Ehret (1987).
63. Morrel-Samuels, Herman, and Bever (1989).
64. Hopkins (in press).
65. Hopkins, Morris, Savage-Rumbaugh, and Rumbaugh (1989).
66. Holloway (1983).
67. Schepers (1950).
68. Tobias (1987).
69. Op. cit. (1987, p. 753).
70. Davidson and Noble (1989).
71. Falk (1980). However Falk does recognize a role for gesture, noting that tool manufacture has at least some of the properties of language.
72. Kimura (1973).

— 8 —

Praxis and the Left Brain

Actions speak louder than words

ABOUT NINE OUT OF EVERY TEN people are right-handed, and the proportion who are left-hemispheric for language is probably even higher. Although there may be weak precursors of handedness as well as a left-hemispheric specialization for vocalization in other species, the combination of the two asymmetries seems to be unique to humans.

Both asymmetries, moreover, apply to activities that are themselves distinctively human. Our handedness is most clearly evident in intricate skills such as writing or throwing. Although other primates, especially chimpanzees, are also capable of quite delicate manual acts, they do not match human dexterity in either sense of the word. In a review of the evidence, Richard E. Passingham concludes that "our superior manual skill is not in doubt."[1] And yet he can find no evidence for any marked difference between humans and chimpanzees in the way in which the motor area of the brain controls the hands or in the size of the motor area itself. The difference, he suggests, must lie in "higher mechanisms in the brain which influence the motor area."[2] This chapter suggests that those mechanisms are normally a function of the left cerebral hemisphere.

Similarly, as we have seen, human language seems quite unlike any other form of animal communication, especially in its flexibility and open-endedness. This specialization, too, is normally left-hemispheric. So what is it that manual skill and language might have in common? On the surface, at least, they are rather different, the one overt and physical, the other symbolic and abstract. This chapter begins to build a more integrated view of the nature of left-hemispheric specialization. First, we will examine the relation between handedness and cerebral dominance for language.

Handedness and Cerebral Dominance for Language

Paul Broca assumed that the side of the brain responsible for language would also be the one controlling the dominant hand, so that in left-handers the *right* hemisphere would be dominant for language. Exceptions to *Broca's rule* were soon to emerge. As early as 1868, the eminent British neurologist John Hughlings Jackson remarked on a right-handed patient who was aphasic but also suffered paralysis of the *left* side, implying damage to the right side of the brain.[3] Here, then, was a case of a right-hander with language apparently represented on the right side.

Rather confusingly, cases like this are referred to as *crossed aphasia*, since the side of the brain that controls the dominant hand appears to be opposite to the one that is dominant for language. Such cases are not uncommon. For example, in their 1959 book *Speech and Brain Mechanisms*, Wilder Penfield and Lamar Roberts reported on 53 cases of right-handers with aphasia following right-hemispheric injury and 66 cases of left-handed aphasics with left-hemispheric injury.

Evidence on the imperfect relation between handedness and cerebral dominance for language has also come from records of epileptic patients under consideration for brain surgery. We have already seen that some patients with intractable epilepsy are treated by commissurotomy, or a splitting of the brain. A more common procedure has been to remove the brain tissue in which the epileptic focus is located. In assessing patients for this operation, it is often important to determine which side of the brain is dominant for language and other mental functions. For this purpose, Juhn Wada, then at the Montreal Neurological Institute (now at the University of British Columbia), developed a procedure known as the *sodium amytal test*.[4] Sodium amytal is injected into the carotid artery on one side of the body, causing a temporary inhibition of function on the same side of the brain. The patient is then asked questions, and if the injection to a given side inhibits the patient's spoken answers, it can be inferred that language is represented on that side. The test is usually given on two separate occasions, with injection to each side in turn.

Data collected from 140 right-handed patients at the Montreal Neurological Institute have shown that 96 percent of them have language represented in the left hemisphere, while 4 percent have language represented in the right hemisphere.[5] A similar study of 74 right-handed patients in Italy put the incidence of left-hemispheric language even higher, at nearly 99 percent, with only one patient showing bilateral representation.[6] It is clear from these data, then, that nearly all right-handers are left-hemispheric for language.

However, this pattern is not simply reversed in left-handers, that is, most left-handers do not simply have their brains in backward.[7] Among 122 left- and mixed-handers in the Montreal study, 70 percent had language represented in the *left* hemisphere, just as most right-handers do. Of the remainder, 15 percent had language represented in the right hemisphere and 15 percent had language represented on *both* sides. Left- and mixed-handers did not differ from each other in these respects. The Italian sample included only 10 left- and mixed-handers, and these also showed a varied pattern: 4 were left-hemispheric, 1 was mixed, and 5 were right-hemispheric.

Similar figures come from the testing of verbal functions following unilateral electroconvulsive therapy (ECT), a technique that is sometimes used for the relief of chronic depression. An electric current is passed through the brain from electrodes attached to the scalp in order to induce convulsions, which are said to have a beneficial effect. The practice at the National Hospital in Queen Square in London has been to deliver ECT to just one side of the brain at a time. The patient's language functions are tested as soon as the patient has recovered enough to tell the questioner his or her name. Again, the evidence suggests that over 95 percent of right-handers are left-hemispheric for language compared with some 70 percent of left- and mixed-handers.[8]

What the data show, then, is that while the great majority of right-handers obey Broca's rule, the majority of left- and mixed-handers do not. Non-right-handers show instead a rather mixed pattern of laterality. This is in fact what we might expect from the genetic theory developed by the British psychologist Marian Annett, introduced in Chapter 4.

Annett's Theory Revisited

To recapitulate, the main idea underlying Annett's theory is that most right-handers inherit a right-shift allele (RS+) that strongly biases their handedness in favor of the right hand. Among those who lack this allele and instead carry two RS− alleles, there is no genotypic bias toward either left- or right-handedness. In the absence of strong environmental pressures, these individuals will be roughly equally divided into left- and right-handers, although many will be ambidextrous.

This theory can be extended to cover cerebral asymmetry for language as well as handedness. We suppose that those inheriting the RS+ allele will generally be left-hemispheric for language as well as right-handed, although some small fraction may deviate from this pattern because of extreme environmental or pathological influences. That is, the right-shift allele truly codes for a *left* shift, since it predisposes a dominance of the left hemisphere; in order not to cause confusion, however, I shall continue doggedly to refer to it as a right-shift allele. Among those carrying two RS− alleles, we assume that the *lack* of bias extends to hemispheric specialization as well as to handedness, and furthermore that these two asymmetries are independent of one another. In these individuals, we therefore expect all possible combinations of handedness and cerebral asymmetry to be about equally frequent.

In fact, the data fit these expectations quite well.[9] Right-handers should consist mostly of individuals carrying the RS+ allele, and the great majority of them should therefore have language represented predominantly in the left cerebral hemisphere. We saw from the studies reviewed above that the proportion of right-handers with left-cerebral representation of language is at least 96 percent. On the other hand, so to speak, left- and mixed-handers include a high proportion of those carrying two RS− alleles, and should therefore show a varied pattern of cerebral asymmetry. The evidence demonstrates this to be true also. However Annett's theory would lead us to expect no overall bias in favor of *either* hemisphere in the representation of language among left- and mixed-handers, whereas in fact there is a preponderance of those with *left*-cerebral representation of language.

The "extra" left- and mixed-handers with left-hemispheric representation of language might be identified as so-called "pathological" left-handers, discussed in Chapter 4; these individuals carry an RS+ allele but are left- or mixed-handed because of some pathology or extreme environmental influence. The figures from the Montreal sample, discussed earlier, imply that something over half of the non-right-handers may have belonged to the pathological group.[10] This estimate may be too high, since it is based on epileptic patients with histories of neurological disorders.

Annett has estimated the number of individuals lacking the right shift from the incidence of aphasia following right-hemispheric damage, regardless of their handedness. In several studies, the incidence was just over 9 percent. Since an equal number of RS− individuals should be aphasic as a result of left-hemispheric damage, this means that the proportion of those carrying two RS− alleles must be about 18 percent. If the total number of non-right-handers is about 12.5

percent (see Chapter 4), then this implies that the proportion of non-right-handers who are "pathological" is about 28 percent.[11]

In the most recent formulation of her theory,[12] Annett argues that the right-shift allele is expressed more strongly in those carrying two RS+ alleles (homozygotes) than in those carrying just one (heterozygotes), which may add to the variation in degree of lateralization. To return to a question that was raised in the previous chapter, then, there may be no "clean" rule as to whether the left-hemispheric dominance for language is relative or absolute. Indeed there are variations ranging from right-hemispheric to left-hemispheric dominance, with some individuals showing bilateral representation. Overall, however, the bias toward left-hemispheric representation is strong and appears to be a distinctive mark of the lop-sided human condition.

In some 18 percent of the population, if Annett's estimate is correct, handedness and hemispheric specialization for language are determined by random influences rather than by a genotypic bias. This does not mean that those without this bias are somehow throwbacks to a more primitive, ape-like form! There are probably advantages to a more symmetrical representation that must be weighed against the advantages of lateralization, and the optimum may be a balance between the two; as we saw in Chapter 4, Annett has argued that adaptation is maximized in those who carry one RS+ allele and one RS− allele, rather than two of the same. In any event, functions such as speech that are normally lateralized need not be handicapped if they are represented bilaterally, as they are in some individuals, although there is some evidence that the lack of consistent lateralization may increase the risk of disorders such as developmental dyslexia and stuttering, as we shall see below.

We have seen, then, that handedness and cerebral asymmetry for language may be controlled by a genetic allele whose presence predisposes the left hemisphere to play the dominant role but whose absence leaves the *direction* of both asymmetries open to random influences. Apparently, handedness and language have enough in common to make it an advantage to have both represented in the same hemisphere, yet they are distinct enough for this not to be obligatory. In what way, then, are the two functions linked?

Disorders of Praxis

Apraxia

We may gain some insight into this from a neurological disorder, or rather a *class* of disorders, known as *apraxia*, which usually results from damage to the left cerebral hemisphere. The term is derived from the Greek word *praxis*, meaning action. The apraxias are disorders of skilled, voluntary movement that cannot be attributed to such things as muscular weakness, lack of understanding, failure to cooperate, or intellectual loss.[13] The patient might have difficulty in demonstrating the use of a comb or toothbrush, even though these objects are recognized and can be named. It may apply also to meaningless actions, where a patient has difficulty imitating random actions performed by another person. It applies to actions involving both hands[14] as well as to those involving a single hand, so it is not simply related to handedness. Also included are actions that do not involve the hands, including for example *buccofacial* actions like blowing a kiss or sticking out one's tongue.

John Hughlings Jackson had described patients with this disorder as early as 1866,[15] but it was the German neurologist Liepmann who provided the first insights into its neurological causes. Based on a number of cases, he argued that apraxia could result from damage either to the left hemisphere or to the corpus callosum (which, remember, is the main tract of fibers connecting the hemispheres). He inferred that the left hemisphere contained *movement formulas* for the organization and execution of actions.[16] Damage to the corpus callosum was thought to produce apraxia of the left hand only, since right-hemispheric control of the left hand was disconnected from the movement formulas, or what are now more verbosely termed *visuokinesthetic motor engrams,*[17] that are resident in the left hemisphere. The evidence on this has been variable; some patients with split or partially split brains show it, while others do not. Just as some people may have bilateral representation of language, so some may have bilateral representation of motor engrams. Despite this variability, evidence from patients with damage to the corpus callosum suggests that in most people the representation is left-hemispheric rather than bilateral or right-hemispheric.[18]

Apraxia is most commonly found following damage to the left hemisphere itself, although in these cases it is often accompanied by aphasia, which complicates the diagnosis. Nevertheless careful analysis has shown that the two disorders are distinct. Kenneth M. Heilman and Leslie J. Gonzales Rothi have argued, in fact, that the visuokinesthetic programs for the execution of skilled actions are stored in the parietal lobe of the left hemisphere, close to Wernicke's area but not overlapping it. Apraxia can result either from damage to this area or from damage to the pathways that connect it to the areas in the frontal lobe that organize and control the movements themselves.[19] Heilman and Rothi also refer to evidence that, in apraxia as in aphasia, there is a receptive as well as an expressive aspect. Damage to the parietal area may result in an impairment in the recognition of skilled movements performed by others, even though the patients themselves can perform these movements.[20]

There is also remarkable evidence that apraxia for nonlinguistic gestures is distinct from aphasia for ASL, even though both involve rather similar movements of the hands and arms. Howard Poizner, Edward S. Klima, and Ursula Bellugi described two patients with left-hemispheric damage and clear aphasia for ASL but no symptoms of apraxia for nonlinguistic gestures.[21] A third patient with extensive left-hemispheric damage showed both apraxia and sign aphasia, suggesting that aphasia and apraxia are nevertheless both mediated by the left hemisphere in the deaf. Although the left hemisphere seems to play the dominant role in both vocal and gestural language, as well as in nonlinguistic praxis, it appears that these various skills are modular in form; that is, they involve different brain circuits.

It has been suggested that apraxia applies particularly to skills that involve *sequences* of action. For example, it has been found that apraxic patients can copy static positions of the hand or arm but cannot copy movements.[22] We shall see in Chapter 10 that rhythm appears to be largely under the control of the left hemisphere. And, of course, spoken language itself is fundamentally sequential, with virtually no spatial component, which may be one reason that it is so overwhelmingly left-hemispheric.

But although sequencing is often a critical ingredient in what we may term *praxic skills*, it may not be a necessary one. Some aspects of ASL are spatial rather than temporal, yet they are disrupted by left-hemispheric damage; conversely, deaf patients with right-hemispheric damage may have severe spatial difficulties, yet their signing remains intact, even the spatial components of it.[23] The important point about praxic skills, therefore, may be that they are independent of spatial features of the environment, rather than that they are necessarily sequential. Making signs is an internally generated act, and although it has a spatial aspect, it does not depend on spatial cues in the environment itself. The spatial environment is more critical, though, in recognizing signs made by others, and there is evidence that perception of some spatial components *is* disrupted by right-hemispheric damage.[24]

Praxic skills may therefore be defined as internally generated, purposive skills that are unconstrained by spatial features of the environment. A prime example, of course, is speech. In seeking the explanation for the asymmetrical representation of language, then, we need not consider the properties of language per se; rather, the explanation may lie in the properties of praxic functions generally. Spoken language may be simply the best example of a function that is wholly praxic, with virtually no dependence on the spatial surround.

In most people, then, praxic skills seem to be represented predominantly in the left cerebral hemisphere, suggesting that lateralization might confer an advantage. If this is so, then we might expect people who are weakly or incompletely lateralized to be deficient in praxic skills. This might apply particularly to left-handers, since a greater proportion of them belong to the group lacking the right-shift allele. Indeed, as we saw in Chapter 4, left-handers are often characterized as clumsy, although this might be attributed in part to the fact that we live in a world constructed largely by and for right-handers, and in part to an age-old prejudice against left-handers simply for daring to be different. We should also remember such outstanding left-handed athletes as John McEnroe, Martina Navratilova, or Babe Ruth, and reflect that *lack* of lateralization might also have its advantages in activities requiring quick reactions to spatial events.

One fairly extensive study suggests that left-handers, as a group, are not deficient in motor skills. Michael Peters and Philip Servos of the University of Guelph compared right- and left-handers on a number of manual test of skill, speed, and strength, as well as on speed of articulation and verbal fluency.[25] They divided the left-handed group into those with consistent and those with inconsistent handedness. In no test were the left-handers, whether of consistent or inconsistent handedness, inferior to the right handers. However, the inconsistent left-handers were better with the left hand on tests of fine motor skill but better with the right hand on tests of strength—a dissociation that might cause problems in activities requiring both strength and skill.

Peters and Servos restricted their study to university students, so that those with problems of language or praxis may have been effectively screened out. There is other evidence, from samples more representative of the general population, that those with mixed or reversed lateralization may have difficulties with at least some aspects of language. Most of the studies have focused on developmental dyslexia and stuttering.

Developmental Dyslexia

Developmental dyslexia, sometimes known as *specific reading disability,* can be distinguished from acquired dyslexia, otherwise known as alexia (discussed in Chapter 7) in that it is not related to any known neurological damage. It refers to difficulties in reading that also cannot be attributed to poor intelligence, sensory problems, motor problems, or lack of motivation. As McDonald Critchley remarked, the problems of the dyslexic are "unlike those met with in the case of a dullard, or a poorly educated person."[26] Indeed, many prominent people are said to have been dyslexic, including the inventor Thomas A. Edison, the brain surgeon Harvey Cushing, the sculptor Auguste Rodin, and President Woodrow Wilson.[27]

Writing is clearly a praxic skill, involving sequential movements of a pen or pencil controlled by the hand. Reading is similarly praxic, requiring sequential movements of the eyes across the printed page. Although it has a spatial component, it is not merely a reaction to spatial events in the environment. Rather, printed words are static, and their spatial scanning is driven by the brain of the reader.

One common view is that reading and writing difficulties result from a confusion of left and right. This view owes its origin in large measure to the American psychiatrist and pediatrician Samuel Torrey Orton, whose influential book *Reading, Writing, and Speech Problems* was published in 1937 but was based on the fruits of research and observation dating from the early 1920s. Orton believed that left-right confusion was caused by poorly established cerebral dominance, which was often manifest in left-handedness or a lack of consistent handedness. Children with poorly established dominance would therefore be especially likely to confuse words like *was* and *saw*, or letters like *p* and *q*, and they might also be likely to write backward. Figure 8.1 shows an example of mirror writing produced by an adolescent dyslexic girl. Such was Orton's influence that confusions and reversals like these are often still regarded as virtually synonymous with dyslexia.

However, Orton's theory was based on outmoded concepts of cerebral dominance and on a curious notion that the two halves of the brain would encode patterns in opposite left-right orientations.[28] It is nevertheless true that a purely symmetrical brain would be unable to tell left from right, and would therefore be unable to read or write any script that depends on a consistent left-right code.[29] Indeed, there is evidence that reading disability is rare in Japan, where script is laid out in vertical columns and symbols have the same meaning if left-right reversed.[30]

Nevertheless, statistical evidence for a relation between reading disability and cerebral lateralization has been fitful, at best. Some studies have shown that mixed- or left-handedness are more common among dyslexics than among normal readers, but others have not, and at least two reviewers have concluded that the relation is

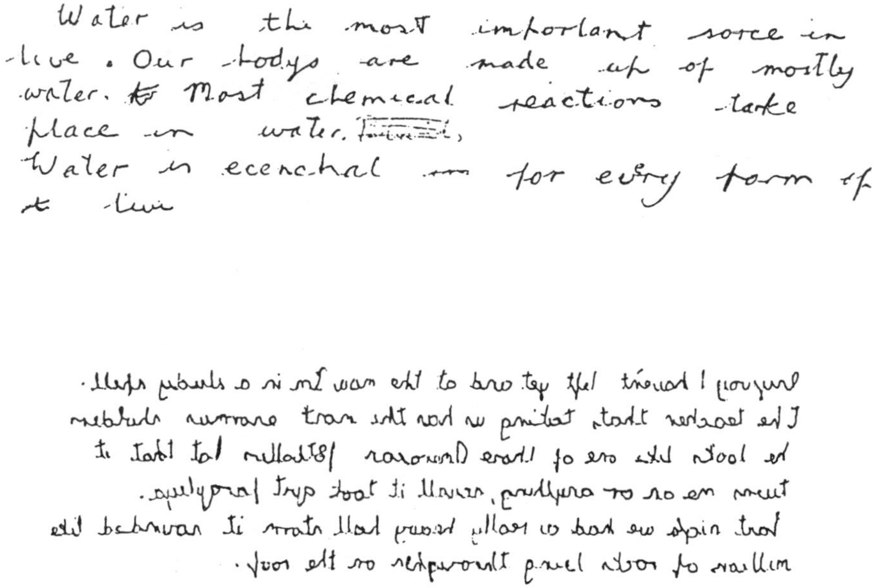

Figure 8.1. Example of normal writing (*top*) and mirror writing (*bottom*) from an adolescent dyslexic girl (from Corballis, 1983).

not convincing.[31] However, Orton himself remarked that many of his dyslexic patients were ambidextrous or showed mixed dominance, and it may be only certain patterns of mixed lateralization that give rise to difficulties. For example, Martha Bridge Denckla of the Harvard Medical School, not an admirer of Orton's theories, was forced to reveal a fact that she has "tried in vain to escape": Nearly two-thirds of the dyslexic children she has studied were right-handed, right-footed, and *left* eye-dominant.[32] The development of reading skills may depend in the first instance on the development of writing, and may be difficult to achieve if the hand and eye are controlled by opposite hemispheres.

There is further evidence that the relation between reading and lateralization may not be a simple one. In a recent study, Marian Annett and Margaret Manning have claimed that reading may be impaired in children at *both* extremes of handedness; that is, in terms of Annett's genetic theory, it is as disadvantageous to reading to carry two RS+ alleles as to carry two RS− alleles.[33] This finding may help explain why previous investigators have not always found a clear relation between reading disability and handedness. Annett and Manning's analysis does not make it clear, however, why a double dose of lateralization would be detrimental to reading.

Of course, handedness is only indirectly related to cerebral asymmetry, and the more direct test of Orton's theory would be to measure cerebral asymmetry itself. In a review of published studies, Hilary Naylor has claimed that there is no convincing evidence that children with reading difficulties differ from normal readers on cerebral asymmetry as measured by visual, auditory, or tactile means.[34] Still, these methods are not especially reliable, as pointed out in Chapter 7, and in fact, Naylor's review seems to provide some evidence that disabled readers *are* less lateralized than normal readers. Out of 39 studies, 15 showed disabled readers to be less lateralized, 21 showed no difference between the groups, and 3 showed the disabled readers to be more lateralized—a decided skew in the direction of reduced lateralization among disabled readers.

Sometimes the dead hand of statistics is no more to be trusted than the embrace of the overenthusiastic theorist. Some case histories of severely dyslexic but otherwise highly intelligent people do suggest that anomalous handedness and left-right confusion may play a role. In her book *Reversals*, the American writer and novelist Eileen Simpson gives a vivid account of her own life as a left hander and dyslexic, but such was her fighting spirit that she largely overcame her disability (but not her bad spelling). Her book reads almost as a testimonial to Samuel Torrey Orton. Another to give testimony on his dyslexia, ambidexterity, and left-right confusions is the British mathematician Kalvis M. Jansons.[35]

Although Orton's theory has lost support since its heyday in the 1920s and 1930s, the idea that dyslexia might be related to reduced cerebral dominance was revived in the 1980s by the late Norman Geschwind of the Harvard Medical School. Along with Peter Behan of Glasgow University, Geschwind maintained that left-handers show not only a higher frequency of reading and learning disabilities than do right-handers, but they also show a higher incidence of disorders of the immune system. These disorders include Crohn's disease, ulcerative colitis, celiac disease, and Hashimoto's thyroiditis.[36] Geschwind and Behan suggested that the male sex hormone testosterone may have an inhibiting effect on the early growth of the left hemisphere, which in some cases results in reading and learning disorders, as well as on the thymus, a gland involved in the development of lymphocytes that recognize self antigens.

Since testosterone is a male hormone, this theory is said to explain why both left-handedness and reading disability are more common in males than in females.[37] Among disabled readers, males outnumber females by as much as four to one,[38] but the difference between the sexes in the incidence of left-handedness is much less pronounced—and some would say nonexistent.[39] Geschwind and Behan's original studies have been criticized on methodological grounds,[40] and empirical support for the relation between left-handedness and autoimmune disorders has been mixed.[41]

Nevertheless, the theory has been developed more fully by Geschwind and Albert M. Galaburda in their 1987 book *Cerebral Lateralization*. They draw attention to several cases of dyslexia in which autopsy revealed abnormalities in brain structure, including an excess of white matter or such abnormalities as "brain warts" on the left side.[42] They also suggest that malformations of the blood vessels (arteriovenous malformations, or AVMs) are more common in males than in females, and refer to six patients with lifelong reading disabilities who had AVMs in the area of the left temporoparietal region (possibly affecting Wernicke's area or the angular gyrus). They relate these anomalies to their theory that testosterone, or some factor related to it, retards growth of the left hemisphere.

They also cite other evidence that anatomical asymmetries normally observed in the human brain are reversed or absent among those with reading or learning disabilities. In the majority of people, the posterior part of the brain is wider on the right than on the left; this applies to some 91 percent of right-handers and 73 percent of left-handers.[43] Brain scans of 24 dyslexic patients revealed that this asymmetry was reversed in 8 of them, with 8 showing the usual asymmetry and 6 showing no asymmetry; note that the group as a whole showed essentially no asymmetry.[44] A follow-up study of 53 children with learning disabilities showed that 22 of them had reversed asymmetry.[45] However, a recent review of evidence of this sort invites continued skepticism.[46]

Reading is one of the most complex and sophisticated skills we possess, involving phonological, spatial, and verbal skills. After many decades of research, the nature of developmental dyslexia remains little understood, perhaps because there are many way in which reading can be impaired. The idea that it may depend, in some cases at least, on anomalies of lateralization is one that has nevertheless persisted over the years, even though the evidence has been inconsistent. It is an idea that may wax and wane until the end of time.

Stuttering

One disorder that seems more purely praxic is stuttering. The writer W. Somerset Maugham was a chronic stutterer, but his easy prose style rules out the suggestion that he had any problems with *language* as such. Another stutterer with highly developed verbal and mathematical skills was Charles Dodgson, better known as Lewis Carroll, and the Dodo in *Alice in Wonderland* was named after the way Carroll stuttered when he tried to pronounce his own surname: "Do-Do-Dodgson." According to one biographer, 8 of the 11 children in the Dodgson family stuttered,[47] suggesting that the affliction was an inherited one. Another of Carroll's biographers, Florence Becker Lennon, has argued that Carroll was born left-handed but forced to use his right hand, which explained not only his stutter but also his obsession with mirrors and left-right reversals, as in *Through the Looking Glass*.[48] As Lennon puts it, he "took his revenge by doing a little reversing himself."[49]

The late King George VI of England was also afflicted with a stutter. He had particular difficulty with the letter *k*, and in his speeches the phrase "His Majesty" was substituted for "King." His stuttering may have been due to weak lateralization. Although a right-handed golfer, he was a left-handed tennis player talented enough to play at Wimbledon.

The idea that stuttering is caused by weak lateralization has nevertheless had a checkered history. One popular theory that goes back at least to the turn of the century is that it is caused by a switch in handedness, and in particular by forcing naturally left-handed children to write with their right hands.[50] The evidence for this is wildly conflicting, with estimates of the proportion of changed handedness among stutterers ranging from 0 to some 70 percent![51] Others have supposed that stuttering is the direct result of weak or mixed cerebral dominance, and not of any enforced switch. This idea seems to have been proposed first in the 1920s by Samuel Torrey Orton, better known for his theories of dyslexia. Orton's colleague Lee Edward Travis at the University of Iowa took up the idea and elaborated the *dominance theory* of stuttering in his 1931 book *Speech Pathology*. It was the springboard for most of the research on stuttering carried out at Iowa in the 1920s and 1930s, but gradually it lost favor. Travis himself changed his view, and in a volume published in 1955 to mark 30 years of research at Iowa, the dominance theory is hardly mentioned.[52]

Wendell Johnson, who followed Travis as director of the speech clinic at Iowa, maintained that the dominance theory was not supported by the evidence. He wrote that "the more skilled and meticulous the investigators became, the more they found stutterers to be, from a neurophysiological point of view, like other people."[53] Yet the change of view might owe something also to a general change in the zeitgeist, one of those shifts in attitude that have characterized the history of science and, more particularly, of psychology. In the years after World War II, organic theories of mental disorders gave way to more functional theories; disabilities were attributed more often to childhood experiences, emotional traumas, or faulty learning than to physical causes. Johnson himself attributed stuttering to the corrective attempts of overly perfectionist parents. If a child's early speech was less than perfectly fluent, an overzealous parent might express dissatisfaction, which only served to increase the dysfluency and create a vicious cycle. Johnson argued that therapy should be directed as much to the parents as to the child.[54]

In more recent years, there has been a return to neuropsychological theories, and something of a revival of the idea that stuttering might related to weak or mixed cerebral lateralization. Although the data are far from consistent, many studies have shown a higher incidence of left-handedness among stutterers than among nonstutterers, but the difference is typically slight; in one large-scale study, for example, 20 percent of stutterers were left- or mixed-handers compared with 14 percent of nonstutterers.[55] However, if it is true that stuttering is at least sometimes caused by a switch from left- to right-handedness, then one might not expect a large difference in handedness between stutterers and nonstutterers. It would be more pertinent to examine the relation of stuttering to cerebral dominance rather than to handedness.

The dominance theory of stuttering was given a boost in 1966, when a Philadelphia neurologist, R. K. Jones, described four cases of stutterers who had been referred for neurosurgical treatment of brain injury in the region of the speech areas.[56] All had stuttered severely since childhood, and three were left-handed, although all four came from families with a high incidence of left-handedness. Their neurological problems were of recent origin and had nothing to do with their stuttering. The sodium amytal test revealed that both hemispheres could control speech. After surgery to the damaged hemisphere each patient was able to speak fluently, without stuttering, and the sodium amytal test now revealed that speech was controlled unilaterally by the hemisphere that had not been operated on.

Another remarkable series of cases has been described by a group of neurosurgeons at St. Anne Hospital in Paris.[57] Several stutterers were reported to have been cured of their stuttering following surgery to one or the other cerebral hemisphere for the relief of epilepsy. The operations were carried out under local anesthesia, and the surgeons were startled to hear two of the patients speaking normally during the operation itself and expressing surprise that they were able to do so!

These observations lend strong support to the idea that bilateral representation of speech may create a conflict of control, resulting in stuttering. Jones' patients, at least, may have lacked the right-shift factor that would have ensured unilateral representation. This does not mean that individuals carrying two RS− alleles will necessarily stutter; the risk may be higher than in those carrying an RS+ allele, but it is probably still low. Moreover there are, no doubt, other causes of stuttering. In two reports that appeared subsequent to Jones', application of the sodium amytal test to stutterers about to undergo surgery revealed the usual left-hemispheric dominance for speech in six of the seven cases.[58]

There has been some speculation as to whether stuttering might be symptomatic of some more general praxic disorder. There is evidence, for example, that stutterers take longer to respond *manually* to a signal than do nonstutterers.[59] William G. Webster of Carleton University in Ottawa has compared stutterers to nonstutterers on the ability to make rapid sequential finger movements according to a prescribed pattern. Although both groups responded at the same rate, the stutterers made more errors and were slower to initiate the sequences. Since both groups were better with the right than with the left hand, Webster concluded that sequencing was controlled by the left hemisphere, but that the stutterers may have been susceptible to interference from competing mechanisms in the right hemisphere. The conflict may be chiefly one of *initiation*; once underway, a sequence may proceed under unilateral control.[60] Stuttering, too, often seems to be a problems of getting started.

In summary, there is some evidence that those lacking the right shift may be somewhat more susceptible to praxic disorders than their more lop-sided friends. The advantages associated with the RS+ allele may therefore have been sufficient to ensure that the majority of individuals carry it, but not sufficient to spread it through the entire human population. Indeed, as suggested in Chapter 4, the optimal condition may be balanced polymorphism, with one allele of each type, and this may have been sufficient to hold the relative proportions of each allele in the population roughly constant at about their present values. Nevertheless, in a diverse world, that minority of individuals who lack the right shift may have special skills of their own, as Leonardo da Vinci, Charlie Chaplin, and John McEnroe can testify.

The Evolution of Praxis

What, then, were the evolutionary contingencies that might have favored the lateralized representation of praxic skills? As pointed out in Chapter 4, for most of our evolutionary history a symmetrical structure has proven more adaptive than an asymmetrical one. Linear motion requires symmetrical limbs, and for animals that move about freely the environment, with its rewards and hazards, is basically the same on the left as on the right. Animals need to be equally alert, sensitive, and reactive to either side. So it is that the limbs and sense organs, and many of the brain mechanisms dealing with them, are placed symmetrically in the body. The advantages of symmetrical structure would apply especially to the sorts of actions that have been called *stimulus-bound*—that is, actions that occur directly in response to environmental events.

Praxic skills no doubt evolved later than stimulus-bound ones. To an extent, they free the organism from immediate environmental constraints, allowing a measure of purpose and planning. Since they are no longer so dependent on environmental constraints, there is no longer any advantage to symmetrical representation. Indeed, bilateral representation may be a disadvantage. If a praxic skill is duplicated in both cerebral hemispheres, there is a risk of conflict; the two hemispheres might be literally at cross purposes. A representation that straddles the hemispheres might be poorly coordinated due to the slowness of neural impulses between the hemispheres. The advantage of asymmetrical representation is summed up by Richard E. Passingham of Oxford University:

> . . . [control by both hemispheres] would not be optimal for the execution of complex sequences of movement as in the production of speech. In such a case we would surely expect the highest skill to be achieved if the sequence was directed by one central programme, located in a single hemisphere, rather than by two separate programmes which must use the long commissural pathways between hemispheres to coordinate their instructions.[61]

The first nonspatial, sequential tasks to be systematically lateralized within a hemisphere may have been vocalizations. As we saw in Chapter 7, even the ultrasonic squeaks of the mouse may be represented in the left cerebral hemisphere, and vocal communication in macaques and certain songbirds also appears to be left-hemispheric. Indeed, vocal communication might be regarded as the purest form of praxis—sequential, purposeful, and with no spatial aspect.

In the case of skills involving the hands, the praxic component must be weighed against the stimulus-bound nature of many manual actions. Indeed, the forelimbs evolved first for locomotion, and so were no doubt symmetrical to begin with. In our own species they continue to play *some* role in locomotion, and do so in symmetrical fashion; imagine the chaos in a swimming final if the competitors pulled harder with one arm than the other, and so swam in curves rather than straight lines. Of course, the forelimbs were exapted for nonlocomotory actions, such as reaching for things, feeding, and manipulation, long before the emergence of the hominids. Even in these cases, however, symmetry may have proven more adaptive than asymmetry; for example, it may have been advantageous to be able to reach for things, or catch things, with equal facility on either side of the body.

The distinction between praxic and nonpraxic functions of the hands is nicely illustrated in a study by Neil V. Watson and Doreen Kimura of the University of Western Ontario.[62] They found that right-handers showed a right-hand superiority in throwing darts, but there was no difference between the hands in blocking (but not catching) table-tennis balls that were launched at them. Throwing has a strong praxic element, and as we saw in Chapter 4, it has even been suggested that handedness originated in throwing. By contrast, blocking a projectile is a direct response to a spatial event, with no praxic component—a response that is best accomplished symmetrically, perhaps, in the interests of self defense.

Although manual praxis seems to be most highly developed in humans, the praxic element may have been present quite early in evolution. As we saw in Chapter 4, even the mouse typically shows a consistent preference for one or the other paw, although as many are left-pawed as are right-pawed. The same is true of other mammals and primates. We have seen that many other animals besides ourselves use primitive tools; for example, chimpanzees use sticks to extract termites from their holes. Even in this case, however, it might be useful to be able to extract termites with either hand, so as to maximize ease of access.

But with the emergence of the hominids, there was a further shift toward the use of the hands for praxis, and this may have been decisive in the evolution of handedness. Our upright stance virtually eliminated the role of the hands and arms in locomotion, and manipulation and manufacture grew more sophisticated. The elements of purpose and planning grew still more prominent. In manufacturing a tool, moreover, one can arrange the layout as one wants, so the spatial constraints imposed by the natural environment become less critical.

So far, I have discussed the relative advantages of symmetry and lateralization in terms of environmental constraints, which tend to favor symmetry, and of the praxic component, which favors lateralized representation. There is another issue here, however, that has to do with storage capacity. It is sometimes suggested that bilateral representation serves an additional adaptive function, that of providing a backup in the event of injury. Many right-handers no doubt curse their very dexterity if injury to the right hand forces them to use the left, whether for writing, eating, or cutting toenails. But they manage; the left hand is not totally useless as a reserve. The same principle may apply to cerebral representation. For example, duplicate memories or skills in the hemispheres may serve as a back-up if one side of the brain is damaged.

The advantage of duplicate representations must be weighed against the disadvantages. William H. Calvin again draws attention to the potential conflict that may be caused by double representation, especially in execution or planning, since the coordination of the two hemispheres depends on "that long, slow trip through the corpus callosum."[63] Norman Geschwind and Albert M. Galaburda suggest that the advantages of duplication may not be so great in any case; "The cat's superior recovery from hemiplegia is of little advantage in the wild, since a few days or even a few hours of disability mean almost certain death."[64] They go on to note that having different skills represented in the two hemispheres increases the diversity of endowment, which may more than compensate for the loss of duplicate representations.

Gordon T. Frost makes this point more specifically in relation to the evolution of the hominids. Double representation, he suggests, would have been wasteful of "neural space,"[65] which may have been especially in demand as the lives of the hominids became more complex. The evolution of the manufacture and use of tools, from *H. habilis* on, was accompanied by an increase in brain size, as we saw in Chapters 2 and 3. However, the size of the birth canal imposes a limit on the size of the brain, and indeed, it seems to have been necessary to resort to a number of "tricks" in order to discover extra space. (Chairpersons of university psychology departments will recognize the problem.) One such trick is the highly convoluted nature of the human brain; it must be folded to fit into the skull, just as one folds an item of clothing to fit it into a suitcase. Another trick was to arrange that human infants be born prematurely, at least relative to other species, so that much of the growth of the human brain could take place after birth. Unilateral representation of praxic skills may have provided a third way of cutting down on the demand for neural space.

Lest we become too enthusiastic over the advantages of asymmetrical representation, it should be remembered that the distinguishing characteristic of human laterality is not so much its presence as its consistency. That is, in most of us, it is the left hemisphere that is dominant for praxis. Other animals may also show dominance, but which side is dominant is a matter of chance. We have seen that other animals show consistent preferences for one or the other forelimb, but there appears to be no genetic control over which side is preferred. In humans, the principle of lateralization seems to have been sufficiently important that it was given the seal of consistency, at least in the majority of us.

In Chapter 7, I noted Dean Falk's theory that right-handedness was derived from the left-hemispheric dominance for vocalization. Since the areas controlling vocalization lie close to those controlling the hands, the left-hemispheric dominance may have spread to control of the hands, favoring the right hand. This may have occurred first in the context of gesture, which would have been closely linked to vocalization. However, praxis goes well beyond gesture and communication, and the more general left-hemispheric dominance for a wide range of praxic skills may reflect a further spread of dominance, or what Falk calls a *field effect*. It may be this spread of dominance that is controlled by the RS+ genetic allele postulated by Marian Annett.

The location of praxic representations in the left hemisphere need not create absolute asymmetry, of course, since each hemisphere has some control over movements on the *same* side of the body, and there is also some exchange of information between hemispheres via the commissures (including the corpus callosum). The bias may nevertheless be sufficient to create a preference for the right hand, but need not prevent right-handed individuals from developing considerable skill with the left hand. That is, right-handedness may be simply a by-product of the left-hemispheric specialization for praxis.

This explanation for handedness explains why handedness seems to have so little to do with the structure of the hands themselves. As I noted in Chapter 4, the hands *look* essentially alike, and the asymmetry resides in their control, not in their architecture. This suggests that handedness did not originate in different functions ascribed to the hands, such as one specialized for holding and one for operating,[66] or in specialization of one hand for such skills as throwing.[67] Rather, handedness may be no more than a secondary consequence of the lateralization of praxic skills, which simply conferred a bias in favor of the contralateral hand.

Falk's theory suggests that the left-hemispheric specialization for both vocal and manual praxis may have been due simply to the field effect. However, more than the mere proximity of the neural areas may have been involved in the two functions that favored their representation in the same hemisphere. Both may have been involved in communication, for example; I have suggested that the early hominids made extensive use of gesture, which were perhaps accompanied by vocalizations that were, at first, of more primitive origin.

There are perhaps other reasons too why it was advantageous to have different functions represented in the same hemisphere. Annett writes that "The main function of the RS+ gene is to ensure that the production and perception of speech sounds depend on the left cerebral hemisphere."[68] Her main reason for this conclusion is that speech depends critically on both sensorimotor feedback from the mouth and auditory feedback from the ear, and coordination of the two is most efficient if they are represented on the same side of the brain. We also saw earlier that there is evidence that opposite dominance of hand and eye might underlie dyslexia in some cases. Although reading is too recent a skill to have been decisive in evolution, consistent handedness and eyedness might have been important in tool making or drawing.

We might ask, why was the *left* hemisphere chosen? But perhaps we should not, for as Thomas Carlyle wrote of handedness: "Why [the right hand] was chosen is a question not to be settled, not worth asking except as a kind of a riddle."[69] I have speculated on this matter elsewhere, although probably at the risk of unsettling the cosmos.[70] It may be that the seed for the choice of side was planted quite early in evolution; the important point, however, lies in the consistency with which the left side appears to have been chosen rather than the fact that it *was* the left.

The Left Brain as Executive

It has often been suggested that the left hemisphere is the executive, the center of intention and will. To some extent, of course, this is already implicit in the idea of praxis, which has to do with purposeful acts that are programmed from within. Nearly 70 years after Liepmann remarked on the specialization of the left hemisphere in purposeful acts, the late Oliver Zangwill, one of the pioneers of neuropsychology, wrote: "there seems little doubt that purposive motor activity proceeds for the most part under left hemispheric control."[71] In similar vein, Roger Sperry, reviewing the evidence from his studies of split-brained patients, concluded as follows:

> The language-dominant hemisphere is . . . the more aggressive, executive, leading hemisphere in the control of the motor system. . . . The mute, minor hemisphere, by contrast, seems to be carried along much as a passive, silent passenger who leaves the driving of behavior mainly to the left hemisphere.[72]

Michael Gazzaniga goes further and regards the left hemisphere as the *interpreter* of action. As we saw in Chapter 7, the left hand of the split-brained patient will often perform actions whose origins are unknown to the articulate left hemisphere. That hemisphere is seldom at a loss for an explanation, however. In one study, Gazzaniga and his colleagues asked a split-brained patient to choose from an array of pictures the ones associated with pictures that were flashed simultaneously to the two hemifields, and so to the two hemispheres. In one example, a picture of the claw of a chicken was flashed to the left hemisphere and a picture of a snow scene was flashed to the right hemisphere. As associated pictures, the patient chose a shovel with the left hand and a chicken with the right hand. When asked to explain these choices, the left brain, oblivious to the snow scene, replied, "Oh, that's simple. The chicken claw goes with the chicken, and you need a shovel to go with the chicken shed."[73]

Gazzaniga suggests that the left hemisphere may play a similar role in "explaining" right-hemispheric dispositions even in normal people. As we shall see in more detail in Chapter 10, the right hemisphere may be the more specialized for emotion, and may bring about shifts of mood for which the left hemisphere may invent plausible (though inaccurate) explanations. We may see here something of Freud's notion of the rational ego trying to explain the aberrations of the emotional id. We shall also see in Chapter 10 that the idea of the rational left hemisphere acting as interpreter and censor of the wayward right hemisphere is an old one, going back to the late nineteenth century. It is in any case difficult to draw inferences about the normal mind from studies of split-brained patients. The left-hemispheric rationalizations of right-hemispheric actions in these patients may be a direct consequence of the disconnecting of the hemispheres, and in normal people the left hemisphere may be much better informed about the springs of action in its silent neighbor.

There are studies of normal people suggesting that the left hemisphere may be dominant simply in making decisions, independently of its verbal specialization or of the programming of sequential actions. Edoardo Bisiach and his colleagues at the University of Milan have demonstrated a right visual-hemifield advantage in making simple binary decisions. In one study, for example, people were asked to press a button if a dot appeared in one or the other hemifield, but to refrain from pressing the button if dots appeared simultaneously in both hemifields. They were quicker to respond when a single dot appeared in the right hemifield, and especially when they also used the right hand to press the button.[74] A follow-up experiment confirmed this and showed that the left-hemispheric superiority was maintained even when the subjects were required to count backward by threes while performing the task. This suggests that the left-hemispheric contribution was unrelated to language.[75]

It has even been suggested that only the left hemisphere is capable of *consciousness*. Echoing Descartes, John C. Eccles has argued that the right hemisphere, by contrast with the left, is a mere "computer," comparable to the brains of lower animals.[76] Zangwill is scornful, however, describing this view as "little more than a desperate rearguard action to save the existence and indivisibility of the soul."[77] The idea is nevertheless expanded by Julian Jaynes in his provocative book *The Origins of Consciousness in the Breakdown of the Bicameral Mind* (1976). As recently as the time of the *Iliad* or the Old Testament, Jaynes argues, the human mind lacked self-awareness and individual purpose, and humans were guided in their decisions and actions by hallucinations, which they interpreted as the voices of the gods. This passive reliance on the inner voices proved inadequate in an age of global catastrophe, however, and there rapidly evolved a new sense of self-awareness, which Jaynes identifies with consciousness and which was mediated by the left cerebral hemisphere. However, the evidence on the evolution of cerebral asymmetry, reviewed in Chapter 7, suggests that the left cerebral dominance for language may go back at least to *H. habilis*.

We can in any case rule out the idea that only the left hemisphere is conscious. There is evidence that patients remain aware of their surroundings when the left hemisphere is incapacitated by the injection of sodium amytal. Studies of split-brained patients have shown that patients can respond intelligently, if nonvocally, to pictures, words, and in some cases even sentences presented to the right hemisphere. The late Norman Geschwind even turned the tables by claiming that the *right* hemisphere is superior in consciousness to the left, on the grounds that it is dominant in the control of attention to spatial events—a topic that will be pursued in Chapter 10.

It has also been argued that the two hemispheres are independently conscious, a possibility that will also be discussed in Chapter 10. An interesting variant on this theme, however, has been developed by the philosopher R. Puccetti. He has argued that the left hemisphere is unaware of the existence of the separate "mind" in the right, which could be why the left hemispheres of Gazzaniga's split-brained patients were compelled to "explain" actions initiated by the right.[78] The silent right hemisphere, by contrast, has known of the existence of a separate consciousness, since from an early age it has heard speech emanating from the mouth of the body it shares with that hemisphere.[79] Only the left hemisphere can speak its mind, but this gives the right a certain superiority in its awareness of self.

The question of whether there are differences between the hemispheres in the awareness of self remains a complex and controversial one. We have seen that Jaynes attributes the awareness of self to the left hemisphere. Roger W. Sperry and his colleagues, however, have shown that the right hemispheres of two split-brained patients were well able to respond to pictures of themselves or their relatives, pets, and belongings.[80] In spite of his characterization of the right hemisphere as the "passive, silent passenger" in the passage quoted above, Sperry remarks in a later article that "after watching repeatedly the superior performance of the right hemisphere in [various] tests . . . , one finds it most difficult to think of this half of the brain as being an automaton lacking in conscious awareness."[81] Even Eccles has subsequently acknowledged that "the minor hemisphere has a limited self-consciousness,"[82] but he still goes on to ask whether it can worry about the future or make decisions based on a value system! In the next chapter, I shall argue that there may well be a sense in which the concept of self does depend upon mechanisms that are primarily left hemispheric.

Conclusions

In this chapter, I have tried to develop an integrated account of the evolution of human laterality, and of the critical role of the left hemisphere. Left-hemispheric specialization probably has to do largely with *praxis*, the organization of purposeful, sequential actions in which spatial constraints imposed by the environment are minimal. In a limited way, this may have emerged early in primate evolution, especially with respect to vocalization.

In hominids, freeing of the hands as a result of bipedalism would have greatly increased the scope for praxis. The role of the left hemisphere in vocal praxis may have been the "seed" that established manual praxis in the same hemisphere—or that favored the selection of the right-shift (RS+) genetic allele that disposed both manual and vocal praxis to be left-hemispheric. To some extent, the unilateral control of nonspatial praxis may have bestowed a general "executive" role on the left hemisphere, although this is probably not absolute. Right-handedness may be a by-product of the left-hemispheric control of praxis.

Praxis itself does not distinguish us from other species. For example birdsong is clearly praxic, intricate beyond the point of human understanding and, in some species, left-hemispheric to boot. Vocal communication may also be left-hemispheric in mammals, including primates. Other species, especially primates, are certainly capable of quite sophisticated manual praxis and show a preference for one hand (or paw). Nonhuman animals, however, seem as likely to be left-handed as right-handed, implying that vocal and manual praxis are unrelated. In most people, by contrast, it is the left hemisphere that appears to be dominant for praxis, whether vocal or manual.

It may be this concentration of praxic control within a single hemisphere that has given rise to our extraordinary ability to execute complex praxic skills—to make speeches, build complex machines, and program computers. But praxis alone may not be sufficient for these activities. The following chapter suggests that out of praxis has emerged a property that may indeed provide the key to human uniqueness.

Notes

1. Passingham (1982, p. 69).
2. Op. cit. (p. 69).
3. Jackson (1868).
4. Wada and Rasmussen (1960). This test is also known as the Wada test.
5. Milner (1975).
6. Rossi and Rosadini (1967).
7. There are case studies of left-handers who do seem to seem to show a cerebral organization that is the left-right reversal of the normal right-handed pattern (e.g., Delis, Knight, and Simpson, 1983; Taylor and Solomon, 1979). Such cases seem to be rare, however.
8. Pratt and Warrington (1972); Warrington and Pratt (1973).
9. Annett (1985); Corballis (1983).
10. An implicit assumption here is that switches in handedness occur more often than switches in hemispheric asymmetry for language (Satz, 1972).
11. Annett (1985).
12. Annett (1985).
13. Heilman and Rothi (1985).
14. Wyke (1971).
15. See Taylor (1958).
16. Liepmann (1908).

17. Heilman and Rothi (1985)
18. Op. cit.
19. Op. cit.
20. Heilman, Rothi, and Valenstein (1982).
21. Poizner, Klima, and Bellugi (1987).
22. Kimura and Archibald (1974).
23. Poizner, Klima, and Bellugi (1987).
24. Op. cit.
25. Peters and Servos (1989).
26. Critchley (1964, p. 11).
27. Thompson (1971).
28. See Corballis (1983, Ch. 8) for a detailed critique of Orton's theory.
29. Corballis and Beale (1970).
30. Makita (1968). Of course, there are other reasons why the reported incidence of reading disability in Japan might be low. Japanese symbols represent either syllables (*kana*) or whole words (*kanji*), and it may be easier to learn these mappings than those that link symbols to phonemes, as in alphabetic scripts.
31. Vernon (1960); Benton (1975).
32. Denckla (1979).
33. Annett and Manning (1989).
34. Naylor (1980).
35. Jansons (1988).
36. Geschwind and Behan (1982). Instead of studying the handedness of groups of dyslexics and matched normal readers, Geschwind and Behan turned the question around and examined reading and learning disabilities in left- and right-handers. They claimed that disabilities were over 10 times more frequent in the strongly left-handed than in the strongly right-handed. However, their definition of learning disabilities was rather loose, and they even included stuttering as one of the symptoms.
37. Geschwind and Behan (1982).
38. Critchley (1975).
39. In a recent survey of over 6000 people, Ellis, Ellis, and Marshall (1988) found that 8.44 percent of males had laterality quotients of less than zero (implying left-handedness) compared with 7.38 percent of females. This difference was not statistically significant. Moreover, there were significant age trends, implying quite a strong cultural component. There is therefore no compelling reason to suppose that the slight gender difference is biologically based.
40. Bishop (1987).
41. From an immunological point of view, it has been argued that there is little evidence for any link between prenatal male hormones and autoimmunity disorders (Wofsy, 1984). Even so, there is at least some evidence of a higher incidence of left-handers than right-handers among those suffering from a variety of disorders related to the autoimmune system: Salcedo, Spiegler, Gibson, and Magilavy (1985) reported no association between handedness and the autoimmune disorder systemic lupus erythe-

matosus (SLE), but Searleman and Fugagli (1987) found find a higher incidence of left-handedness among people with Crohn's disease, ulcerative colitis, and type I diabetes, and Smith (1987) observed a relation between left-handedness and diabetes. Evidence on the role of sex hormones in modulating handedness or cerebral asymmetry is similarly mixed; for instance, Hines and Shipley (1984) found that prenatal exposure to the masculinizing hormone diethylstilbestrol did not alter the normal pattern of cerebral asymmetry in women, but Nass, Baker, Speiser, Virdis, Balsamo, Cacciari, Loche, Dumic, and New (1987) were apparently unanimous in stating that there was an increased incidence of left-handedness in women exposed to an abnormally high androgen level due to congenital adrenal hyperplasia. However, Rich and McKeever (1990) found no evidence for a relation between autoimmune disorders and what they called *anomalous dominance,* and van Strien, Bouma, and Bakker (1987) found no relation between handedness and autoimmune disease in a large group of university sudents, although they did find evidence that the left handers had been slightly more prone to birth complications. Finally, Bishop (1987) examined records from over 10,000 children and found no relation between handedness and the following immune diseases: allergy, eczema, asthma, psoriasis, diabetes, and migraine.

42. Geschwind and Galaburda (1987).

43. LeMay (1977).

44. Hier, LeMay, Rosenberger, and Perlo (1978).

45. Rosenberger and Hier (1981). A further study of 26 dyslexic boys by Haslam, Dalby, Johns, and Rademaker (1981) produced only 3 with reversed asymmetry, although 11 showed no asymmetry. However 13 of these boys revealed so-called soft neurological signs, and 12 had records indicating abnormal pregnancy or delivery, suggesting that their reading dificulties may have been of pathological origin.

46. For a recent review, see Hynd and Semrud-Clikeman (1989).

47. Wood (1966).

48. The theme of left-right reversal is engagingly documented, along with other historical and philosophical aspects of Carroll's work, in Martin Gardner's excellent book, *The Annotated Alice* (1965).

49. Lennon (1945), cited by Gardner (1965).

50. See Harris (1980).

51. Van Riper (1971).

52. Johnson (1955).

53. Op. cit., p. 9.

54. Johnson (1959).

55. Morley (1957).

56. Jones (1966).

57. Mazars, Hécaen, Tzavaras, and Merreune (1970).

58. Andrews, Quinn, and Sorby (1972); Lussenhop, Boggs, LaBorwit, and Wallc (1973).

59. For example, Bordon (1983); Starkweather, Franklin, and Smigo (1984).

60. Webster (1986).
61. Passingham (1982, p. 103).
62. Watson and Kimura (1989).
63. Calvin (1987, p. 267).
64. Geschwind and Galaburda (1987, pp. 20–21).
65. Frost (1980).
66. Bruner (1968).
67. Calvin (1983).
68. Annett (1985, p. 400).
69. Quoted by Glezer (1987, p. 273).
70. Corballis (1983, Chapter 9).
71. Zangwill (1976, p. 305).
72. Sperry (1974, p. 11).
73. Gazzaniga and LeDoux (1978).
74. Bisiach, Mini, Sterzi, and Vallar (1982).
75. Vallar, Bisiach, Cerizza, and Rusconi (1988).
76. Eccles (1965); Popper and Eccles (1977).
77. Zangwill (1976, p. 304).
78. Puccetti (1981).
79. This is well understood by New Zealanders, who also share a hemisphere with a more talkative neighbor.
80. Sperry, Zaidel, and Zaidel (1979).
81. Sperry (1984, p. 666).
82. Eccles (1981).

— 9 —

The Generative Mind

> . . . but it is a melancholy of mine own, compounded of many simples, extracted from many objects. . . . [1]

IN THIS CHAPTER we reach the crux of my argument for a discontinuity between humans and other species. In Chapter 8, I argued that the left hemispheres of most humans are specialized for what I have called praxis. This may not be unique to humans, although we have almost certainly developed praxic skills to a level well beyond that reached by other animals. In this chapter I will focus on the way we represent things in our minds, and argue that out of praxis there has emerged a special form of representation whose most important property is *generativity*. This form of representation, I suggest, *is* unique to humans.

We have already seen that human language is generative, in the sense that we can create an unlimited number of sentences from a small number of phonemes. Phonemes are combined according to the rules of grammar, in hierarchical fashion, to form morphemes, words, phrases, and sentences. These are representational in that they can stand for real and possible states of the world. In these respects, language appears to be unlike any other form of communication between animals.

Generativity also characterizes forms of representation other than linguistic ones, including in particular the mental representation of objects and scenes. I will argue that this was an adaptation to the open-ended nature of the manufactured human environment, with its unlimited variety of shapes and combinations of shapes. There is evidence, moreover, that this generative form of representation is also left-hemispheric, at least in the majority of people. That is, the left hemisphere of most humans—that magic carpet of tissue—may be uniquely equipped with what I shall call a *generative assembling device*, or *GAD*. This hypothetical device, which is related to the device known as the LAD that we encountered in Chapter 5, is responsible for constructing representations in generative fashion from small vocabularies of primitive units.

The Representation of Objects

Marr's Theory

The idea that the representation of objects is essentially hierarchical and combinatorial was developed by the late David Marr in his influential 1982 book, *Vision*.[2] Marr gave an account of how our brains process visual information, from the point at which light enters the eye to that at which we recognize the array of objects in the world around us; Figure 9.1 presents an overview. This account was *computational* in the sense that the processes of vision were conceived as a series of computations of the sort that might be carried out on a digital computer. It was also based on what was known of the neurophysiology of vision. The way in which we perceive and represent the visual world is far from fully understood, however, and Marr's theory is in many respects incomplete. Nevertheless it is a useful starting point for a discussion of how objects are represented in the brain.

I shall skip fairly quickly over the early stages of visual processing, as these are not especially relevant in this context. It all begins, of course, at the eye, which acts as a lens system that focuses light on the retina, the network of light-sensitive cells at the back of the eyeball. The array of objects in the real world is therefore projected as an image on the retina. If we ignore variations in color, this can be conceived most simply in terms of variations in the intensity of light across the retinal surface. From this retinal image, the nervous system takes over, and the first stages of visual analysis result in what Marr called the *primal sketch*. This consists mainly of a map-like record of changes in intensity. Such changes signal the boundaries between objects or between parts of objects, so that the retinal image is reduced to something like a line drawing.

This set of local descriptions makes up the *raw primal sketch*. They are then grouped by a series of recursive operations into larger-scale elements to form the *full primal sketch*. To take an example from Marr, the raw primal sketch of a close-up view of a cat consists mainly of descriptions of the individual hairs, but the next level

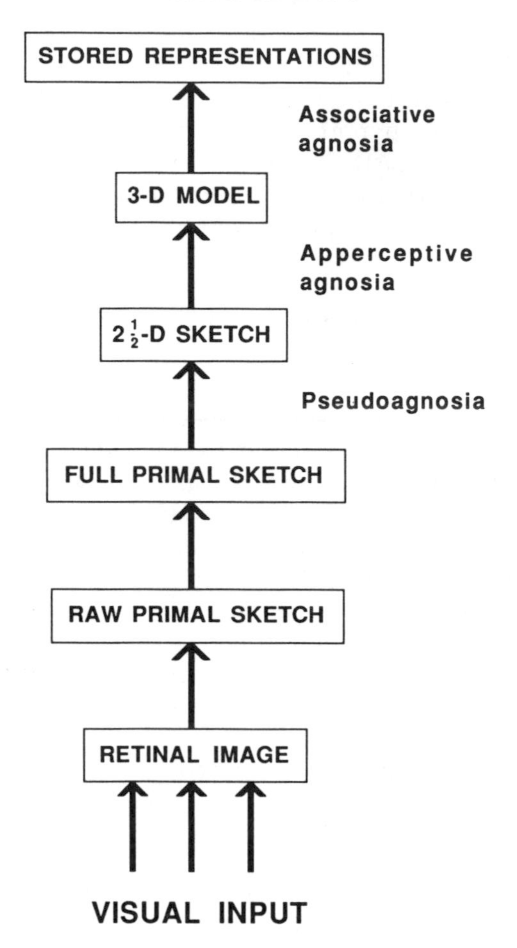

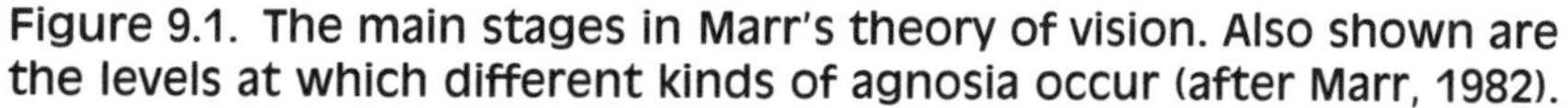

Figure 9.1. The main stages in Marr's theory of vision. Also shown are the levels at which different kinds of agnosia occur (after Marr, 1982).

might comprise the markings on the surface of the coat. At a still higher level, there might be elements that respond to the parallel nature of these markings, and so on.

From the full primal sketch, the visual system next computes what Marr oddly called the *2½-D sketch*. This includes information about the depth of each point on all surfaces visible to the observer. One cue to depth is supplied by stereopsis, which is derived from the fact that each eye has a slightly different vantage point and so receives a slightly different image of the world. Other cues for variations in depth include shading, gradients in texture, surface contours, and so on.[3] By this stage, then, the observer is informed about the geometry of visible surfaces, including their edges, orientations, and distances from the observer. Objects are not yet represented as *volumes,* however, so the representation is not yet a full three-dimensional one, which explains its odd name.

So far, the processing simply represents routine computations on the visual input. There is no need to suppose that there are fundamental differences between humans and other mammals in the mechanisms involved, and indeed, much of the basic neurophysiological information has come from the cat and monkey. The processing required to form the 2½-D sketch may also be described as *bottom-up,* in that it is driven by the input itself and requires no conceptual knowledge on the part of the observer. Bottom-up processing may be contrasted with *top-down* processing, in which interpretation does depend on what we know about objects, and may involve the testing of hypotheses about what a particular object is.

The next stage in Marr's theory is what he calls the *3-D model.* At this, the highest level of visual processing, the visual array is conceived as comprising objects that occupy volume in space. The 3-D model is also *object-centered* rather than *viewer-centered* in that the shape of an object is understood in terms of its own internal axes rather than in terms of the viewer's axes. That is, one might now perceive a bottle, say, in terms of its own top and bottom rather than in terms of the top and bottom of one's own vantage point. This allows the description of the object to be independent of its orientation in space, so that it can be recognized regardless of how it is viewed.[4]

The representation of objects in the 3-D model is critical to their *recognition.* That is, each object must be represented in such a way that it can be matched with information that is stored in memory. If we are to recognize a bicycle, say, we must extract a description from the image of a bicycle that can be compared with what we know about the shape of a bicycle, and if the two descriptions match, we can assert that the thing "out there" is indeed a bicycle. It is at this stage that Marr presents us with a theory that begins to resemble language in its generative structure.

One set of primitive units that Marr uses to construct representations consists of what he calls *generalized cones.* Figure 9.2 shows how cylinders (which belong to the class of generalized cones) of varying size and length can be arranged to represent the human body. The representation is hierarchical. At the crudest level, the body is just a single cylinder; at the next level, it is broken down into six connected cylinders representing the head, torso, arms, and legs. The arm can be broken down further into two cylinders representing

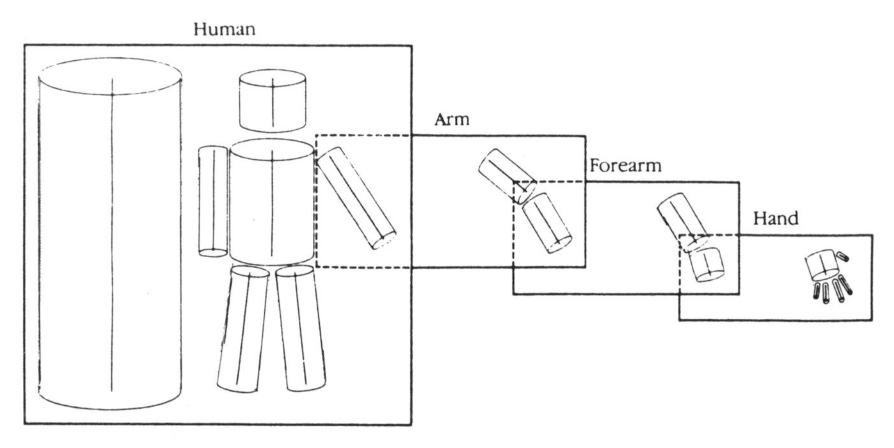

Figure 9.2. Diagrams showing how cylinders can be arranged to represent the human body at increasing levels of precision (from Marr and Nishihara, 1978).

the upper arm and forearm, and so on. By this process, one can achieve an increasingly fine description that can be matched against stored information at whatever level of precision is required.

This method of representation has much in common with the construction of a sentence. It is recursive in that routines are embedded in other routines. It is no doubt rule-governed, although its grammar is even less well understood than is that of language. Most important, it is generative in that there is essentially no limit to the number of different representations that might be formed. That is, many objects can be represented, albeit crudely in many cases, by an assemblage of cylinders. Figure 9.3 shows a number of examples.

Biederman's Theory

The notion that objects might be understood in terms of a set of idealized primitives, such as cubes, blocks, and so on, is actually an old one. It was part of Plato's theory of ideals and was the important principle of the Cubist school of painting.[5] Most shape primitives may be said to belong to the class of generalized cones, which may be defined as the three-dimensional volumes swept out by moving two-dimensional shapes along a path in the third dimension. For example, a rectangle moved in a straight line might create a brick-like shape, while a circle would create a cylinder. The two-dimensional shape can be altered in size as it sweeps out its path, and the path may be curved or straight. Some examples of generalized cones formed in this way are shown in Figure 9.4.

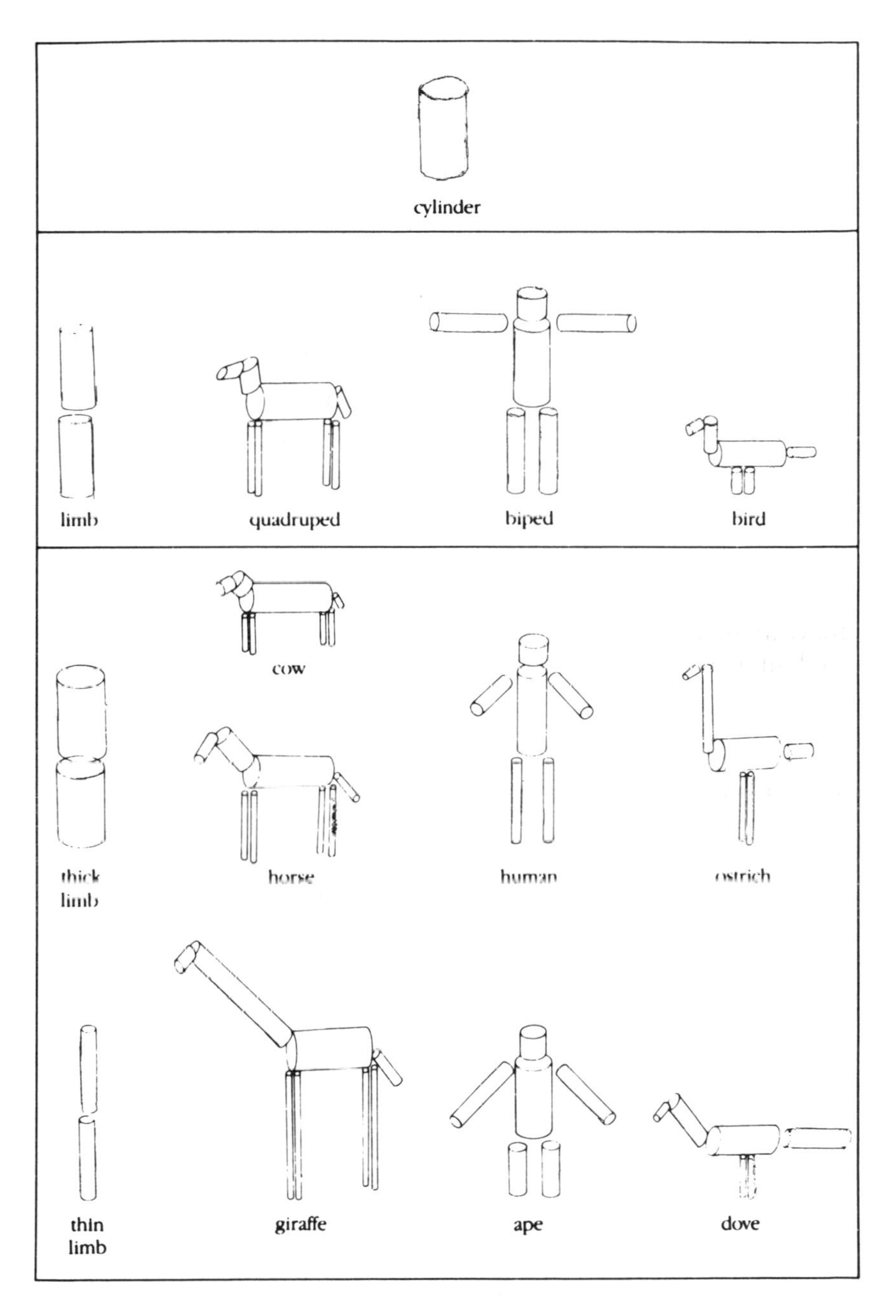

Figure 9.3. Representations of various objects constructed from cylinders (from Marr and Nishihara, 1978).

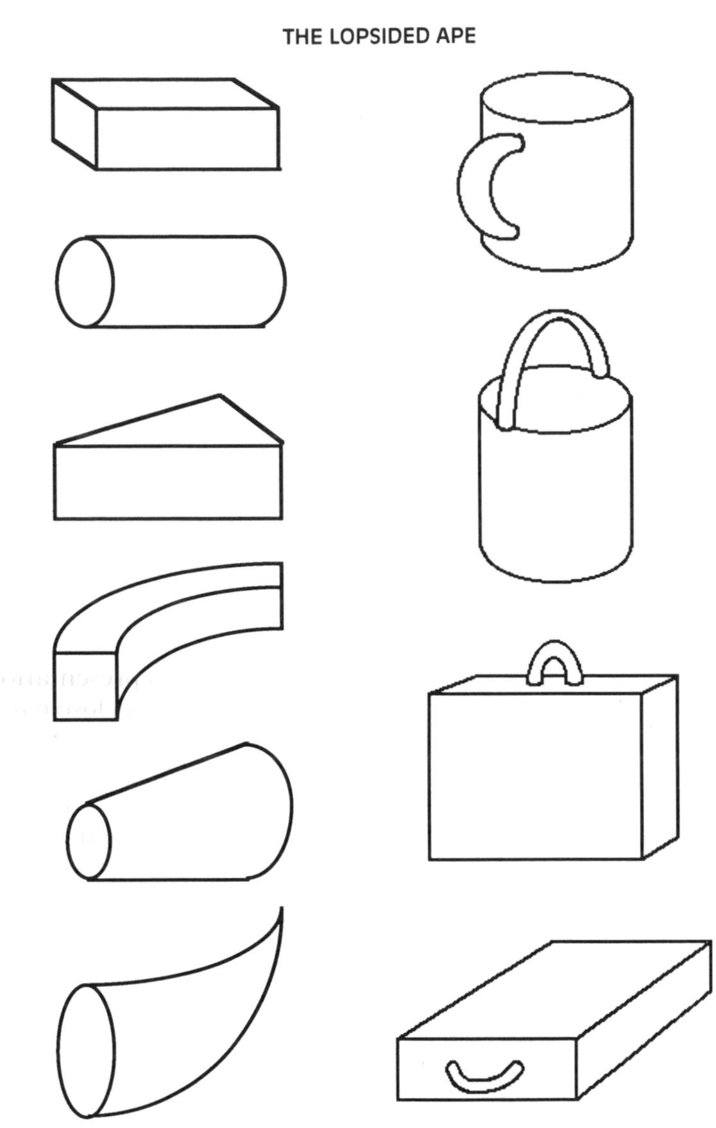

Figure 9.4. Examples of geons (*left*) from the set proposed by Biederman and examples of two-geon objects (*right*) (after Biederman, 1987).

Irving Biederman of the University of Minnesota has proposed a set of 36 generalized cones that he calls *geons*, which is short for "geometric ions."[6] He suggests that this basic vocabulary is sufficient to generate virtually all of the shapes we know, at least to the point at which they can be recognized. The geons are chosen, moreover, so that they are as discriminable from one another as possible, even when rotated into different orientations. All of them are basically convex in shape, so that any region of concavity in a more complex shape can be regarded as defining the boundary between geons, which helps the observer to parse the shape into its component geons. Figure 9.4 includes some examples of simple two-geon shapes, illustrating that the same geons can be aranged in different ways to represent different shapes.

Biederman has applied his geon theory, or *recognition-by-components* (*RBC*) theory, principally to the recognition of objects. That is, he gives an account of how visual objects are parsed into their component geons and then matched with geon-based representations that are stored in memory. In support of this idea, he has shown that recognition of mutilated pictures is generally better, other things being equal, if individual geons are preserved in the picture than if they are mutilated. He has also shown that objects drawn with only two or three of their geons can be identified almost as rapidly as fully colored, detailed, textured photographs of the same objects.

Biederman also makes the parallel between the representation of objects and human language more explicit. There is a close parallel between the geons that make up shapes and the phonemes that constitute spoken language, and indeed the number of geons proposed by Biederman is very similar to the number of phonemes, which ranges from about 16 to about 44 in different languages.[7] The assembling of geons, like that of phonemes, is presumably hierarchical: Just as phonemes are combined to form words and then words are combined to form sentences, so geons are combined to form objects and objects to form scenes. Here, too, the number of meaningful combinations is essentially unlimited. Indeed, in imagination we can combine objects to form scenes in ways that need not even conform to practical reality, just as we can compose nonsense sentences, although even imagined scenes probably do conform to basic rules that govern the actual locations and movements of objects in space.[8] We may therefore imagine such events as a cat playing a fiddle or a cow jumping over the moon—events that would never actually occur but that are nonetheless physically coherent. Like the rules of grammar, these rules are invoked unconsciously, and indeed, our explicit knowledge of them is fragmentary, at best.

It is possible to quibble, or course, with the actual vocabulary of geons proposed by Biederman. Perhaps the geons vary from culture to culture, just a phonemes vary from language to language. But the power of a part-based, syntactic system lies in its combinatorial properties rather than in the precise nature of the parts. This is illustrated by construction toys for children, such as Lego; even with a limited range of blocks or component parts, it is possible to build a wide variety of objects in recognizable imitation of real objects.

Pros and Cons of a Vocabulary-Based System

A geon-based system is especially well adapted to a technological environment, in which objects are often in fact composed of standard parts and proliferate in unlimited fashion. Think of the number of objects that include knobs, handles, or wheels, among their parts. Biederman calculates that, with a vocabulary of just 36 geons, one can construct about 154 *million* three-geon objects! Such a system is therefore effectively an open-ended one, readily adaptable and expandable to meet the demands of an ever-changing environment.

This generative aspect may also aid the process of recognition itself. That is, we may recognize an object in part by generating an image from memory and matching it against an image formed from the input,[9] perhaps at the level of the 3-D model itself. One of the problems in recognizing a familiar entity, such as a chair, is that the image on the retina that is formed by the chair is infinitely variable, depending on its location, orientation, color, type, and so on. This problem may be solved in part by a bottom-up analysis of the input itself, and in part by the top-down process of generating images of various candidate objects and transforming them so as to discover the best "fit" to the input. The same basic idea underlies Liberman's motor theory of speech perception, in which speech sounds are recognized in terms of how they are uttered (top-down) as well as how they sound (bottom-up).

A vocabulary-based system makes for effective storage and retrieval in other ways. The basic vocabulary of geons are idealized, or "customized," so that a good deal of unnecessary information may be discarded. For example, a cylinder may suffice for the leg of a chair, the arm of a person, or the handle of a frying pan. Idealized geons may also serve as a basis for cross-referencing, and for retrieving information from memory. If objects are simply represented as unrelated entities, like a stack of photographs, we have no system for quickly finding the ones we want. However indexing by parts can provide a ready means of locating what we are looking for.[10] To illustrate this point, the ease of finding a word in a dictionary derives from the fact that words are made up of parts (letters) that can be systematically ordered. Imagine how difficult it would be to find a word if it comprised a single, holistic entity, like a Chinese logogram.

But for all that, any system of representation based on a small vocabulary of building blocks has limitations. Such a system cannot capture the full extent of human knowledge about the shapes of objects, just as not everything we know can be put in words. Consequently we may have access also to a mode of representation that is holistic rather than assembled from parts and that appeared much earlier in evolution. This mode may be better adapted to the recognition of patterns in the natural environment, such as individual faces, or the forms of animals, or the texture of vegetation. It may represent nuances of shape and texture that cannot be captured easily in a geon-based construction.

Accordingly, two sorts of descriptions may be extracted from the 3-D model sketch for matching against stored information, one generative and geon-based, the other holistic and nongenerative. I now document evidence that the geon-based form of representation is primarily left-hemispheric, while the holistic form of representation tends to be right-hemispheric.

GAD and the Left Hemisphere

Recognizing Objects

Evidence that the two cerebral hemispheres might be specialized for different kinds of representations comes from a neurological disorder known as *agnosia*. Patients with this disorder have difficulty recognizing objects, despite normal intelligence and otherwise normal vision. Often the disorder is highly specific, so that the patient has difficulty with only certain classes of objects, such as animals, faces, or vegetation.

In 1890, Heinrich Lissauer distinguished between two levels of agnosia. The first he called *apperceptive agnosia*, referring to a disturbance at the level at which the three-dimensional structure of an object is understood but prior to the level at which meaning is assigned. In Marr's terminology, the patient with this disorder might be considered to have difficulty constructing the 3-D model from the 2½-D sketch. The second he called *associative agnosia*, which refers to the failure to recognize what some object is, even though the three-dimensional structure may be correctly perceived.[11] It is this second level of agnosia that is of primary concern here, since it has to do with the descriptions that are extracted from the 3-D model for matching against what is stored in memory. Figure 9.1 shows the levels, in Marr's system, at which these agnosias are thought to arise.

Associative agnosia can occur following damage to either side of the brain. However, in a recent study of 99 published cases, Martha J. Farah of Carnegie-Mellon University suggests that there may be a link between the *kind* of associative agnosia and the side of the brain damage.[12] She first contrasts two extreme forms of associative agnosia. One, which we encountered in Chapter 7, is *alexia*, or the inability to read (or recognize printed words), and this is nearly always the result of left-hemispheric damage. The other is *prosopagnosia*, or the inability to recognize familiar faces.[13] This involves damage to the right hemisphere, sometimes exclusively,[14] although the more general pattern is that of damage to both hemispheres.[15]

Farah notes that agnosia for objects other than faces or words always seems to occur in association with either prosopagnosia or alexia, or both. Prosopagnosia and alexia, by contrast, may each occur in isolation, and they do not seem to occur together unless there is also agnosia for other objects as well.[16] Farah concludes that prosopagnosia and alexia are "markers" for different kinds of representation. One, whose marker is prosopagnosia, is holistic and represents objects as unified "pictures" rather than as assemblages of parts. The other, whose marker is alexia, has to do with the partwise representation of shapes documented above. This is left-hemispheric and may be uniquely human. This distinction between holistic and partwise representations is, of course, essentially the one drawn in the previous section.

Farah points out that, for the most part, the agnosias associated with prosopagnosia have to do with natural objects, such as animals[17] or foodstuffs,[18] or with discriminations that involve subtle distinctions of shape, such as the discrimination between different makes of automobile,[19] rather than classifications that depend on a partwise analysis. The agnosias associated with alexia, by contrast, tend to involve everyday objects, most of which are manufactured or are clearly composed of parts.[20]

It is unlikely, though, that specific objects or shapes are represented exclusively in one way or the other. For example, animals might be represented both in holistic fashion and as assemblages of parts; Irving Biederman gives several examples of how representations of animals might be constructed out of geons, and yet if the present account is correct, we might also expect animals to be represented in the holistic mode, since the recognition of animals undoubtedly preceded human evolution. The identification of faces may involve different kinds of contributions from the two hemispheres.[21] The right hemisphere might contribute more to the distinction of one individual's face from another's, while the left might code a partwise representation that enables one to recognize a face simply as a face. Even printed words may be represented in the right hemisphere in a holistic fashion.[22]

Overall, then, the evidence is consistent with the idea of a partwise, generative mode of representation that is predominantly left-hemispheric and that may be uniquely human. This may be contrasted with an earlier, more holistic mode that may be biased toward the right hemisphere—a topic to be discussed more fully in the next chapter. It should not be thought, however, that this mode is inferior; it is the product, after all, of millions of years of evolution and is tuned to the most subtle aspects of the natural environment. The left-hemispheric mode has the advantage of open-endedness and flexibility, but in other respects, GAD may be a crude adaptation to human profligacy.

Generating Images

Further evidence on GAD and the left hemisphere comes from a rather unlikely source: mental imagery. Most of us experience visual imagery, as when we dream or simply fantasize about our favorite places, people, or pictures. The conjuring of images of objects in one's mind may be regarded as the converse of recognition. Whereas recognition begins with the object itself and is completed when the input is matched with information stored in memory, visual imagery begins with stored knowledge of what an object or shape is like and constructs from it an internal "picture." This constructed image has much in common with the perception of an actual object.[23]

Over 100 years ago, John Hughlings Jackson wrote that "the posterior lobe of the right side [of the brain] . . . is the chief seat of the revival of images."[24] Indeed, imagery is generally regarded as right-hemispheric, since it is analogue and picture-like, as opposed to language, which is symbolic and propositional. In 1983, however, H. Ehrlichman and J. Barrett published a review of published cases in which patients with brain damage reported a loss of visual imagery and concluded that "reports of loss of imagery are not contingent on damage to the right hemisphere. If anything, it is damage to the posterior areas of the left hemisphere which is more often associated with reported loss of imagery."[25] In a more extensive review, in which she was careful to distinguish between different aspects of imagery, Martha J. Farah made the stronger claim that "the critical area for image generation may be close to the posterior language centers of the left hemisphere."[26]

The specific role of the posterior part of the left hemisphere, and particularly the left occipital lobe, has been confirmed in several more recent studies. For example, in a study of freehand drawing in patients with brain damage, Murray Grossman observed that one patient with damage to the left posterior part of the brain seemed unable to generate the mental images required for drawing.[27] An experiment on brain activity in normal people, measured by the pattern of blood flow in different parts of the brain, showed also that the left occipital lobe was activated when the subjects were asked questions whose answers required visual imagery.[28] Farah herself, along with several colleagues, has also recorded electrical potentials from the scalps of people while they formed images of words and has demonstrated increased activity over the left occipital lobe.[29] There is also evidence that those who experience vivid imagery show more activity in the posterior part of the left hemisphere than those who do not experience vivid imagery.[30] It has even been suggested that loss of dreaming is associated with damage to the posterior part of the left side of the brain.[31]

The specialized role of the left hemisphere in generating images was rather strikingly demonstrated in studies of a split-brained patient, known as J.W. He was given tasks that required him to envisage the shapes of lowercase letters given only their uppercase versions,[32] or to envisage uppercase letters given only their lowercase equivalents.[33] He was unable to carry out these tasks when the letters were flashed to the left of a fixation point, and so to the right hemisphere, but had little difficulty when the letters were flashed to the right of fixation, and so the left hemisphere. His right hemisphere also proved incapable of performing a task requiring him to imagine some named animal and decide whether its ears protruded above the skull (as on a cat or mouse) or not (as on a sheep), while his left hemisphere had no trouble with this.[34] A second split-brained patient known as V.P. was also tested, although less extensively, and gave broadly similar results.

It is, in a way, not surprising that the generation of images should be controlled by the side of the brain responsible for language, since one of the functions of language is to evoke images, and it is indeed this aspect that gives language its quality of vicariousness. We can talk about other places and other times precisely because our words are born of images and convey those images (or something like them) to others.[35] In all of the experiments on image generation described above, the stimulus for the evocation of an image was verbal—a letter or a word. Image generation is presumably more efficient, therefore, if its mechanism is located in the same hemisphere as the mechanisms for language.

There are, however, some discrepant findings. Another split-brained patient, L.B., seems quite well able to generate images of lowercase letters from their uppercase equivalents, and vice versa, regardless of which hemisphere the input is projected to.[36] Similar studies with normal people fail to show an evidence for a right-hemifield advantage.[37] Moreover, the split-brained patient J.W. was able to perform certain imagery tasks quite well when the input was to the right hemisphere. One such task required him to judge whether named objects, such as a book or a nose, were taller than they were wide. Another was to judge which of two named animals, such as a goat or a hog, was larger. In these cases, the left and right hemispheres proved equally adept.

Stephen Kosslyn of Harvard University, who was involved in some of these experiments and has made extensive studies of the nature of imagery, has argued that the critical left-hemispheric component is the generation of images *by parts*.[38] He quotes from the experience of a patient with left-hemispheric damage who seems clearly to be having trouble with GAD:

> When I try to image a plant, an animal, an object, I can recall but one part, my inner vision is fleeting, fragmented; if I'm asked to imagine the head of a cow, I know that it has ears and horns, but I can't revisualize their respective places. In the same way, I cannot determine how many fingers a frog paw has, even though I have manipulated this animal every day in the laboratory. . . .[39]

Farah has concurred that the deficit may be specifically one of constructing images by parts. She and her colleagues describe the case of a patient with left posterior brain damage who appears to be unable to generate the appearances of objects but is able to perform other spatial tasks. For example, he can describe how to get from his home (in France) to Paris. Farah concludes that the evidence from this patient is consistent with "the hypothesis that image generation deficits after left posterior brain damage involve an inability to assemble mental images from their parts."[40]

Kosslyn suggests that the right hemisphere is able to construct an outline image, sufficient for judgments about size or outline characteristics but not for judgments about the arrangement of component parts. He also describes a study suggesting that some people, at least, can create images of letters in holistic rather than partwise fashion, and that this ability is right-hemispheric rather than left-hemispheric.[41] This might explain why some studies have failed to show a left-hemispheric advantage in the generation of images of letters.

The idea that the two hemispheres may contribute differently to the generation of images is supported by another study of the effects of brain injury on the ability to draw, carried out by Andrew Kirk and Andrew Kertesz of the University of Western Ontario. Overall, drawing was impaired more by damage to the left than to the right side of the brain. However, patients with left-sided damage tended to oversimplify their drawings, leaving out details, although the overall organization was generally accurate; this is consistent with the idea that the left-hemisphere is specialized for partwise representations. Patients with right-sided damage showed a converse pattern of deficits. They produced detailed drawings but had some difficulty with spatial organization and synthesis.[42]

In summary, the evidence suggests a left-hemispheric specialization for a partwise, grammatical process in both the recognition of objects and the generation of images. These parallel the left-hemispheric specialization for the understanding and production of sentences. Yet the left-hemispheric component in object recognition and image generation has nothing to do with language. This suggests that GAD's benevolent influence extends beyond language, at least to the representation of objects.

It is important to distinguish the generative function of GAD from creativity. Although we may construct sentences or imagined scenes that we have never constructed before, these activities need not be particularly creative in the usual sense of that term. Both language and visual imagination are generally effortless and relatively automatic—to the point where we ourselves cannot discern the rules that govern them. Creativity, by contrast, is usually considered a more effortful process, perhaps often going beyond the accepted rules and even establishing new rules. Of course, we may use language and imagery creatively, but this usually involves the addition of features or properties that go beyond the medium itself.

The Evolution of GAD

Representation of Objects by Nonhuman Species

I have reviewed the evidence on language in nonhuman species in Chapter 6 and concluded that only humans are capable of truly generative language. The question now is whether animals other than humans represent objects in a generative, geon-based fashion. Since geon theory is relatively new, it has yet to be tested explicitly in other animals, at least to my knowledge.

There is evidence, though, that other animals can form quite sophisticated representations of shapes. These include even the pigeon, one of experimental psychology's favorite subjects. In a classic experiment, Richard J. Herrnstein and D. H. Loveland of Harvard University showed that pigeons could learn to discriminate slides that contained photographs of humans from those that did not and could generalize this knowledge to slides they had not seen before.[43] This suggests that the birds could form representations of humans sophisticated enough to pick out exemplars that differed substantially in posture, orientation, and even clothing. Since then, pigeons have been taught a variety of discriminations based on natural categories, such as pictures that do or do not depict water, or trees, or fish, or birds.[44] Herrnstein has claimed that pigeons cannot discriminate the presence or absence of artificial objects, such as bottles or chairs;[45] this is consistent with the idea that geon-based representation may have evolved with the manufacture of objects and is therefore unique to humans.

More recent studies have suggested, however, that pigeons *can* categorize manufactured objects. For example, it has been shown that they can classify pictures into *four* categories, such as cat, flower, car, and chair, that include natural as well as artificial objects[46]—and, indeed, the artificial categories seem to cause no more problem than the natural ones. One might therefore be tempted to think that even the pigeon has found GAD. But in all these studies the number of presented categories is small, even though different categories might be selected on different blocks of trials. Moreover the pigeon is notoriously clever at discovering ways of discriminating patterns that do not require a high level of understanding. For example, it has been suggested that pigeons always discriminate pictures on the basis of small, local features rather than global perceptions.[47]

There is also evidence that nonhuman primates can categorize objects. Viki, the chimpanzee reared by the Hayeses, readily learned to choose between pairs of pictures, each drawn from a different class. For example, each pair might consist of a house and a dog, each one depicted differently on each trial, and Viki might be required to choose the dog.[48] The categories were changed each day, so that on another day Viki might be shown a cat and a telephone on different trials. Nevertheless, Viki was typically able to learn the task after a single trial. Other more recent investigators have also demonstrated that primates can learn such categories.[49]

We also saw in Chapter 6 that chimpanzees and gorillas can learn symbols numbering at least in the tens, if not in the hundreds. Of course, many of these represent actions or qualities rather than objects, but even so, these studies indicate that nonhuman primates can use symbols to discriminate perhaps tens of natural or artificial objects. Where the symbols themselves constitute shapes, as in the case of David Premack's Sara, they must themselves be added to the repertoire of known shapes—although they are two-dimensional and do not really lend themselves to geon-based representation. To keep matters in perspective, however, Biederman has estimated, probably conservatively, that humans can discriminate some 30,000 different shapes.[50] This provides a reasonable basis for supposing, then, that humans have evolved a special generative mode of representation.

Nevertheless, it can be anticipated that research on the ability of other animals to represent shapes will grow more sophisticated and will parallel research that has aimed to demonstrate language in nonhuman animals. For the present, we can probably rest reasonably safely with the claim that a generative, geon-based form of representation is uniquely human. This claim is, of course, reinforced by the evidence that it has many of the properties of language and that it appears to be primarily left-hemispheric.

Let me now consider, then, when and how GAD might have developed in hominid evolution.

GAD in Hominid Evolution

The proliferation of objects in the human environment is, of course, due largely to manufacture, although we may also have evolved an unusual ability to identify and classify natural objects, such as plants or other animals. Since tools appear early in the record of *Homo*, the line that led to ourselves, it is possible that GAD began to emerge over 2 million years ago with *H. habilis*.

One whose views are consistent with this theory is Ralph L. Holloway. Although he does not focus specifically on generativity, as I have done, he sees both language and the manufacture of objects as critical in the evolution of human thought. He writes that "any theoretical model that describes language *also* describes stone tool making."[51] However, as we saw in Chapter 3, Holloway does not see either of these activities alone as critical. He stresses rather that such activities as tool making, hunting, and gathering were fundamentally social processes in which the emphasis was on cooperation rather than aggression. This new social environment, in turn, favored the development of communicative skills. Tool making may have provided the kick for generativity, but hominid evolution thereafter depended upon a positive-feedback system between hominid activities and the environments created by them. The unique feature of human evolution was that the hominids created their *own* environment, so that the selective pressures of evolution were in effect under their own control.

Although this positive-feedback system would have accelerated the evolutionary process, Holloway argues for a gradual rather than a punctuate evolution of language and associated manual skills from the australopithecines on. Part of the evidence for this is that the characteristic pattern of brain asymmetry dates back to the earliest hominids. In keeping with the theme of this book, Holloway writes that

> fossil hominid endocasts, from *Australopithecus* on, do show a typical *Homo* pattern of left-occipital, right-frontal petalial asymmetry, which has been strongly correlated with right-handedness in modern humans. It is the combination of the two asymmetries which is striking both in modern *Homo* and in the fossil hominids.[52]

Holloway also suggests that the same combinatorial principles underlie both language and tool making; both of these, he writes, "utilize a limited number of basic units that can be combined in a *finite* number of ways."[53] But note the word "finite" here. True generativity, as we have seen, allows units to be combined in an essentially *unlimited* number of ways. In fact, generativity in tool making does not seem to have begun until well *after* the emergence of *H. sapiens sapiens*. Although brain size increased systematically from the australopithecines through *H. habilis* and *H erectus* to *H. sapiens*, tools themselves changed only slowly from their first appearance in the historical record some 2.5 million years ago until the emergence of blade technology in Africa about 70,000 years ago and the evolutionary explosion in Europe and Asia some 35,000 years ago. That is, some major change seems to have occurred *after* the brain reached its present size, but it is associated only with *H. sapiens sapiens*.

In opposition to Holloway, then, some authors have suggested that language and representational skills emerged late in human evolution. An extreme position is taken by Iain Davidson and William Noble, who focus on the cave drawings of the Upper Paleolithic a mere 32,000 years ago. They argue that the cave drawings were outward manifestations of a newfound inner process of *depiction*, or the ability to generate an image and reflect upon it. Depiction may have had its origins in gesture:

> A hominid confronting a bison while in the presence of another hominid might have had gestural communication, developed through prior selection for increasing tool making and use, by which to indicate to the other the presence of the bison—by gesturing in the air its distinctive humped outline. Repetition of this act in a situation in which the gesture was fixed by making a trace of the outline in the mud on a cave floor would indicate "a bison" because of the mimicry that formed the gesture. This "frozen" gesture would remain to be seen in the absence of a bison. It would thus afford the materials for communicating about a "bison" in the absence of a bison.[54]

According to Davidson and Noble, this ability to depict was critical to the evolution of language. Since the depiction of an object persists in the absence of the thing depicted, communication can achieve the property of displacement; we can talk about things that are not physically present or that occurred at some other point in time. The notion of depiction also has implications for visual perception. Davidson and Noble refer to the theories of the late James J. Gibson, who argued that perception was "direct" and required no internal representations.[55] According to Davidson and Noble, this may be true of nonhuman species, but humans evolved internal representations, or depictions, of objects that have radically altered the way we perceive the world.

Davidson and Noble suggest, then, that the marks of language and depiction did not appear until a mere 32,000 years ago, well after the appearance of *H. sapiens sapiens*. However, it is difficult to accept that true language emerged quite as late as this. For one thing, language is universally human, and by 32,000 years ago, the various populations of *H. sapiens sapiens* were widely scattered. It follows that language must have evolved independently in different populations, which is on the face of it unlikely. It is conceivable, though, that hominids made drawings much earlier, but that no traces remain.

We also saw in Chapter 6 that at least one species of ape, the pygmy chimpanzee, seems to be capable of using symbols to refer to objects or places that are not immediately present. Although symbolic representation may come more naturally to humans than to other species, it does not seem to be uniquely human. If the pygmy chimpanzee is anything to go by, one would guess that a representational capacity was present early in hominid evolution, and possibly before the split between the apes and hominids.

A Rapprochement

In Chapter 6, I outlined two scenarios for the evolution of language. In one, generative language appeared in *H. sapiens sapiens* as a punctuate evolutionary event, perhaps due to some random genetic reshuffle. In the scenario that I prefer, however, generative language evolved gradually, although probably in accelerating fashion, in the 2 million or so years of evolution from *H. habilis* to *H. sapiens sapiens*, much as Holloway has suggested. What distinguished *H. sapiens sapiens*, though, was the switch from a means of expression that was primarily gestural to one that was primarily vocal.

According to this scenario, then, GAD evolved prior to Eve, but Eve was able to exapt it for purposes other than language. Hominids had long been social creatures, and efficient communication may have been more important to their survival than the making of tools. Nevertheless, when vocal speech took over from gesture, the freeing of the hands meant that they could be dedicated increasingly to manipulation and manufacture, and indeed to such frivolous activities as drawing and the making of ornaments. There was idle work for idle hands to do.

Besides the hands, the eyes would also be freed by a conversion to vocal language. Instead of having to watch their companions for signs and process those signs visually, our ancestors would have been able to pay more attention to the objects they made and used and to the world about them. They could also talk as they watched and manipulated, and so share ideas and insights. Generative principles, previously associated with language, might then have been applied to the manufacture of objects, and indeed to their perception. Gradually, objects made by humans would have begun to outnumber naturally occurring ones and to place a burden on memory. By reducing objects to their parts and to rules governing their combination, our ancestors may have bestowed an open-endedness on their ability to process objects and scenes that approached the open-endedness of language.

Contrary to Davidson and Noble, then, I suggest that the appearance of drawings and ornamentation in the Upper Paleolithic was not critical to the development of language, but was rather a release of manual and visual creativity brought about by an earlier conversion to vocal language. It was probably cultural, moreover, rather than biological. However, it may well have been critical to the subsequent development of writing and eventual literacy, which, in turn, have had a profound effect on technology and control over the environment. The societies of Europe and Asia seem to have had something of a head start in these developments, although not without cost to the natural environment. This is not to attribute any biological superiority to those who descended from the cave artists of the Upper Paleolithic; rather, this was merely a cultural development that took off (and eventually landed us on the moon).

The switch from a gestural to a vocal form of language would also have increased the intensity of Holloway's positive feedback loop. Manufacturing and communication could then function cooperatively rather than competitively. The output of manufactured objects would have increased, making the environment more complex, and so adding to the pressure for communication and further industry.

The freeing of the hands may have had consequences for recreation as well as for industry. The appearance of cave drawings may have been essentially recreational in spirit—an outlet for a new-found generative and representational skill. Even today, art is not essential to survival, although it does make the world a more agreeable place to live in. Moreover, different societies may have found different outlets for those idle hands and eyes, which could be why the appearance of cave drawings in the Upper Paleolithic was apparently restricted to parts of Europe and Asia.

After the Evolutionary Explosion

Following the evolutionary explosion, manufacturing continued to develop at an ever-increasing pace. It is not clear precisely when it achieved a level of generativity comparable to that of language. However, there can be little doubt of the generativity of modern technology, which is relentlessly combinatorial: The same units—cogs, wheels, axles, nails, screws, handles, microchips, etc—recur in an enormous variety of manufactured objects.

One of the most significant developments was not in manufacture itself, but in the development of writing systems. We saw above moreover that alexia, or the inability to recognize printed words, may serve as a marker for associative agnosia for objects that are constructed of parts. Indeed, the evolution of writing systems may be said to have begun with the cave drawings of 32,000 years ago and to have largely paralleled the evolution of manufactured objects, as well as of our ways of representing them. Let me therefore briefly outline how writing systems developed from a holistic to a partwise form of representation.[56]

The Development of Writing

From the earliest cave drawings, then, depictions of objects first developed standard forms, known as *pictograms*, that were designed expressly for the purpose of communication. The earliest known pictograms came from Sumeria and spread to surrounding areas about 5000 years ago. Pictograms were widespread in the ancient world, and are still to be found among the American native peoples and in Africa, as well as in international road signs or labeled instructions on clothes and other goods.

Pictograms, in turn, evolved into *ideograms*, in which the symbols stood for abstract as well as concrete ideas. For example, a pictogram of the sun could be taken to refer to heat, or an arrow could indicate the verb "to go." Ideograms are also constructed from combinations of pictograms, as in the combination of Chinese pictographs for "eye" and "water" to produce the ideogram for "tear."

Ideographic symbols also became more stylized, and at some point in their development achieved the status of *logograms*, in which each symbol stood for a word and could be read as such. Ancient Mesopotamian cuneiform inscriptions, Egyptian hieroglyphics, and Chinese characters are all logographic.

Logograms remain essentially pictorial, however, and make little or no reference to the *sounds* of speech. For example, the numerals 1, 2, and 3 contain no information as to how they should be sounded, and are in fact spoken differently by speakers of different languages. Historically, though, sound-based elements began to emerge in some writing systems, especially as some concepts could not be represented pictorially. Among the Sumerians, for example, it was necessary to develop sound-based symbols to represent people's names, which were seldom based on picturable objects. Indeed, Sumerian cuneiform writing was purely sound-based in that each symbol represented a particular word *as sounded*. Homophonic words, which sound the same but have different meanings (as in "see" and "sea" in English), were always written the same way. However, the basic unit was still the whole word.

The next important development was the *syllabary*, in which individual symbols stood for syllables rather than whole words. The first syllabary seems to have been developed by the Semites and Phoenicians about 1700 B.C. and adapted in Old Hebrew, Cypriote, and Persian scripts. Two surviving syllabaries are the Japanese scripts *Katakana* and *Hiragana*, which were formalized in about the ninth century. The Japanese have also retained a logographic system known as *Kanji*, which was borrowed from the Chinese. It has been suggested that *Kanji* remains important because of the large number of homophones, or words that sound the same but have different meanings, in spoken Japanese. For example, the word sounded as *ka* has over 200 meanings, but each has a different logograph so that these meanings can be distinguished.

The final phase in the evolution of writing systems was the development of *alphabets*, in which the symbols stand for phonemes. The beauty of representing language at this level is that relatively few phonemes are required to generate every word in a language, so that only a small number of symbols is required. As we saw earlier, for example, American English is based on just 44 phonemes, whereas there are hundreds of thousands of words. Of course, alphabets typically do not map precisely onto phonemes in one-to-one fashion; for instance, there are just 26 letters in the English alphabet, and some phonemes, such as the *ch* sound in "cheat," are based on combinations of letters. Even so, alphabetic systems are far more practical than ideographic systems or even syllabaries.

Although spoken language probably consisted of phonemes at least from the emergence of *H. sapiens sapiens*, the explicit notion of the phoneme appears to have come about gradually, and even today it is difficult to define objectively. In Egypto-Semitic scripts, for example, the symbols represented consonants only; vowel markers were added later as an afterthought. The first fully alphabetic system was the Greek alphabet, representing the birth of the idea that the phoneme, rather than the syllable, was the basic element of spoken language.

With the arrival of alphabetic scripts, then, we have full recognition of the partwise nature of vocal language and its mapping onto a visual representation. This may have reinforced a geon-based system for representing (and perhaps even constructing) objects; recall that agnosia for the partwise representations of objects seems always to be associated with alexia. Of course, it may be noted that objects are three-dimensional and have a more complex spatial structure than words, in which the parts are simply strung together in a line. Nevertheless, both may depend on similar combinatorial principles.

Alphabetic writing, of course, exhibits the full generativity of spoken language, since the one maps fairly directly onto the other. Insofar as one form of associative agnosia for objects—and especially for objects that are represented in partwise fashion—is associated in turn with alexia, we may perhaps infer that the partwise, generative representation of objects was derived, in turn, initially from the partwise, generative nature of language.

Beyond Writing and Common Objects

It might be said that written forms of expression have evolved beyond the mere representation of spoken language to encompass mathematics, itself a generative, rule-governed form of representation. I know of no evidence on the representation of such abstract systems of mathematics as algebra or formal logic in the brain, but neurologists have long been aware that brain damage can cause impairments in calculation, a condition known as *acalculia*.[57] This is usually associated with damage to the left side of the brain, although right-sided damage can also cause difficulties, especially when there is a spatial component (as in placing numbers in columns for adding).[58] Of course, developments in mathematics are essential to advances in manufacture; no motor vehicle, harbor bridge, airplane, or modern high-rise building could be constructed without the aid of mathematics. The power of rule-governed generativity grows ever stronger, distinguishing the human environment more and more from the habitats of other species—although many species are forced to share the human environment with us.

GAD can be regarded as both an assembling and an analytic device. Not only do we construct more complex objects by assembling parts in more complex ways, we also analyze existing objects more and more into component parts. The "models" that we construct are also built from primitive building blocks, again often resembling models made from children's construction toys. Models of molecules consist of colored balls and connecting rods; models of atoms resemble our models of planetary systems; even models of the mind consist of boxes and arrows in varying combinations! The influence of GAD may therefore be seen not only in our perceptions of common objects but in our theories of mind and matter, ranging from subatomic particles to our conceptions of the universe.

There are, of course, differences between human cultures in these applications of GAD, but we must resist the temptation to conclude that they reflect biological differences. The exploitation of generative principles has depended on culture, and is embedded in culture rather than in the genes. The generative principle itself is no doubt biologically programmed but its vast applications must be taught. Even specific languages must be taught, but the child's readiness to learn them is biological; similarly, I know of no evidence that children of different cultures are not equally ready to learn to write, or do mathematics, or build things, given equal opportunity and equal motivation.

It may be recalled from Chapter 1 that A. R. Wallace, Darwin's contemporary, was puzzled by the differences among races in how "civilized" they were. He could not understand how it was that different races were all equally human, yet their accomplishments were so widely varied. Wallace may have been biased in his appreciation of what it means to be human or civilized, but there is no denying the extraordinary range of human activities. Wallace thought that the answer to this conundrum lay in God, who created humans to possess powers beyond their immediate concerns. We now see that it was not God but GAD that bestowed on us a potential that is surely never fully realized in any one individual.

Is There a Universal GAD?

In this chapter, I have suggested that the generative, rule-governed nature of human thought and representation may be largely left-hemispheric and uniquely human. These properties are usually associated specifically with language rather than with other aspects of thought, such as object recognition or imagery. Consequently the ideas developed in this chapter may seem to run somewhat counter to the Chomskian idea that language is not only peculiar to our species but also unique *within* the human repertoire.

The issue here is whether GAD is a general device applying to different domains, such as language and the representation of objects and scenes, or whether there are different GAD-like devices for each separate domain. The issue is in fact part of the more general question as to whether language is a special system with properties of its own, or whether the properties of language are general properties of the human mind. This question was recently debated by two of the most influential figures of our time, Noam Chomsky and the late Jean Piaget, the Swiss developmental psychologist, along with other commentators and camp followers.[59] It was not resolved, although several commentators (such as Jean-Pierre Changeux) suggested compromise solutions.

Nevertheless the ideas developed in this chapter may be said to lie closer to Piaget's than to Chomsky's. To Piaget, the important distinction is not between language and other aspects of thought but between two kinds of thought. One has to do with what Piaget calls success, or the direct adaptation to the world; this is immediate and nonreflective, and is sometimes called *practical intelligence.* The other has to do with *understanding,* and involves the conceptualization of the world.[60] This is sometimes called *representational intelligence,* and is concerned with the disinterested acquisition of knowledge and reflection upon it. Although language is part of this second system, GAD may be a more general property of it.

Piaget's distinction is very close to the one I drew earlier between the holistic, largely right-hemispheric mode of representation and the partwise, generative-cum-analytic, left-hemispheric mode. Indeed, essentially the same dichotomy recurs frequently in the history of ideas and has led to speculation that the brain is a dual organ, with each side representing a different mode of thought and even of consciousness. The dual brain is explored more fully in the next chapter.

GAD, Recursion, and Self-Knowledge

The notion that GAD might correspond to some general, representational intelligence brings us back to a topic raised in the previous chapter. Could it be that GAD underlies the concept of self? One of the properties of language, as we saw in Chapter 5, is *recursion*, the embedding of structures within structures. Recursion seems to be a property not just of language but of other forms of representational thought as well—that is, of GAD itself. It may be this recursive property that allows us to turn our thoughts inward, to introspect, and indeed to know ourselves. Without recursion, a person or an animal may know something, but recursion permits one to *know that one knows*. As Gordon G Gallup, Jr., puts it: "The basic distinction is between having an experience and being aware of having an experience."[61]

Recognition of oneself may not be uniquely left-hemispheric or even uniquely human. As we saw in the previous chapter, Roger W. Sperry and his colleagues have shown that the right hemispheres of two split-brained patients were clearly capable of recognizing pictures of themselves, or of relatives and personal belongings.[62] Similarly, Gallup has shown that chimpanzees can apparently recognize themselves in a mirror. For example, unknown to the animals, he painted part of one eyebrow ridge and the opposite ear bright red. When they viewed themselves in a mirror, the chimpanzees quickly learned to examine themselves for clues to the disfigurement. Although orangutans seem similarly capable of self-recognition, other primates, including rhesus monkeys, java monkeys, macaques, gibbons, and baboons, do not show self-directed actions when they see themselves in a mirror; rather, they treat their mirror images as though there was some different animal "behind" the mirror. Gallup suggests that the ability of chimpanzees and orangutans to identify themselves in a mirror does not depend simply on figuring out how a mirror works, because even monkeys can do this in other contexts. Rather, they readily learn to identify themselves in a mirror because they already have a concept of self.[63]

These studies suggest that self-recognition may not be uniquely human, although it is not widely shared with other animals or even with other primates. However, recognition of a photograph or a mirror image of oneself is not the same as awareness of one's own *thoughts* and does not seem to capture the recursiveness of thinking about oneself. It is possible that only humans have the facility to use the concept of self recursively. As Phillip Johnson-Laird points out, one can decide *whether to decide*, or decide whether to make a decision oneself or seek advice from someone else. Indeed, one can even proceed to a three-level hierarchy by, say, *knowing that* one can decide whether to make a decision oneself or seek advice elsewhere—a form of knowledge that can be comforting (and therefore adaptive) in cases of difficult decisions. While this sort of embedding may suggest the possibility of infinite regress, in practice one's ability to generate recursive structures is limited by short-term memory.[64]

I suggest, then, that this recursive use of self-knowledge may be a GAD-given power, unique to humans and normally a property of the left cerebral hemisphere. To this extent, then, there may well be some truth to the claims of authors such as John C. Eccles and Julian Jaynes that self-consciousness is uniquely human and uniquely left-hemispheric. But we are on dangerous ground here, because our knowledge of our own self-consciousness depends largely, if not wholly, on subjective evidence, and other animals are barred from giving testimony. Perhaps our self-important notions of self-consciousness are just part of the conspiracy to reserve for ourselves a special niche. It is time, therefore, to move to another chapter.

Notes

1. More wisdom from Jaques in Shakespeare's *As You Like It*.

2. Marr (1982). See also the introductory articles by Marr and Nishihara (1978) and by Pinker (1984).

3. For more details, the reader should consult Marr (1982) or any other basic textbook in vision. However, the precise mechanisms by which the brain computes these early sketches or maps of the visual world are not of concern to the major points I wish to make in this chapter.

4. Marr argues that objects must be described relative to their own axes if they are to be recognized independently of their orientations in space. I have argued elsewhere that this need not be so; that is, there may be descriptions that are independent of *any* reference axes, and that suffice to describe most objects uniquely and independently of orientation (Corballis, 1988).

5. Johnson-Laird (1988).

6. Biederman (1987).

7. There is also a deeper sense in which geons are analogous to phonemes. Phonemes themselves are characterized in terms of dichotomous contrasts, such as whether they are voiced or unvoiced, nasal or oral, plosive or not, and so on. Biederman suggests that geons can be similarly characterized in terms of primitive visual contrats, such as whether they are symmetrical or not, have straight or curved edges, have a straight or curved axis, or whether they expand, contract, or remain constant in size along the axis.

8. See Shepard (1984).

9. Kosslyn (1980).

10. Pylyshyn (1973) makes essentially this point in arguing that images are not stored as picture-like entities.

11. A more elaborate classification of agnosias has been suggested by Humphreys and Riddoch (1987).

12. Farah (in press).

13. For an insightful account, see the title chapter in Oliver Sack's (1985) entertaining book, *The Man who Mistook his Wife for a Hat*.

14. De Renzi (1986); Landis, Cummings, Christen, Bogen, and Imhof (1986).

15. See Young (1988) for a review.

16. Two cases among the 99 reviewed by Farah were equivocal in this

respect. Both patients suffered from alexia and prosopagnosia, and in both cases there was some reason to believe that the patients also suffered from object agnosia. However, both reports were rather contradictory about this.

17. For example, Gomori and Hawryluk (1984).

18. Damasio, Damasio, and Van Hoesen (1982).

19. Op. cit.

20. For example, McCarthy and Warrington (1986). Still, there are exceptions; in one case, for example, a patient with a left-hemispheric stroke was selectively impaired in the naming of fruits and vegetables but had little difficulty with other sorts of objects. In general, fruits and vegetables are not the sorts of objects that one might associate with a geon-based representation.

21. For a discussion of the factors influencing the relative contributions of the two hemispheres to the processing of faces, see Sergent (1988). These factors probably reflect purely perceptual aspects of processing, as well as those to do with representation.

22. Coltheart (1980)—but see Shallice (1988) for critical commentary. Ogden (1984) has reported a case of a right-handed man with all the signs of left-hemispheric speech who suffered a form of dyslexia following damage to the posterior *right* hemisphere. She concludes that the nature of his disorder "is consistent with other evidence that the right hemisphere mediates the holistic, nonphonological processing of printed words" (p. 277).

23. Shepard (1978); Farah (1988).

24. Quoted by Ley (1983, p. 252).

25. Ehrlichman and Barrett (1983, p. 61).

26. Farah (1984, p. 268).

27. Grossman (1988).

28. Goldenberg, Podreka, Steiner, Willmes, Suess, and Deecke (1989).

29. Farah, Perronet, Weisberg, and Monheit (1989).

30. Marks (1980).

31. Greenberg and Farah (1986).

32. Farah, Gazzaniga, Holtzman, and Kosslyn (1985).

33. Kosslyn, Holtzman, Farah, and Gazzaniga (1985).

34. Op. cit.

35. Johnson-Laird (1988) has argued that we make use of images even in the processing of quite abstract sentences, as in assessing the truth of syllogisms. For example, given the premises:

> None of the archeologists is a biologist.
> All the biologists are chess-players.

one may be required to assess the truth of such conclusions as:

> None of the archeologists is a chess-player.

for

> Some of the chess-players are not archeologists.

Johnson-Laird suggests that most people proceed, not by applying the rules of formal logic, but by setting up images of some arbitrary number of archeologists, biologists, and chess players, according to the specifications of the premises, and then evaluating the conclusions according to the images so formed.

36. Corballis and Sergent (1988, and unpublished experiments).

37. Sergent (1989).

38. Kosslyn does not refer explicitly to Biederman's geons when discussing partwise representation.

39. From Deleval, De Mol, and Noterman (1983), quoted in translation by Kosslyn (1988, p. 1625).

40. Farah, Levine, and Calvanio (1988).

41. Kosslyn (1988).

42. Kirk and Kertesz (1989); see also Gainotti and Tiacci (1970).

43. Herrnstein and Loveland (1964).

44. For example, Herrnstein, Loveland, and Cable (1976); Poole and Lander (1971).

45. Herrnstein (1985).

46. Bhatt, Wasserman, Reynolds, and Knauss (1988).

47. Cerella (1979). It might be thought that this implies partwise representation, of the sort that I have claimed to be uniquely human. However, the pigeon's ability to discriminate pictures by focusing on specific features does not prove that it has formed a coherent representation. More likely, it has discovered a feature, or set of features, that sorts the slides into the required categories at a better than chance level, and so secures a fair return of rewards.

48. Hayes and Hayes (1953).

49. Davenport and Rogers (1971); Gardner and Gardner (1985); Schrier, Angarella, and Povar (1984).

50. Biederman (1987).

51. Holloway (1981, p. 290; his italics).

52. Op. cit. (pp. 291–292).

53. Op. cit. (p. 290; my italics).

54. Davidson and Noble (1989, p. 130).

55. Gibson (1966, 1979).

56. For more detailed accounts, see Gelb's (1952) classic work or the more recent book by Gaur (1987). Henderson (1982) also provides a useful summary.

57. Henschen (1919); see Boller and Grafman (1983) for a review.

58. Rosselli and Ardila (1989).

59. Piattelli-Palmarini (1980).

60. Piaget (1978).

61. Gallup (1977, p. 329).

62. Sperry, Zaidel, and Zaidel (1979).

63. This work is summarized by Gallup (1977).

64. Johnson-Laird (1988).

—10—

The Duality of the Brain

> Farewell, dear task! Thou often has beguiled
> Pain from my heart, and sickness from my brow —
> For other eyes than those which wept or smiled
> O'er the progressing page will meet thee now!
> Farewell—farewell! my weak and friendless child!
> Thy parent's love can no assistance lend
> To thy young dawn upon a waste so wild,
> As this gay world may be without a friend!
> Yet I will crave for thee whate'er may bend
> Thy timid footsteps through its wintry waste,
> That thy hard lot (if hard betide thee) end
> In thy first day—that thou shalt not be traced
> With the slow slime of sorrow from the hour
> Thou has dared, like me, at things beyond thy pow'r.[1]

Historical Background

With this sonnet, published in 1844, the English physician Arthur C. Wigan bade farewell to the manuscript of his book *The Duality of Mind*, which to his undoubted relief was published later the same year. The theme of the book was that the two sides of the brain function as separate "minds," each with its own consciousness. The idea was not entirely new, and indeed, there has long been fascination with the fact that the brain is essentially duplicated, in mirror-image fashion, about its midline. Wigan gave the idea vivid expression, however, by providing examples of what he saw as a corresponding duality of *mind*. The theme is one that has persisted, in various guises, to the present day. We shall see, in fact, that

Joseph E. Bogen, one of the surgeons who carried out the split-brain operations in California in the 1960s, has also suggested that the two sides of the brain are independently conscious and has referred to this view as *neo-Wiganism.*[2]

Wigan thought that in order to achieve a normal life, the two hemispheres had to be coordinated through "exercise and moral cultivation." His main interest, however, was in the breakdown of coordination, which could result in a variety of pathological conditions. He cited the observation of the French physician E.L. Georget that there was often an asymmetry between the two halves of the skull in the insane. He suggested that the presence of two minds, each separately conscious, might explain the *déjà vu* experience, the sense that something has happened before even though it is not explicitly recalled. One mind has no recollection of the happening, while the other supplies the feeling that it is familiar. Wigan also referred to cases of motiveless crime, suggesting that one side of the brain might commit a crime in innocence, with the motive supplied by the other side.

The duality of the brain also offered a natural explanation for cases of double personality, in which an individual might exhibit quite different "personae" at different times, with one apparently having no knowledge of the other. Such "Jekyll and Hyde" cases attracted considerable interest in the late nineteenth and early twentieth centuries, culminating perhaps in Morton Prince's book *The Dissociation of a Personality* (1906), a study of a certain Miss Beauchamp who occasionally appeared to adopt the personality of an altogether different person known as Sally B. The idea that the two hemispheres were responsible for such cases may have lost favor, however, with later evidence that the personality could be split into more than two parts.[3]

Although cases of double personality suggest different *kinds* of consciousness in the two hemispheres, this was not part of Wigan's theme. The only systematic difference between the hemispheres that he mentioned was a general superiority of the left hemisphere, which he inferred from the dominance of the right hand. He wrote that

> when one brain is decidedly superior in power (as I believe to be generally the case with the left), in this case the right brain aids the left and corroborates its fellow as an assistant aids a workman; and more is done by the two directed as one will, than both could have executed separately.[4]

Even this passage, however, contradicts Wigan's more general theme that the two hemispheres function independently.

Wigan, of course, wrote before Broca's discoveries of the lateralized representation of speech in the brain, and he was unaware of Dax's earlier report. To Wigan's successors, one consequence of Broca's work was simply to reinforce the idea of cerebral dominance, with the right hemisphere being generally regarded as merely the nondominant or "minor" hemisphere. Even as late as 1962, the British zoologist J.Z. Young wondered if the right hemisphere might be merely a "vestige," although he wisely allowed that he would rather keep his than lose it.

Although the right hemisphere was generally regarded as inferior, at least until the 1960s, there was still considerable speculation about the special contribution it might make. In 1864 the perspicacious John Hughlings Jackson suggested that if "expression" resided in one hemisphere, then one might raise the question of whether "perception—its corresponding opposite," might reside in the other. A few years later, the French neurologist Armand de Fleury[5] and the Austrian physiologist Sigmund Exner[6] independently made the similar suggestion that motor functions were represented more widely in the left hemisphere, while sensory representation was more widespread in the right hemisphere.

But the main speculation about the right hemisphere, especially in the latter part of the nineteenth century, had to do with its contribution to emotion. In France, Jules Bernard Luys was struck by personality differences between those with left- and right-brain damage. Those with right-brain damage tended to be more manic and hyperemotional than those with left-brain damage, who were generally passive and apathetic. Luys argued for an "emotion center" in the right hemisphere.[7] This idea was supported by the observation that patients with hysterical disorders tended to show their symptoms on the *left* side of the body—an asymmetry that had been noted prior to Broca's discoveries about the left-hemispheric representation of speech.[8] This asymmetry featured in theories of hysteria that were prevalent in France in the latter part of the nineteenth century. It was attributed, for example, to Jean-Martin Charcot, famous for his work on hysteria and hypnosis; indeed, it was even dubbed *Charcot's rule* by Paul Richer in a work published in 1881, although Charcot himself was later skeptical of it.[9] Pierre Janet, the other influential theorist on the nature of hysteria, also made much of the preponderance of left-sided symptoms,[10] although he later expressed general doubts about a neurological approach to hysteria.[11]

The idea of hemispheric duality led to some bizarre developments in the treatment of hysteria in the latter part of the nineteenth century. One of these had to do with the application of metal discs to the afflicted side of the body, a technique known as *metallotherapy*. The idea was developed by a French doctor named Victor Burq and later taken up by Charcot himself. Different patients were said to be sensitive to different metals. Magnets were also used to control patients' thought processes and even to transfer thoughts from one patient to another.

Another technique, developed by one of Charcot's assistants, Gabriel Descourtis, was *hemihypnosis*, in which the two sides of the brain and body were separately hypnotized. There were remarkable stories and even photographs of patients in a double trance, with one side of the face smiling and the other side showing fear. The technique was simply to hypnotize the patient with each eye open in turn, on the mistaken belief that each eye projected to the opposite hemisphere. It may be recalled from Chapter 7 that each eye in fact projects to *both* hemispheres, and such disregard for the anatomical facts may be one reason why hemihypnosis eventually fell into disrepute.[12]

These extraordinary studies were also accompanied by much speculation, especially in France, about the wider significance of hemispheric duality. The emotional right hemisphere was generally seen as inferior to the rational left and was associated with an inferior stage of evolution. In keeping with the prejudices of the time, the right hemisphere was linked with animality, instinct, the female, nonwhite inferiority, and madness, while the left stood for humanness, volition, the male, white superiority, and reason. Many of these ideas were developed by the French physician Gaetan Delaunay, an entirely respectable figure at the time.[13] There is a remarkable similarity between these associations and the linking of various polar opposites to the left and right hands, documented in Chapter 4, except that the left and right labels have been reversed in deference to the neurological facts.[14]

Ideas based on the double brain rapidly lost popularity after the turn of the century. The historian Anne Harrington states that she could find almost nothing written on the topic between 1920 and 1960.[15] One reason may be that the experiments on such techniques as metallotherapy and hemihypnosis simply lost credibility. This was also the time during which psychiatry emerged as a discipline distinct from neurology, and adopted an approach that was more functional and less tied to brain function. It may be recalled, too, that this was the age of behaviorism, when psychologists also adopted a purely functional approach, eschewing references to both mind and brain. In any event, this period of silence on the double brain seems to have been sufficient to cause amnesia for the developments of the late nineteenth century, and were it not for the diligence of Anne Harrington, we might still not know about them.

Neo-Wiganism

From the 1960s on, however, the double brain has made a remarkable comeback. To some extent, this may have been due to a revival of interest after World War II in the psychological consequences of brain damage, often the result of injuries sustained in the war itself. But what really seems to have turned the tide were the studies of the split-brained patients in the 1960s. In these patients, as explained in Chapter 7, it is possible to test each hemisphere of the brain independently of the other. These studies, as we have seen, confirmed the dominance of the left hemisphere for speech. However, it soon became apparent that the right hemisphere could perform some tasks as well as the left hemisphere, or even better, even though it was incapable of speech.

For the first time since the forgotten days of hemihypnosis, the right hemisphere was again (and perhaps more legitimately) able to find expression independently of its talkative neighbor. The results had a dramatic quality that may, in retrospect, have given an exaggerated sense of their importance. Michael S. Gazzaniga, a pioneer in the testing of these patients, writes:

> It is difficult to overdescribe the riveting experience of observing a right hemisphere perform or fail to perform many of the tasks that I have related. There is a certain eerie quality to watching a hand draw or point to places when the left brain of the patient does not in fact know under what command the left motor system is responding.[16]

In any event, this work led to a form of neo-Wiganism in which not only were the two sides of the brain considered capable of independent thought, but also each side was held to represent a *style* of consciousness that was complementary to the other. Leading the way, Joseph E. Bogen characterized the right brain as *appositional,* in contrast to the *propositional* mode of the left brain.[17] The idea was pursued by Robert E. Ornstein in his popular book *The Psychology of Consciousness* (1972) and now permeates contemporary folklore. The left hemisphere is portrayed as linear, rational, analytic, and fundamentally Western in its style of thought; the right, by contrast, is seen as divergent, intuitive, holistic, and representative of the Eastern way of thinking.

As Bogen recognized, the opposition between these modes of thought is actually an old one, underlying such dualities as the pre-Confucian concepts of *yin* and *yang*, the Hindu notions of *buddhi* and *manas*, and Levi-Strauss' distinction between the *positive* and the *mythic*. The polarity also has much in common with that underlying the nineteenth-century concept of the double brain, with the right hemisphere representing the more intuitive, emotional aspect of consciousness. There is one important difference, though: In the new Wiganism, the right-hemispheric contribution is not generally regarded as inferior. Indeed, it is often romanticized to the point of being superior to that of the left hemisphere.

In particular, the right hemisphere is often described as the source of creativity. One author, for example, has deplored the overemphasis on left-hemispheric values in education, noting "the tragic lack of effort to develop our children's right-brain strengths. That potential—a source of equally essential creative, artistic, and intellectual capacity—is at present largely unawakened in our schools."[18] The message is conveyed in the very title of an influential book, *Drawing on the Right Side of the Brain* (1979), by Betty Edwards, an art teacher; she writes that "a new way of seeing will be developed by tapping the special functions of the right hemisphere."[19] Some authors have nevertheless retained something of the darker, more negative aspect. In his entertaining book *The Dragons of Eden* (1977), Carl Sagan characterizes the right hemisphere as intuitive, emotional, and paranoid, often seeing patterns or conspiracies where they do not exist. In the pursuit of knowledge and truth, the left hemisphere is needed in order to submit proposed patterns to critical analysis, although the right hemisphere is equally important in detecting patterns in the first place.

With hindsight, it is clear that the polarities linked to the two sides of the brain owe as much to the temper of the times as to the neurological evidence. In the late nineteenth century, the duality gave expression to the common view that women and the people of other races were inferior to the white European male. These undesirables, along with the insane, were relegated to the right hemisphere. The 1960s, by contrast, were times of protest against racism, sexism, and war. This was the age of gurus and members of Eastern religious sects parading in Western streets. Consequently the right hemisphere was romanticized and stood for creativity, pacifism, the flower people, and the exploited peoples of Asia. The left hemisphere, by contrast, stood for the Western military-industrial establishment.[20] The cause of the right hemisphere was neatly captured in the catch-cry, "Make love, not war."

In the meantime, in the conservative 1980s, there has been something of a reaction against this rather facile linking of the extremes of a dichotomy with the two hemispheres.[21] Gazzaniga has suggested that the fascination of witnessing the right hemisphere at work in the split-brained patient has led to an exaggeration of its capabilities. In a provocative article, he goes so far as to write, "it could well be argued that the cognitive skills of a normal disconnected right hemisphere without language are vastly inferior to the cognitive skills of a chimpanzee"[22]—a remark that takes us back to the nineteenth-century view of the right hemisphere as inferior and brutish. Nevertheless, Gazzaniga's skepticism is a necessary corrective to the extravagant theorizing that has dominated the doublethink of the past quarter of a century.

In what follows, I shall attempt a more cautious review of the functions in which the right hemisphere seems to play a major role. This will lead to a reappraisal of the nature of hemispheric specialization.

Right-Hemispheric Functions

Perception

Some years after he had speculated about right-hemispheric specialization for perception, John Hughlings Jackson described a patient with a tumor in the right hemisphere who suffered from what he called *imperception*.[23] The patient failed to recognize familiar people and places and was spatially disoriented. Since then, there has been a good deal of evidence that the right hemisphere is indeed the more specialized for a wide variety of aspects of nonverbal perception. Some of these have more to do with the *recognition* of objects than with their actual perception and will be discussed in the next section. However, these is also evidence that patients with right-sided brain damage perform more poorly than patients with comparable left-sided damage on such low-level perceptual tests as differentiating figure from ground,[24] localizing points in space,[25] judging stereoscopic depth,[26] judging the directions of lines,[27] and so on.

Studies of split-brained patients also suggest a superiority of the right hemisphere in perceptual tasks. For example, the right hemisphere seems to be better than the left at matching arcs of circles to whole circles of the same diameter,[28] in matching fragments of shapes to whole shapes,[29] in judging the alignment of dots,[30] and in certain geometric judgments.[31] Similarly, comparisons between the visual hemifields of normal people have shown a left-hemifield (or right-hemispheric) advantage in such tasks as judging stereoscopic depth,[32] localizing dots,[33] judging the curvature of lines,[34] judging color,[35] and judging lightness.[36] These and other results[37] complement the right-hemifield (left-hemispheric) advantages for verbal tasks discussed in Chapter 7.

Hemispheric differences in perception are not confined to vision. Studies of dichotic listening in normal people have shown a left-ear advantage in perception of environmental sounds,[38] sonar sounds,[39] and nonverbal vocalizations,[40] as well as various aspects of music, to be discussed later. There is also evidence that the left hand is more accurate than the right in discriminating between nonverbal shapes.[41] This evidence comes from the *dichhaptic* technique, an analogue of the dichotic technique in which shapes are presented simultaneously to the two hands. Under this condition, the left-hand advantage is thought to reflect right-hemispheric superiority.

The right-hemispheric advantage in these perceptual skills, some of them quite elementary, is not absolute, however, and is in most cases probably quite slight. In testing the differences between the visual hemifields in normal people, it will be recalled, it is necessary to flash the visual information very briefly, so that the lateralized presentation is not disrupted by an eye movement. It is probably only under restricted conditions of this sort that differences between the hemifields emerge. Under conditions of normal viewing, there is probably very little difference between left and right hemifields in our ability to judge stereoscopic depth or the curvature of a line. Gazzaniga has also pointed out that the hemispheric differences on perceptual tasks in split-brained patients tend to be slight and variable.[42]

It is possible that these findings are not so much the result of a right-hemispheric specialization as of a left-hemispheric weakness. This weakness may have arisen simply because of the invasion of language and praxic representation. In the testing of split-brained patients, for example, Gazzaniga and his colleagues observed that the right-hemispheric superiority was especially marked in what have been called *manipulospatial skills,* such as arranging blocks according to a visually displayed pattern,[43] or drawing a picture of a three-dimensional cube.[44] J. E. LeDoux has claimed that these skills are mediated by the area of the right hemisphere that is opposite Wernicke's area in the left. He concludes that the manipulospatial skills served by this area are bilaterally represented in nonhuman primates, but are strongly biased in favor of the right hemisphere in humans because the left-hemispheric region has been taken over by language.[45]

Gazzaniga has pointed out that not all split-brained patients show this extreme right-hemispheric advantage for manipulospatial skill.[46] A deficit in manipulospatial skill is sometimes known as *constructional apraxia* and in contrast to other apraxias (see Chapter 8) has generally been associated with right-hemispheric damage. In reviewing the evidence, however, Enrico De Renzi has noted that, while earlier studies linked it with right-hemispheric damage, more recent studies have shown no hemispheric bias.[47] The early investigators may simply have selected right-hemispheric cases as being somehow more interesting or revealing.[48] The evidence suggests that manipulospatial skills may involve both spatial and sequential skills, and may be dependent on both right—and left-hemispheric specializations.[49]

However, these considerations need not invalidate LeDoux's general point. The left-hemispheric specialization for language and other praxic skills does seem to be a strongly established dominance, taking precedence over other functions that may have been bilaterally represented, or only weakly lateralized, in our primate forebears. Many right-hemispheric advantages may well be consequences of invasion of the left hemisphere by praxic and language skills. Other functions may have to accommodate themselves to the predominantly left-hemispheric representation of language and praxis, and the way in which this is done may be subject to wide variations.

Visual Recognition

Chapter 10 outlined the different stages involved in the analysis of visual scenes and introduced the phenomenon of *agnosia*, or the inability to recognize familiar objects as a result of brain injury. In particular, Lissauer's distinction between apperceptive and associative agnosia was noted. Some authors also refer to a condition known as *pseudoagnosia*,[50] so called because it is not specific to objects as such; rather, it is a result of disruption to the early stages of visual processing (see Figure 9.1). As we saw above, it is more pronounced following damage to the right than to the left hemisphere.

At a higher level of processing, but still prior to the level at which meaning is extracted, an object may be perceived as a three-dimensional structure. As we saw in Chapter 10, Lissauer called disruption to this stage *apperceptive agnosia*. Studies of groups of patients have suggested that agnosias generally thought of as apperceptive are more common following right- than left-hemispheric damage.[51] This is borne out by studies of individual cases. For example, Elizabeth K. Warrington and Merle James of the National Hospital at Queen Square, London, have recently described three patients with apperceptive agnosia, all with damage to the posterior right hemisphere.[52] These patients scored normally on tests of sensory and perceptual discrimination and had a normal appreciation of the meanings of objects. For example, when given a choice of pictures of objects, they could correctly choose the heaviest, or the most dangerous, or the one most likely to be found in a kitchen. Their difficulty lay in identifying objects that were depicted under various degraded

Figure 10.1. Examples of overlapping figures used to test for apperceptive agnosia.

conditions. For example, they had difficulty recognizing objects when they were viewed from unusual angles, or when they were represented as silhouettes or as projected shadows.[53] These patients also had difficulty in picking out objects drawn in overlapping fashion, as in Figure 10.1.

The right-hemispheric bias in apperceptive agnosia may again simply reflect that hemisphere's greater involvement in perceptual analysis, although at a higher level than that of the sorts of elementary sensory and perceptual processes described in the previous section. Again, this bias seems to be relative rather than absolute. Patients with apperceptive agnosia *can* see and understand objects, provided that the viewing conditions are favorable; it is only when viewing is degraded in some way that difficulties arise.[54]

As we saw in Chapter 9, apperceptive agnosia may be contrasted to *associative agnosia,* in which objects are correctly seen as three-dimensional objects but their meaning is not understood. Patients with this disorder have difficulty in naming objects or describing their functions, or in picking out an object from an array according to some quality (e.g., dangerousness), or in demonstrating how an object is used. However, they seem to have no difficulty at the apperceptive level. For example, they can color in the individual objects of an overlapping drawing, as in Figure 10.1, even though they cannot name the objects. Aside from their difficulty in naming objects, these patients are not aphasic. One patient, for example, had difficulty naming objects or pictures, but no trouble naming familiar noises;[55] another had difficulty naming pictures, but no difficulty naming objects from a verbal description.[56]

I argued in Chapter 9 that there are two kinds of associative agnosia. One has to do with the representation of shapes in a partwise, grammatical fashion, is associated with alexia, and usually results from left-hemispheric damage. The other has to do with the holistic representation of shapes, is associated with prosopagnosia, and usually results from right-hemispheric damage. Again, the rightward bias in the holistic form of representation may be at least partly due to the left-sided representation of the generative component, or GAD. Note, however, that the recognition of objects must often depend on contributions from both hemispheres. Even if GAD is employed, the earlier analyses might have called upon the superior perceptual abilities of the right hemisphere. For subtle aspects of recognition, moreover, the partwise representations available in the left-hemisphere may have been insufficient, and it may have been necessary to call upon the more holistic, picture-like representations stored predominantly in the right hemisphere.

As we saw in Chapter 9, prosopagnosia (or the inability to recognize familiar faces) is associated more with damage to the right than to the left side of the brain. This asymmetry may be manifest in the normal perception of faces. For example, it is claimed that the right side of a face looks more like the whole face than does the left side! This is demonstrated by taking a picture of a face viewed from the front, cutting it in half down the middle, and then completing each half with its own mirror image to form a full face that is symmetrical about its midline, as shown in Figure 10.2. Most people judge the full face constructed from the left side of the picture (or the right side of the *face*) to be more like the original than the face constructed from the right side of the picture. The suggested explanation is that the left side of the picture usually falls in the left visual hemifield and is therefore projected to the right side of the brain, which is superior to the left side in forming representations of faces.[57]

Spatial Transformations

Most of us possess an ability to make imaginary transformations of the world about us. We can imagine, for example, how the living room might look if the furniture were rearranged. This skill may play a role in ordinary visual perception, enabling us to anticipate scenes and so perceive them more rapidly and efficiently when they actually occur. However, it goes beyond perception and is an impor-

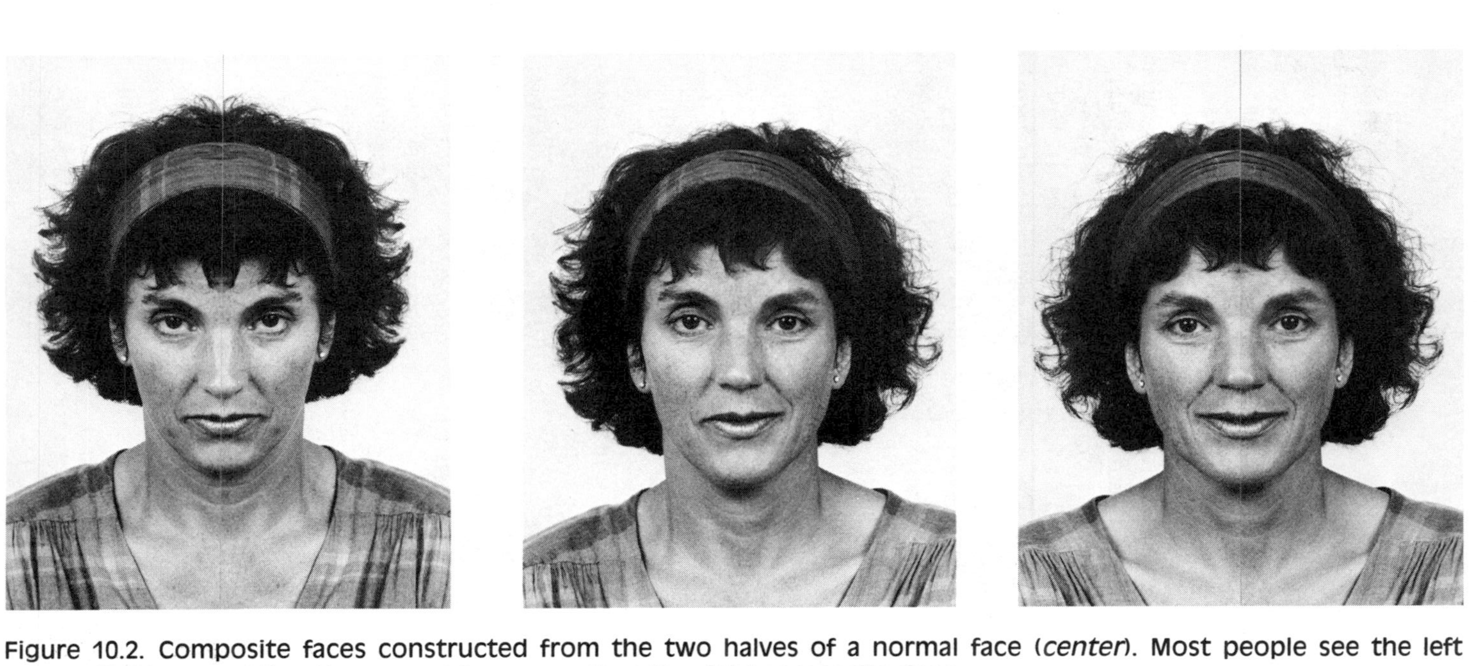

Figure 10.2. Composite faces constructed from the two halves of a normal face (*center*). Most people see the left composite as resembling the normal face more than the right composite does.

Figure 10.3. In order to determine whether these letters are normal or backward, most people first rotate them, physically or mentally, to the upright.

tant aspect of what is called *spatial ability,* useful in professions such as the visual arts or architecture.

The spatial transformation that has been most studied is *mental rotation,* the act of imagining an object turning around so that it is pictured in a different orientation. The study of mental rotation was pioneered by Roger Shepard and his colleagues at Stanford University. One of the techniques they use is to present patterns such as letters or digits in varying orientations and time people as they decide whether they are normal or backward (mirror-reversed). Most people mentally rotate the pattern to its normal upright position before making the decision. Examples are shown in Figure 10.3; in order to decide whether the letter R as shown in the figure is normal or backward, most readers will be tempted to rotate it to the upright, either by turning the book around (which is cheating) or by performing a mental act. By plotting the time it takes to make the decision as a function of the orientation, it is possible to estimate the rate of mental rotation.[58]

There is evidence that mental rotation is primarily right-hemispheric. Justine Sergent and I have tested each cerebral hemisphere of a split-brained man on the mental rotation of letters and other figures by flashing them briefly in the left or right hemifield. His performance was quite normal when projection was to the left hemifield, indicating that the right hemisphere was capable of the task. When projection was to the right hemifield, and so to the left hemisphere, his performance was at first quite random. With repeated trials his left hemisphere mastered the art of mental rotation, but it remains slower and less accurate than the right hemisphere.[59] This is striking evidence of right-hemispheric superiority in a person who, on other tasks, has typically not shown clear right-hemispheric advantages.

There is also evidence that patients with damage to the posterior part of the right cerebral hemisphere have difficulty with mental rotation, while those with corresponding damage to the left cerebral hemisphere do not.[60] The study of patterns of blood flow in the brain also suggests that the right parietal lobe is activated during mental rotation.[61] Curiously, however, studies of normal people in which the patterns to be rotated are flashed in the left or right hemifield have not shown consistent hemifield differences. A possible reason for the inconsistency of results is that the patterns must presumably be recognized before they are rotated, and this may confer a left-hemispheric advantage that acts in opposition to the right-hemispheric advantage for the rotation itself.[62]

In any event, the right-hemispheric specialization is probably not absolute. The split-brained patient that Sergent and I studied is now able to carry out mental rotation of patterns projected to the left hemisphere, although he retains a right-hemispheric dominance. By contrast, his right hemisphere still cannot produce speech, even nearly 30 years after his operation.[63] Again, these observations suggest that the right-hemispheric advantage for mental rotation may be secondary to the left-hemispheric specializations for language, praxis, and generativity.

Spatial Attention and Awareness

Evidence that the right hemisphere might be more specialized for spatial attention comes from a remarkable disorder known as *hemineglect*. Patients suffering from this show a tendency to ignore one side of space. For example, they may eat from only one side of the plate, or dress only one side of the body, and if asked to copy a drawing they may produce only one side of it.

Most cases of hemineglect result from damage to the right side of the brain and affect the left side of space. Figure 10.4 shows some examples of left-sided neglect in drawings done by patients with right-sided brain injury. However, examples of right-sided neglect following left-sided damage are not unknown, and may be even be as common as left-sided neglect if patients are tested shortly after sustaining the injury; left-sided neglect may simply be more persistent.[64] There is also evidence that right-sided neglect is associated more with damage to the forward portions of the left hemisphere, while left-sided neglect is generally associated with the rearward regions of the right hemisphere.[65]

There has been considerable debate as to the nature of hemi-

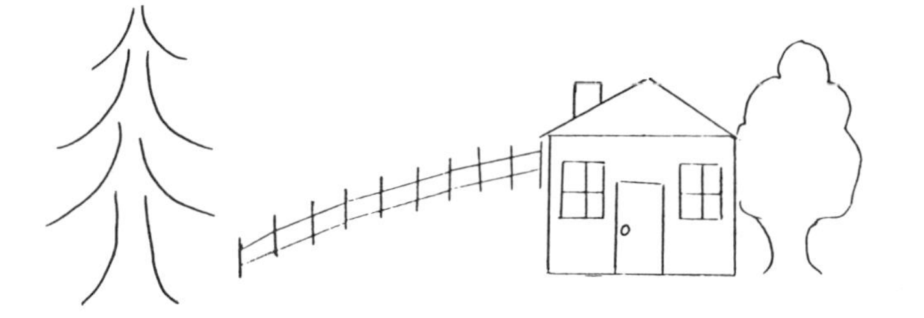

COPY THIS DRAWING IN THE SPACE BELOW

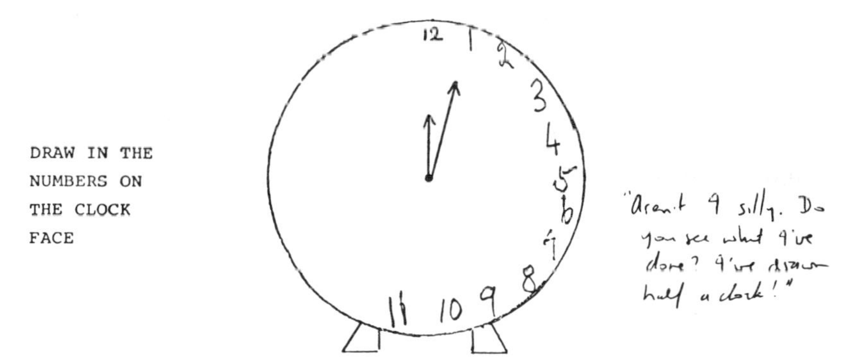

Figure 10.4. Drawings by a patient with right-sided brain damage, illustrating left hemineglect (samples donated by J.A. Ogden, from Corballis, 1983).

neglect. One thing that is clear, however, is that it cannot be attributed simply to the inability to actually *see* things on one side of the visual field. Patients with hemineglect tend to ignore one side of space more or less regardless of where they are *looking*, so it is not simply a matter of where images fall on the retina. Indeed, Edoardo Bisiach and Claudio Luzzatti of the University of Milan have demonstrated hemineglect in visual *imagery* in the absence of any visual input at all. Two patients with right-sided damage were asked to imagine the Piazza del Duomo (the cathedral square) in Milan and to describe various landmarks. They systematically omitted those on the left side. But when asked to imagine the square from the opposite end, they proceeded to describe all the landmarks, now on the right, that they had previously omitted![66]

Most authors seem to agree that hemineglect is a disorder of attention rather than of perception.[67] If pressed, patients often *can* name objects on the neglected side of space. In a follow-up study of neglect of one side of the image of the cathedral square in Milan by Bisiach and his colleagues, patients were able to describe both sides of the square when asked specifically for the left and right sides in turn.[68] The comment written beside the clock in Figure 10.3 was recorded from the patient and indicates some appreciation of what he had done. There is often an emotional component, too, as though the patient knows about the objects on the neglected side but wishes to deny their existence. In the earlier study by Bisiach and Luzzatti referred to above, the authors note that the patients sometimes did refer to landmarks on the neglected side, but that they did so "in a kind of absent-minded, almost annoyed tone."[69]

If hemineglect is indeed attentional, then it implies a dominance of the right hemisphere, since the disorder is more marked and more persistent following right-sided than following left-sided damage. One explanation is based on the ideas of Marcel Kinsbourne.[70] He suggests that each hemisphere controls attention across space in a graded fashion; accordingly, the left hemisphere has a bias in favor of the right side, and the right hemisphere a bias toward the left side. These biases are relative rather than absolute; given two objects, the left hemisphere will favor attention to the rightward one, and the right hemisphere will favor attention to the leftward one, regardless of the absolute positions of the objects.

In humans, the gradient of bias may be steeper in the left than in the right hemisphere. Alternatively, the attentional profile of the left hemisphere may simply be lower overall than that of the right, again perhaps as a consequence of the invading presence of language and praxic functions.[71] Consequently, neglect of the left side is greater following left-hemispheric damage than is neglect of the right side following right-hemispheric damage. Again, then, what is seen as a specialization of the right hemisphere may be instead a diminution of left-hemisphere functions. To my knowledge, there is no evidence of a comparable hemispheric asymmetry in nonhuman primates.

There are still unresolved questions as to the nature of hemineglect, and indications that it can be manifest in different ways. It may depend, for example, on the kind of material that is viewed. In one intriguing case of a right-handed patient with damage to both sides of the brain, there was a *left*-sided neglect in the reading of words (*neglect dyslexia*) but a *right*-sided neglect on tests that did not involve language.[72] This suggests that the left hemisphere may be more dominant for reading, while the right hemisphere is the more dominant for other, nonlinguistic spatial activities.

Besides hemineglect, there are other more general disorders of awareness that seem more often related to right-hemispheric than to left-hemispheric damage. One striking example is the unawareness of disease, for which M.J. Babinski in 1914 coined the term *anosognosia*. For instance, patients are often unaware that they are suffering from hemiplegia, or paralysis of one side of the body; this is much more common when the left side of the body is affected, implying damage to the right side of the brain. Brain-injured patients may even be unaware of being blind, a form of anosognosia known as *Anton's syndrome*.[73] Anosognosia is not always related to right-hemispheric damage; for example, patients with aphasia following left-hemispheric damage are often unaware of their deficits. Here for example is one description of patient suffering from jargon aphasia:

> It is amazing to see such a patient uttering in a confident and natural way utterly meaningless words or extraordinary sentences. For instance one of our patients called on by a neighbour who wore splendid new shoes told her admiringly: "Oh what beautiful chemists you have." This interchange of words which, of course, surprised the neighbour, was the beginning of a paraphasic jargon and the first symptom of a left temporal tumor.[74]

Even so, anosognosia is most common following right-hemispheric injury and suggests that the left hemisphere does not have a monopoly on conscious awareness.

Emotion

In hemineglect, as we have seen, and indeed in other disorders of awareness, there often seems to be an element of denial, suggesting an emotional component. Indeed, it has long been observed that damage to the right hemisphere produces what Janet called *"la belle indifférence"*[75]—a general loss of emotion or a denial that anything is wrong. By contrast, damage to the left hemisphere often results in what has been termed the *catastrophic reaction,* with outbursts of weeping, swearing, despair, guilt, or aggression.[76] Some of the evidence suggests that positive emotions are housed in the left hemisphere and negative ones in the right. For example, it has been observed that patients undergoing the sodium amytal test often show a catastrophic reaction when the left hemisphere is anesthetized but a euphoric or maniacal reaction when the right hemisphere is anesthetized.[77]

There are still those who link mental illness to hemispheric asymmetry. For example, it has been claimed that epilepsy originating in the left temporal lobe of the brain is associated with schizophrenia, while epilepsy originating in the right temporal lobe is associated with manic-depressive illness.[78] As ever, these associations remain controversial.[79]

There is nevertheless a good deal of evidence from studies of normal people that the two sides of the brain contribute differently to emotion. Dichotic-listening experiments have shown a left-ear (or right-hemispheric) advantage in identifying emotional intonation in speech[80] and in recognizing nonverbal emotional sounds such as laughing and crying.[81] The left-hemifield (right-hemispheric) advantage in the perception of briefly flashed faces is enhanced when the faces expressed emotion, compared with when their expressions were neutral.[82] Several investigators have argued that the right-hemispheric advantage in perception of facial emotion is distinct from the right-hemispheric component in the actual perception of faces themselves.[83]

Most right-handed people also differ in the extent to which they can express emotion on the two sides of the face, and in fact have better control over the facial muscles on the left than on the right.[84] You can check this by trying to wink, grimace, or smile as broadly as possible with each side of the face in turn and watching the results in a mirror—but do remember that the mirror reverses left and right![85] Quite apart from the matter of muscular control, there is evidence that emotions are expressed more intensely on the left than on the right side of the face.[86]

There is also evidence that emotion expressed on the left side of the face is perceived as more intense than that expressed on the right side. Composite pictures can be made by reflecting each half of a picture of a face, as described earlier; if the original face depicts negative emotions, such as sadness, fear, disgust, or anger, the com-

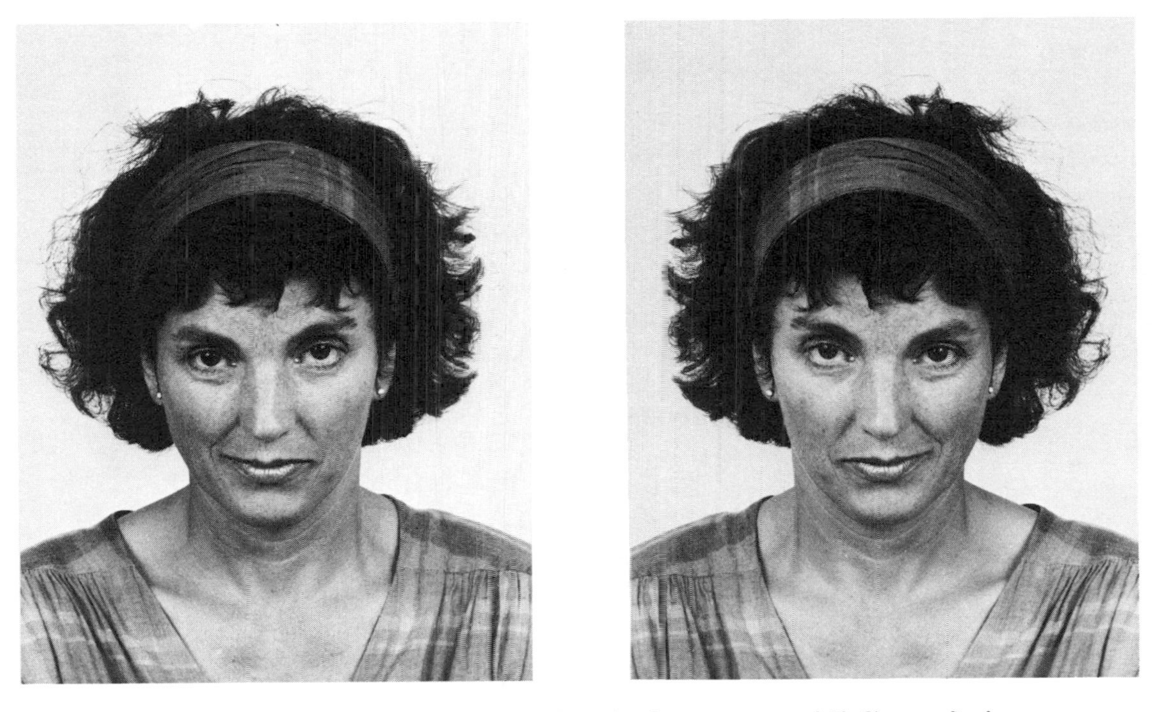

Figure 10.5. Normal (*right panel*) and mirror-reversed (*left panel*) pictures of a half-smiling face. Which one looks happier?

posite made from the left half seems to express these emotions more strongly than the composite made from the right half.[87] Even when composites are made from halves of a relaxed face, the left-half composite seems sadder than the right![88]

However, the interpretation of emotion in composite faces may depend also on asymmetry in the observer. If composites are made of faces that are half smiling and half neutral, the face is seen as more expressive if the smile is seen on the left side than on the right.[89] This is illustrated in Figure 10.5. Note that in this case it is the *right* side of the depicted face that is seen on the left, and that has the more powerful influence on perceived emotion, presumably because it is projected to the observer's right hemisphere.

Bernard B. Schiff and Mary Lamon of the University of Toronto have provided a new twist on the differences between the sides of the face.[90] They found that if people were required to pull back one corner of the mouth and lift it, and to hold this awkward grimace, they experienced different emotions, depending on which side was involved. Left-sided grimaces tended to make them feel sad, while right-sided grimaces induced a more positive emotion, although one that was difficult to characterize. These emotions spilled over into the stories they told about neutral pictures. Here, then, is a technique you might try if you want to cheer yourself up.

Since the left side of the brain controls the right side of the face, and vice versa, these results are again consistent with the idea that the two sides of the brain control complementary emotions, with the right side tending to evoke negative emotions and the left side positive ones. Overriding this complementarity, however, the right side seems to have the stronger overall influence on emotional state.

Music

It has long been known that damage to the brain can cause what is known as *amusia*, or an impairment of musical ability. This is sometimes associated with aphasia, but often not. There have even been rather striking cases of musicians who have become aphasic but have nevertheless retained their musical skills. For example, Maurice Ravel, the famous musician and composer, suffered sensory aphasia as a result of an automobile accident at the age of 57, but was still able to recognize tunes and identify musical errors at a high level of accuracy.[91] The Russian composer V. B. Shebalin suffered severe sensory aphasia following a vascular lesion to the left hemisphere at the age of 51, but his musical skills remained intact and he even continued to compose some notable pieces of music.[92]

Cases such as these suggest that musical ability depends more on the right than on the left hemisphere. This conclusion receives some support from the literature on amusia,[93] although, as Robert J. Zatorre remarks, "this literature is disappointing in its lack of firm conclusions."[94] Cases of amusia no doubt often pass unnoticed, since musical abilities vary widely in the normal population and are not always tested following brain injury.

There have, however, been experimental studies of the musical abilities of patients following surgery to the left or right side of the brain. Most of this work has focused on the role of the temporal lobe, since this includes the main areas for the analysis of sounds. Using a test of musical abilities known as the Seashore test, Brenda Milner of the Montreal Neurological Institute found that patients with left-sided removals showed no musical deficits, while those with right-sided removals were impaired in timber and tonal memory, which involved comparisons of two short melodies.[95] Others have reported basically similar results, although some have found that damage to both sides produces greater impairment than damage to the right side alone.[96] Dichotic-listening studies have generally revealed a left-ear advantage for aspects of musical perception, such as the discrimination of melodies,[97] musical pitches,[98] timbres,[99] and harmonics.[100] Such studies confirm a general right-hemispheric advantage, although there is also some conflicting evidence.[101]

One aspect of music that may depend more on the left than the right hemisphere is rhythm. For example, although there seems to be no difference between the hands in tapping in time to a metronome, the right hand is generally better at tapping a simple rhythm. Since this appears to be true of the majority of left-handers as well as of right-handers, it presumably relates to a left-hemispheric dominance rather than to handedness *per se*.[102] In a Japanese study, it was found that two patients with damage to the corpus callosum had lost the ability to tap given rhythmic patterns with the left hand, while performance with the right hand was unimpaired; the authors concluded that the "right hemisphere is not qualified to voluntarily control tempi of repetitive movements."[103]

Most people also find it fairly easy to tap a rhythm (e.g., *dah*-dit-dit-*dah*-dah) with the right hand while tapping the beat (as italicized) with the left hand. Reversing this arrangement, however, proves much more difficult—the reader should try it. Again, this is partially independent of handedness itself.[104] In keyboard instruments, it is, of course, the right hand that usually carries the melody and the intricacies of rhythm, while the left hand carries the beat.

Much of the evidence discussed so far has had to do with relatively superficial aspects of music. Could there be some deeper level at which music is generative in the way that language is, and might be expected to call upon the services of GAD? Fred Lerdahl and Ray

Jackendoff have suggested that this is indeed so. They have developed a generative theory of tonal music (*GTTM*) that is surely at least a close relative of GAD.[105] Like language, music depends on culture, and just as there are thousands of different languages, so there are thousands of musical idioms, and different styles within these idioms are analogous to the different dialects of language. There may even be a critical period for acquiring musical skills. For example, some people possess what is known as *absolute pitch,* the ability to name pitches of individual sounds or to sing named pitches accurately. There is evidence that almost all musicians who began their training before the age of 6 possess absolute pitch, compared with none of those who began after the age of 11.[106]

Yet although music appears to have many of the properties of language, including generativity, it does not seem to have the *representational* quality of language. According to one group of authors:

> Music does not exist to convey meaning in the same way [as language]. It exists to provide listening satisfaction, to accompany rituals, and so on. . . . In the case of music, we are in a similar position to someone who is attempting to judge the grammatical correctness of a language he or she does not understand.[107]

If this is so, then the generativity of music may be formal rather than psychological; that is, to the average listener at least, music is perceived in holistic, emotional fashion, not as an explicit, propositional representation of something. It has been described as a "language" of the right hemisphere, but one that is emotional, nonverbal, and holistic.[108]

Jackendoff makes a similar point.[109] He suggests that in music, unlike language, all levels of analysis are important; we listen to the sound quality of the instruments, the rhythm, the melody, and perhaps only if we are trained in music do we begin to appreciate the deeper meaning of a given piece. In language, by contrast, the meaning is of paramount importance, and there is less emphasis on the surface form unless one is listening to a reading of poetry, say, where the more "musical" aspect may gain in importance. As we have seen, the perception even of elementary speech sounds seems to strip them of all *perceptual* quality, so that they are heard as linguistic units rather than as sounds.

A number of investigators have found that trained musicians are more likely than the musically naive to show a right-ear advantage in processing melodies.[110] However this, in turn, may depend on the kind of music. For example, among musicians familiar with Western tonal music, there is a left-ear advantage for music that does not conform to this type of musical system.[111] We might expect those with musical training to experience music at a deeper, more generative level, so these results might be taken as evidence that this deeper level is fundamentally left-hemispheric, but that there is a right-hemispheric bias at the more superficial perceptual levels.

What of the evolution of music? Ray Jackendoff suggests that it may have predated language:

> From an evolutionary point of view, there is no reason to think that musical structure came into being in splendid isolation, as a structure sui generis that somehow came to be linked to the affective response by brand-new pathways. More plausible is that musical perception, a highly specialized cognitive activity, is linked to some phylogenetically older cognitive representation that in turn has links to the affective response.[112]

He goes on to note the intimate relation between music and dance, which is a bodily expression of emotion.

Another early manifestation of music is singing. Charles Darwin suggested that singing has its origin in mating.[113] Singing may also play a role in cementing social cohesion and is a feature of almost every social gathering, whether for celebration, religious observance, or watching rugby in Wales. Parents sing to their children, and the intonational contours of baby talk also often have a song-like quality that may be important in expressing attitude or in controlling approach and withdrawal.[114] Given the rather diffuse yet pervasive quality of music in human society, it may well have been a precursor to language, perhaps even providing the raw stuff out of which generative grammar was forged.

But music is an enigma. Although it is pervasive in human life and experience, we do not really know what it is for or where it came from. Because it seems in some respects complementary to language, there has been a tendency to regard it as right- rather than left-hemispheric. On the other hand, it has a generative structure similar to that of language, which suggests that it should be left-hemispheric. However, music is perceived, processed, and appreciated at different levels, and the relative contributions of the two hemispheres probably differ from one level to the next. Moreover, whereas all normal people acquire language, not all people are musically sophisticated; indeed, those who acquire music at the level analogous to that of generative representation in language may be only a small minority. These considerations make it extremely difficult to fit music into the general theme of this book, and I leave readers to make of it what they will. Only enjoy.

Evolution of Hemispheric Duality

As recently as 1980, Stephen F. Walker concluded a review of cerebral asymmetries in the animal brain as follows:

> On the basis of reports of hemispheric asymmetries . . . it is difficult to reject the null hypothesis that the vertebrate nervous system is an entirely symmetrical device, with the possible exception of the brain of humans and canaries. In only these two species is there strong evidence that damage to one side of the brain has behavioral effects different from those which result when the other side suffers similar injuries.[115]

Since that time, there have been many reports of asymmetries in the brains of nonhuman animals, to the point where some have denied any discontinuity between humans and other species. In the final chapter of his book on cerebral lateralization in nonhuman species, for example, the late Norman Geschwind wrote:

> It is still commonly believed that the human has some completely distinctive characteristics, for example, the possession of language and a high level of artistic and musical abilities. It is my belief that the discovery of dominance in animals will play a major role in removing these last barriers to the special position of humans.[116]

One of the difficulties in evaluating these claims is that everything, even a billiard ball, is asymmetrical if examined closely enough or with sensitive enough measuring instruments. There is little doubt that individual animals have asymmetrical brains. The real question is whether there are systematic asymmetries that might be regarded as precursors to those of the human brain. I think that there probably are, but I am not willing yet to go as far as Geschwind in arguing that they remove the barriers to the special position of humans.

One likely precursor is a right-hemispheric specialization for emotion. Lesley J. Rogers of the University of New England in Armidale, Australia, has observed that in domestic chicks the right hemisphere is dominant for attack and copulation.[117] There is also evidence for a right-hemispheric bias in emotional reactivity in rats, as expressed in killing mice. However, this seems to depend on whether they have been handled or not; according to Victor H. Denenberg, early handling reduces the emotional reactivity of the left hemisphere more than that of the right.[118] Denenberg also reports that the left hemisphere has an inhibitory effect on the right, so that the incidence of muricide (mouse killing) is increased in rats with split brains.

There is also evidence for a right-hemispheric specialization for emotion in primates. For example, split-brained monkeys were more responsive to videotaped recordings of monkeys, people, animals, and scenery when these were exposed to the right hemisphere than when they were exposed to the left.[119] The late Stuart Dimond and Rashida Harries of University College, Cardiff, noted that people, gorillas, chimpanzees, and orangutans touched their own faces more often with the left hand than with the right, while monkeys showed no such asymmetry (and in fact rarely touched their own faces).[120] They related this to a right-hemispheric specialization for emotion. Since apes appear to show no consistent handedness in manipulation, these observations suggest that the right-hemispheric specialization for emotion may have preceded the left-hemispheric specialization for praxis—or at least for *manual* praxis.

There is also evidence that nonhuman animals may show a right-hemispheric bias in spatial representations. Denenberg notes that rats appear to have a bias to turn to the left, implying a dominance of the right hemisphere, but this seems to depend on the sex and strain of the animals;[121] in a different strain, Stanley D. Glick and his colleagues have reported a right-turning bias.[122] In monkeys, Gregor W. Jason, Alan Cowey, and Lawrence Weiskrantz have found a right-hemispheric advantage in spatial discrimination involving the location of a dot in a square; four animals with right-sided removal of the occipital lobe performed worse than five animals with left-sided removal.[123] Charles R. Hamilton and Betty A. Vermeire have recently reported a right-hemispheric advantage in the perception of faces in split-brained monkeys,[124] an asymmetry that resembles that in humans but is independent of handedness or language.[125] And in unpublished work, Robin D. Morris and William D. Hopkins of the Yerkes Regional Primate Research Center at Emory University showed composites of human faces like those of Figure 10.5 to three chimpanzees and found that they tended to choose as "happiest" those in which the smiling half was on the left, again implying a right-brained advantage.[126]

There is also evidence, however, that certain visual discriminations are *left*-hemispheric. In domestic chicks, Rogers found a left-hemispheric specialization in the discrimination of pebbles from grains of food, and Hamilton and Vermeire found that their split-brained monkeys were better able to discriminate tilted lines when they were projected to the left hemisphere. These are the sorts of discriminations that are normally associated with the right hemisphere in humans. Hopkins and his coworkers have also claimed evidence for a right hemifield (or left-hemispheric) advantage in the discrimination of visual forms in two chimpanzees, whereas there was no such asymmetry in either humans or monkeys.[127]

Although the evidence is patchy, it is probably fair to conclude that there are genuine precursors of hemispheric duality in humans. We have seen that there is evidence for left hemispheric specialization for communication in other species, including birds, mice, monkeys, and perhaps dolphins and chimpanzees. On the basis of research carried out mainly in the Soviet Union, the Russian biologist Vsevolod Bianki has recently claimed that in rats and mice, there is a fairly general dominance of the left hemisphere for motor control.[128] Bianki also claims that there are right-hemispheric specializations comparable to those in humans, and argues for a duality of the brain that applies across a wide range of species.

Bianki suggests that this duality may be characterized in terms of a left-hemispheric specialization for induction, which proceeds from the particular to the general, and a complementary right-hemispheric specialization for deduction, which proceeds from the general to the particular. Denenberg, on the other hand, suggests that "the left hemisphere is preferentially biased to receive and transmit communication; the right is selectively set to deal with spatial and affective matters."[129] But it is perhaps too early to tell whether the various systematic asymmetries that have been observed in different species can truly be captured in any simple duality. Denenberg's claims, for example, drew the following response from one group of commentators:

> Denenberg makes this case from a dozen studies showing lateralization of several specific functions, but he does not mention the hundreds of papers that demonstrate the symmetry of the brain for the majority of behavioral, electro-physiological, metabolic, and morphological indices.[130]

My own view is that the most likely precursors of hemispheric asymmetry in humans are a right-hemispheric bias for emotion and perhaps for spatial representation and a left-hemispheric bias for praxis. The biases are probably fairly slight. In the human brain, however, the left-hemispheric specialization for generativity (or GAD) may have been superimposed on this earlier scheme and may have thrown an additional right-sided bias on those functions that are not involved in praxis, generative language, or other generative functions. To this extent, at least, many of the specializations of the right hemisphere may have come about by default, as a secondary consequence of the occupation of the left hemispheric by GAD.[131] There are three reasons for believing this to be so.

First, the compendium of so-called right-hemispheric skills is much more heterogeneous than that of left-hemispheric skills, ranging from elementary perceptual functions to more complex skills such as mental rotation, the recognition of faces, the perception of music, and emotion. This suggests a rather general biasing of functions to the right, although some of these functions, such as emotion, may have been lateralized prior to the invasion of GAD. Second, left-hemispheric specialization seems more pronounced and more unitary in character than right-hemispheric functions, implying a wholesale invasion—although again, there may have been a prior left-hemispheric specialization for praxis. Third, the functions associated with GAD appear to be uniquely human, and therefore much more recent in evolution, than the functions associated with the right hemisphere. GAD may have been as predatory in its occupation of brain space as in its effects on the environment, and so may have thrown a more marked lopsidedness into our human brains than exists in any other species.

I do not mean to imply, however, that those functions loosely described as right-hemispheric remain as they were in our primate forebears. In the passage quoted above, Geschwind notes that humans seem to have special talents for music and art, and these depend at least in part on skills that are biased toward the right hemisphere. Chapter 9 suggested that these talents may have flowered as a consequence of the freeing of the hands and eyes when *H. sapiens sapiens* switched from gestural to vocal language. The generative aspect of these activities may well depend on GAD in the left hemisphere, but the emotional, spatial, and melodic aspects involve aptitudes that are fundamentally right-hemispheric, but are probably uniquely human in their sophistication and subtlety.

New Myths for Old

In summary, the picture of hemispheric duality in humans that emerges is not very different from the popular account of the left hemisphere as more analytic, rational, and propositional, and the right as more holistic, intuitive, and appositional. The GAD in the left hemisphere does indeed have an analytic quality, since it involves representations that are constructed of parts. These representations have a rule-governed, or propositional, format. However, I have argued that the important property of GAD is not its analytic quality, but rather its generative aspect. It is GAD that has given humans the unlimited possibilities of verbal expression, and of the manufacture, use, and representation of objects.

By contrast, the earlier forms of representation that are, in humans, biased toward the right hemisphere are more holistic, and their intuitive quality may derive from the fact that they are preverbal. Indeed, many of the functions that seem to be associated more with the right hemisphere, such as emotion, the appreciation of music, or the perception of faces, are precisely the sorts of functions for which words often fail us. They may also be the sorts of functions that cannot be adequately simulated on a digital computer.[132] The computer is itself a product of GAD (a GADget), and its powers are therefore merely GAD-like.

Although there is a certain complementarity between the functions of the two sides of the brain, the evidence suggests that they do not function independently. Most functions, in fact, seem to depend on the specialized capacities of *both* hemispheres. To recognize an object may require the spatial skills of the right hemisphere if it is presented in some unusual orientation, or under otherwise degraded conditions, but it may require the parsing operations of left-hemispheric GAD in order to then determine what it actually is. Even at the final stage of matching the input against stored information, both hemispheres may play a role; the left hemisphere may be primarily involved in determining that some large object is a car, but it may require the more holistic descriptions of the right hemisphere in order to determine the make of the car. We also saw that different aspects of music may engage the two hemispheres to different degrees.

In this respect, spoken language may be more or less unique as a complex skill in tapping the skills of the left hemisphere, with essentially no right-hemispheric component, except perhaps in prosodic or emotional aspects of speech. However, both sign language and written language have spatial components, and there is correspondingly more of a right-hemispheric component. Similarly, we saw that constructional apraxia may occur following either left- or right-hemispheric damage, depending upon whether the generative or spatial aspects are critically involved.

It is, of course, easy to criticize the wilder extremes of Wiganism and neo-Wiganism, of the conceptions of hemispheric duality that prevailed at the turn of the century or that are still paraded in the popular press. To some extent, as pointed out earlier, the double brain has served as a vehicle for the prejudices of the age. However, I do not think that these characterizations of cerebral duality are all wrong; the fact that rather similar characterizations have emerged seemingly independently at different periods in history suggests some measure of truth.

But neither do I think that any revision of these ideas can be free of bias or contemporary mythology. As we move into the 1990s, it may be fitting that the left hemisphere is identified as the vehicle of GAD, the generator, at once populating and polluting the natural environment with words and objects. For the compensatory forces of nature, and for the more holistic, "natural" side to our biological makeup, we might indeed look more to our right minds. And in the new mythology, we might replace the idea of two independent minds with that of a single brain that can draw on different forms of representation. Cooperation rather than conflict between the different aspects of our makeup might at least offer some hope for the future.

Perhaps, though, we should leave the last word to Rudyard Kipling, in his remarkable poem entitled *The Two-sided Man*:[133]

> Much I owe to the Lands that grew—
> More to the Lives that fed—
> But most to Allah Who gave me two
> Separate sides to my head.
>
> Much I reflect on the Good and the True
> In the faiths beneath the sun
> But most upon Allah who gave me two
> Sides to my head, not one.
>
> *I* would go without shirt or shoe,
> Friends, tobacco or bread,
> Sooner the lose for a minute the two
> Separate sides of my head!

Notes

1. Published in *The Illuminated Magazine* of October 1844.
2. Bogen (1969b). Bogen also arranged to have Wigan's book republished in 1985.
3. The *Three Faces of Eve*, by Thigpen and Cleckley (1957), was later made into a film. A more recent study by Confer and Ables (1983) documents the split of a personality into *five* distinct entities.
4. Wigan (1844, p. 313).
5. de Fleury (1872).
6. Exner (1881).
7. Luys (1881).

8. Briquet (1859).
9. Charcot (1878).
10. Janet (1899).
11. For a thorough and entertaining history of the "double brain" movement in the latter part of the nineteenth century, see Harrington (1987).
12. The rise and fall of metallotherapy and hemihypnosis are again documented in Harrington's (1987) book.
13. See Harrington (1987) for documentation.
14. Deference to the facts was not, however, a conspicuous feature of the ideas of the time.
15. Harrington (1987, p. 248).
16. Gazzaniga (1988, pp. 446–447).
17. Bogen (1969a,b); Bogen and Bogen (1969).
18. Garrett (1976, p. 244).
19. Edwards (1979, p. vii). For further examples and a critique, see Harris's (1988) excellent review.
20. Corballis (1980).
21. For a useful critique, see Harris (1988).
22. Gazzaniga (1983, p. 536).
23. Jackson (1876).
24. Russo and Vignolo (1967); Teuber and Weinstein (1956).
25. Hannay, Varney, and Benton (1976).
26. Carmon and Bechtold (1969).
27. Benton, Hannay, and Varney (1975).
28. Nebes (1971).
29. Nebes (1972).
30. Nebes (1973).
31. Franco and Sperry (1977).
32. Durnford and Kimura (1971).
33. Bryden (1976).
34. Longden, Ellis, and Iverson (1976).
35. Davidoff (1976).
36. Davidoff (1975).
37. See Bryden (1982) for a review.
38. Curry (1967).
39. Chaney and Webster (1966).
40. King and Kimura (1972).
41. Bradshaw, Burden, and Nettleton (1986); Cohen and Levy (1986).
42. Gazzaniga (1983, 1988).
43. For example, Gazzaniga, Bogen, and Sperry (1965).
44. LeDoux, Wilson, and Gazzaniga (1977).
45. LeDoux (1983).
46. Gazzaniga (1988).
47. De Renzi (1982).
48. Shallice (1988).
49. There is evidence that left-sided damage disrupts the sequential aspect of construction (Hécaen and Assal, 1970), while right-sided damage

disrupts the spatial aspect (Blakemore, Iverson, and Zangwill, 1972). In one constructional apraxic with left-sided damage, the difficulty seemed to lie primarily in telling left from right (Riddoch and Humphreys, 1988).

50. Shallice (1988).

51. For example, De Renzi, Scotti, and Spinnler (1969); Milner (1958).

52. Warrington and James (1988).

53. These deficits can be related to the theory of object perception developed by Marr and Nishihara (1978). As we saw in Chapter 9, this theory has to do with how the perceiver makes a structural description of an object from low-level perceptual information. In the case of objects depicted from an unusual angle, this involves assigning internal axes to the representation. Processes such as this may be involved in apperception and may be largely right-hemispheric. For further discussion, see Humphreys and Riddoch (1984) and Shallice (1988).

54. The nature of the visual degradation necessary to induce apperceptive agnosia has been a matter of some controversy. It has been suggested that the condition is most apparent when an object is presented in such a way that its principal axis cannot be easily determined, as when a flashlight, for example, is viewed end-on rather than from the side (Humphreys and Riddoch, 1984; Warrington and Taylor, 1973). This has been taken to support the view of Marr (1982) that, in order to construct a representation of an object that can be recognized, it is necessary to discover its internal axes, including most critically the principal axis. Other evidence has suggested, however, that forms of degradation not critically involving the principal axis are equally effective in inducing apperceptive agnosia (Warrington and James, 1986; and see Shallice, 1988, for further review). This, incidentally, supports my own view that the specification of internal axes is not critical to the identification of an object (Corballis, 1988).

55. Hécaen, Goldblum, Masure, and Ramier (1974).

56. McCarthy and Warrington (1986).

57. Gilbert and Bakan (1973).

58. Cooper and Shepard (1973). The average rate of mental rotation is roughly one revolution per second. Perhaps this is the origin of the *second* as the unit of time; it may be defined as the time it takes to imaging a wheel turning through one revolution!

59. Corballis and Sergent (1988, 1989a,b).

60. Ratcliff (1979).

61. Deutsch, Bourbon, Papanicolaou, and Eisenberg (1988).

62. See Corballis and Sergent (1989b).

63. For a conflicting opinion, see Johnson (1984), who showed that, under certain restricted conditions, L.B. can sometimes vocalize information received by the right hemisphere. Our own observations (Corballis and Sergent, 1988), as well as those of Myers and Sperry (1985), suggest that he does this by transferring the information to the left hemisphere. This is accomplished either by external *cross-cueing*, or by a crude transfer at a subcortical level, i.e., below the level of the corpus callosum and the other sectioned commissures.

64. Ogden (1985a).
65. Op. cit.
66. Bisiach and Luzzatti (1978); see also Bisiach, Luzzatti, and Perani (1979) and Ogden (1985b) for further demonstrations of the neglect of a visual image.
67. See De Renzi (1982) for a review. Some authors, such as Bisiach and Vallar (1988), have claimed that neglect can also arise from a one-sided degradation of *representational space.*
68. Bisiach, Capitani, Luzzatti, and Perani (1981).
69. Bisiach and Luzzatti (1978, p. 132).
70. Kinsbourne (1987).
71. Corballis (1983, p. 52).
72. Costello and Warrington (1987).
73. For a review of this and other forms of anosognosia, see McGlynn and Schacter (1989).
74. Alajouanine (1956, p. 23).
75. Janet (1907).
76. Gainotti (1972).
77. Terzian (1964).
78. Flor-Henry (1983, 1985).
79. See Meyer-Bahlburg (1983).
80. Haggard and Parkinson (1971).
81. Carmon and Nachshon (1973).
82. Suberi and McKeever (1977).
83. Ley and Bryden (1979); McKeever and Dixon (1981); Strauss and Moscovitch (1981).
84. Chaurasia and Goswami (1975).
85. It only seems to, actually, but that is another story.
86. Borod and Caron (1980); Campbell (1978).
87. Sackeim, Gur, and Saucy (1978).
88. Campbell (1978).
89. Levy, Heller, Banich, and Burton (1983).
90. Schiff and Lamon (1989).
91. Alajouanine (1948).
92. Luria, Tsvetkova, and Futer (1965).
93. See Benton (1977) for a review.
94. Zatorre (1984, p. 197).
95. Milner (1962).
96. Zatorre (1984, 1989).
97. Kimura (1964).
98. Blumstein and Cooper (1974).
99. Kallman and Corballis (1975).
100. Gordon (1970).
101. See Peretz and Morais (1988) for a review.
102. Wolff, Hurwitz, and Moss (1977).
103. Kashiwagi, Kashiwagi, Nishikawa, and Okuda (1989).
104. Ibbotson and Morton (1981).

105. Lerdahl and Jackendoff (1983).
106. Sergeant (1969).
107. West, Howell, and Cross (1985).
108. Falk (1987b).
109. Jackendoff (1987).
110. Bever and Chiarello (1974); Wagner and Hannon (1981).
111. Johnson, Bowers, Gamble, Lyons, Presbey, and Vetter (1977).
112. Jackendoff (1987, p. 237).
113. Darwin (1901—originally published in 1871).
114. Papousek and Papousek (1981).
115. Walker (1980, p. 361).
116. Geschwind (1985, p. 271).
117. Rogers (1980).
118. Denenberg (1981).
119. Ifune, Vermeire, and Hamilton (1984).
120. Dimond and Harries (1984).
121. Denenberg (1981); Denenberg and Yutzey (1985).
122. Glick and Shapiro (1985).
123. Jason, Cowey, and Weiskrantz (1984).
124. Hamilton and Vermeire (1988).
125. However, Overman and Doty (1982) found no evidence for hemispheric asymmetry in macaques in the analysis of faces.
126. Morris and Hopkins (1989).
127. Hopkins, Washburn, and Rumbaugh (1990).
128. Bianki (1988). Much of this work was hitherto unknown to Western readers.
129. Denenberg (1981, p. 20).
130. Bures, Buresova, and Krinavek (1981).
131. Corballis and Morgan (1978).
132. It is true, of course, that little progress has been made in simulating human language or the parsing of visual scenes. But *some* progress has been made, and these problems do at least lend themselves to the propositional language of computers. In the case of emotion or musical appreciation, it can be argued that computer scientists do not know even where to begin. Computers can *make* music quite well, of course, but is there any sense in which they can be said to appreciate it?
133. Kipling (1927, p. 568).

—11—

The Plastic Brain

> Every baby has a month of heaven and a month of hell before birth, so that it may make its choice with its eyes open.
>
> SAMUEL BUTLER[1]

SO FAR, MY ARGUMENT for a discontinuity between humans and other animals has rested on functions normally associated with the left side of the human brain. The most notable of these is language, whose properties of generativity and open-endedness seem to be unique to humans. Chapter 9 suggested that we may also use a left-hemispheric, generative mode to represent shapes, derived from the fact that we live in an ever-changing sea of manufactured objects. Manufacture is itself relentlessly generative, to the point where it has taken over the human environment and increased our ability to know and recognize the objects around us. A generative mode of representation, based on a fixed vocabulary of primitive units of shape and on rules for their combination, provides a rapid and ready (though sometimes rough) means of adapting to the pace of environmental change. It has been suggested that generativity in language, manufacture, and representation might be attributed to a generative assembling device, or GAD, that is normally represented in the left cerebral hemisphere.

This chapter reviews evidence that our characteristic lopsidedness, or precursors of it, is present very early in life, probably even before birth. The evidence is consistent with one of the themes of this book, namely, that human laterality is fundamentally biological rather than cultural. But it would be wrong to conclude from this that laterality is a fixed entity, inflexibly wired into the system as part of our human heritage.

One of the awkward facts of the human condition is that not all of us are right-handed or left-hemispheric for language. By implication, GAD is not necessarily resident in the left hemisphere. As we have seen, those who deviate from the lopsided norm are often dismissed as being somehow inferior, and we might be tempted even to declare them GADless. Such a temptation must surely be resisted. The world is full of talented left-handers, and as we saw in Chapter 4, there is no evidence that left-handers, as a group, are in any way inferior to right-handers in intelligence or ability. Nor is there any evidence that those otherwise normal individuals with language represented predominantly in the right cerebral hemisphere are in any way linguistically inferior, although there is admittedly little direct information on this point.

Although the nature of the functions that are normally left-hemispheric may provide an insight into what is special about humans, then, left-hemispheric representation may not itself be critical. This chapter also reviews evidence that the two hemispheres may in fact have equal *potential;* that is, both may be equally capable of mediating language, or GAD, or perhaps any of the functions normally biased toward a particular hemisphere. Chapter 9 suggested that the left hemisphere may be a "magic carpet" whose warp and woof are especially designed to contain those forms of representation that make us uniquely human. That conclusion must be qualified in this chapter. The allocation of functions to the hemispheres seems to be somewhat flexible, permitting some variation in precisely how and where the magic carpet is laid down.

But this raises a problem. On the one hand, we have evidence that lateral asymmetries are already present very early in life. On the other hand, there is evidence for equipotentiality, such that either hemisphere is potentially capable of the specialization normally associated with the other. A final objective of this chapter is to suggest how this problem might be resolved. It begins, then, with evidence of asymmetries that can be detected very early in life and suggests that laterality unfolds according to a fixed biological program.

Evidence of Asymmetries in Early Infancy

Asymmetrical Reflexes

First, there are a number of asymmetrical reflexes that are apparent in infancy, or even before. One of the most striking is the *tonic neck reflex,* which appears as early as the 28th week after conception and usually disappears by some 20 weeks following birth. It is induced by gently turning the head to one side, whereupon the arm and leg on that side extend, while the arm and leg on the opposite side flex, producing what is sometimes called a *fencer's posture.* The majority of infants show the reflex to rightward rather than leftward turns of the head.

There is evidence that the tonic neck reflex shows some relation to subsequent handedness. Those infants who show a rightward tonic neck reflex nearly all become right-handed, while those with no consistent bias or with a left tonic neck reflex show mixed handedness later on.[2] This may again be interpreted as consistent with Annett's theory if we assume that the RS+ allele causes a right shift in the reflex as well as in handedness. Consequently, the majority of those with a rightward reflex carry an RS+ allele, while those with a leftward reflex or with no consistent bias belong to the minority carrying two RS− alleles.

After birth, most infants tend to sleep or lie with their heads turned to the right,[3] an asymmetry that is readily noticeable in any hospital ward of healthy babies. It cannot be attributed to the way babies are placed by their nurses or parents, since it persists despite the common practice of shifting the head in an attempt to ensure equal turns to either side. It is probably related to the rightward bias in the tonic neck reflex. Jacqueline Liederman has claimed, in fact, that there is no intrinsic bias in the tonic neck reflex at all; rather, the appearance of bias is an artifact of the infants' tendency to lie with their heads to the right.[4]

Another reflex that is present immediately after birth is the *stepping reflex,* which is elicited when the infant is lowered onto a flat surface. When contact occurs, the legs make coordinated stepping movements. According to one report, the leading leg is the right leg in most infants. Although I know of no studies in which this asymmetry has been correlated with subsequent handedness, the proportion of infants leading with the right leg is reported as about 88 percent,[5] which is in close agreement with the proportion of right-handers in the normal adult population.

Postural Asymmetries

Even before birth, during the last month or so of pregnancy, the human fetus is asymmetrically placed in the mother's womb. In most cases the fetus is placed head down, with the back to the mother's left side, as illustrated in Figure 11.1. It has been suggested that this positioning in the womb might be responsible for right-handedness, since the left arm is jammed against the mother's pel-

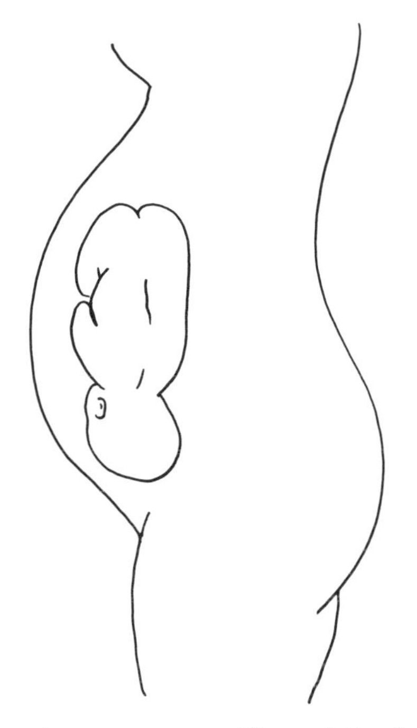

Figure 11.1. The most common position of the human fetus in late pregnancy.

vis and backbone, while the right arm is allowed more freedom to move.[6] However, the evidence for this is only indirect and is related to asymmetries in the posture that infants show at birth.

About two-thirds of infants emerge from the womb head first with their backs to the mother's left side, probably as a fairly direct consequence of fetal position prior to birth,[7] although it has been claimed that over 50 percent of fetuses change their positions in the 24 hours before birth.[8] Nevertheless those born with their backs to the mother's left show a preference after birth for lying with their heads turned to the right, while those born with their backs to the mother's right show no systematic preference for head turn after birth. This head-turn preference, in turn, is strongly related to hand preference in infancy, with the right-turn preference leading to right-handedness and the left-turn preference leading to left-handedness.[9]

In Chapter 3 it was suggested that human children are born prematurely, at least by comparison with the offspring of other primates. Conceivably this may hold part of the key to handedness. That is, the human newborn, unlike the newborn chimpanzee or gorilla, may still be influenced by fetal asymmetries, which then become part of early reaching and grasping, and so lead to consistent handedness. The main difficulty with this account is that the characteristic fetal asymmetry applies to only two-thirds of babies, whereas the proportion of right-handedness is close to 90 percent. Moreover one recent study based on a fairly small sample claims that there is no relation between fetal position and subsequent handedness.[10]

Handedness

Handedness itself may be detected early in infancy, although not, of course, in all of its manifestations. There is evidence that most 2-month-old infants will hold a rattle longer in the right than in the left hand,[11] and that the majority also grip a rattle more strongly in the right hand.[12] These asymmetries might be regarded as indicating differences in *performance* between the hands.

The *preference* for one or the other hand in reaching for or picking up objects seems to emerge a little later and to show some fluctuation. In one classic study carried out in the Yale Clinic of Child Development in the 1940s, Arnold Gesell and Louise B. Ames took moving films of groups of children ranging in age from 8 weeks to 10 years. The children were filmed in settings contrived to make them respond to a variety of objects, such as pencils, paper, cubes, and construction toys, and the preferred hand in reaching for or manipulating these objects was noted. Gesell and Ames found that the first clear preferences, at about 16 to 20 weeks, were usually for the *left* hand.[13] Although others have reported the same phenomenon,[14] still others are skeptical,[15] with Marian Annett calling it "a hardy perennial in the literature [that] should be rooted out as thoroughly misleading."[16]

Be this as it may, Gesell and Ames also noted that there were fluctuations in hand preferences in the first year, and that cyclic fluctuations persisted until the age of about 8 years. One hand, usually the right, gradually became dominant despite these fluctuations. From the orderly way in which handedness unfolded, Gesell and Ames inferred that handedness was not a matter of learning or imitation. Rather, they concluded, "handedness is a product of growth."

Other studies, focusing on performance rather than on preference, show an even more remarkable consistency in the development of handedness. For instance, Marian Annett has studied the development of differences in skill between the two hands on a peg-moving task, in which ten pegs must be moved from one row of holes to another with each hand in turn. On average, both children and adults are faster with the right hand, and the difference in average time remains roughly constant from the age of 3.5 years to over 50 years! This is so despite the fact that the time taken by each hand decreases sharply during childhood, but is roughly constant from about 12 years of age. The proportion of people who are faster with the left hand than with the right also remains roughly constant, at somewhere between 10 and 20 percent.[17]

Cerebral Asymmetry

We saw in Chapter 5 that children are sensitized to critical features of speech sounds very early in life. There is also evidence that very young babies can discriminate faces from nonfaces within the first few hours, and possibly even minutes, following birth.[18] Since these and other perceptual skills are lateralized in the adult, it is reasonable to inquire whether they are already lateralized in infancy.

For her Ph.D. thesis at McGill University, Anne Entus neatly adapted the dichotic-listening technique (discussed in Chapter 7) for testing cerebral asymmetries in infants. She placed the nipple of a bottle in the mouths of babes and sucklings aged from 3 weeks to 3 months, and whenever the infants sucked they heard two sounds (say, *da* and *ba*), one in each ear. They received no milk, alas, but the sounds proved just as rewarding, and the infants sucked eagerly to hear more. After a while, however, the rate of sucking dropped, and when it fell below a critical level, Entus changed the sound in just one ear; for example, *ba* in the right ear might be changed to *ga*. This produced an increase in sucking as the infant's interest was regained. Later, when sucking in response to the changed sound had dropped, Entus changed the sound in the *other* ear (e.g., *da* might be changed to *ma*).

Entus found that when the sounds were syllables, as in the examples given above, most of the infants showed greater recovery of sucking when the change occurred in the right ear than when it occurred in the left ear.[19] This could not be attributed simply to better hearing in the right ear, since a *left*-ear advantage was found when musical sounds were used instead of speech sounds. The right-ear advantage for speech sounds was taken to reflect a left-hemispheric specialization for speech, while the left-ear advantage for music was taken to reflect a right-hemispheric advantage for nonverbal sounds.

Catherine T. Best and her colleagues at Wesleyan University found similar results using heartbeat as the critical measure. Heartbeat slows down following any change in sound, and 3-month-old infants showed more deceleration with dichotic speech sounds when the change was in the right ear compared to the left ear. The opposite was the case with musical sounds.[20] In one later study, the right-ear advantage for speech was found in 3- and 4-month-old infants but not in 2-month-old infants, while the left-ear advantage for musical sounds was present even in the 2-month-olds.[21] Best has suggested that right-hemispheric specialization develops slightly earlier than left-hemispheric specialization.[22] However, a more recent study suggests a right-ear advantage and therefore a left-hemispheric specialization for speech sounds in infants a mere 4 *days* old.[23]

Another group of investigators developed a visual technique for measuring hemispheric asymmetry for the perception of speech in infants.[24] They showed video recordings of a woman speaking pairs of syllables, like *mama, bebi,* and *lulu*. On one side of a split screen, the woman was shown with her lips moving in synchrony with the sound of the utterance, while on the other side she was shown with her lip movements corresponding to a different syllable, and therefore out of synchrony with what was heard, as in a poorly dubbed movie. Infants aged 5 to 6 months spent more time looking at the synchronized speech than at the unsynchronized speech, but this difference was reliable only when they were looking at the right side of the split screen. This was taken as evidence for a left-hemispheric specialization in linking the visual and acoustic aspects of speech. The message for new parents is that it might be better to speak to the infant from the right side of the crib but to play music from the left side.

Using a rather similar technique, Scania de Schonen and her colleagues in Marseille showed that in infants as young as 4 months there was a *left*-hemifield (or right-hemispheric) advantage in recognizing faces.[25] This was true both of the recognition of faces the babies already knew, such as those of their mothers or caregivers, and also of faces learned in the course of the experimental session. If just gazing fondly at your infant, then, it might be advisable to do so from the left side of the crib, scurrying to the other side only if you wish to talk.

More evidence on early lateralization comes from electrical recordings taken from the scalp. In one, the response to speech sounds was larger over the left than over the right temporal lobe of the brain. This was true in infants ranging in age from 1 *week* to 10 months (which again does not square with the age trend in the heart-rate study described above), as well as in a group of adults.[26] In another study, electrical signals were recorded continuously from 6-month-old infants as they listened to natural speech or music, and systematic differences between the two sides of the brain were again observed, consistent with a left-hemispheric involvement in speech and a right-hemispheric involvement in music.[27]

There are also anatomical asymmetries between the two sides of the brain that have been related to hemispheric specialization and that can be detected in infancy. For example, in the majority of people the temporal planum, an area in the temporal lobe that on the left comprises part of Wernicke's area, is larger on the left than on the right.[28] This asymmetry is present also in newborns,[29] and can even be detected in the human fetus as early as the 31st week of pregnancy.[30]

So far, the studies reviewed suggest that asymmetries observed in adulthood, or precursors of them, may be present in early infancy. However, there is one curious asymmetry that seems to show a reversal between infancy and adulthood. We saw in the previous chapter that most people express emotion more intensely on the left than on the right side of the face. Studies of infants in the first year of life have shown just the opposite, with more intense emotional expression on the *right* side of the face.[31]

Summary

The evidence shows that the human brain is markedly lopsided from infancy. Moreover, the asymmetries observed in infancy, with the exception of the about-face just mentioned, are in most cases clearly related to asymmetries present in adulthood. This is not to say, of course, that the adult pattern is present in the infant; most language functions, for instance, do not emerge until well beyond infancy and could not possibly be lateralized before they appear. Nevertheless, in those functions that can be measured from infancy on, the degree of lateralization seems to remain remarkably constant. There is little evidence that lateralization actually develops, or increases over time, once it is established.[32]

Equipotentiality

The idea that the two hemispheres of the brain might have equal *potential*, at least for the representation of speech, goes back to the French physician Pierre Marie, who proposed it simply on the grounds that the two hemispheres of the brain are mirror images:

> The inborn centers which we know (and they are not numerous) are always bilateral and very clearly symmetrical. The motor centers of the limbs, the centers of vision, have their locations in each of the two hemispheres in symmetrical regions. . . . How can we admit the existence of an inborn center for speech which would be neither bilateral nor symmetrical?[33]

We now know, of course, that the brain is not quite so structurally symmetrical as Marie thought it was, but the high degree of symmetry still lends some force to his argument.

Marie's argument was essentially an inference based on symmetrical structure, although he was no doubt aware that in some individuals speech was represented primarily in the right side of the brain. There is also evidence, however, that even in those destined to become left-cerebrally dominant for language the right hemisphere can take over, with little or no impediment. Indeed, it is rather striking that the best evidence for equipotentiality has to do with language, since language is normally the most *lateralized* of functions. We shall see later that it is difficult, if not impossible, to demonstrate equipotentiality for spatial skills.

The most convincing demonstration that the right hemisphere can take over language comes from studies of patients who have had one or the other cerebral hemisphere surgically removed. This rather drastic operation is known as *hemispherectomy*,[34] and it is sometimes necessary for the treatment of brain disorders that would otherwise prove fatal. Following removal of the left hemisphere, the extent to which the right hemisphere can take over seems to depend on the age at which the left hemisphere became incapacitated, whether through the operation itself or because of some prior condition.

According to a review of evidence published in 1962,[35] if the left hemisphere is incapacitated before the age of 2 years, the right hemisphere can take over language more or less completely. Thereafter, the recovery of language declines, so that by the age of puberty or thereabouts, incapacitation of the left hemisphere results in more or less permanent aphasia. Patients who undergo left hemispherectomy in adulthood typically have great trouble with expressive speech, although they may retain the ability to swear (which is understandable), or to recite automatic phrases, or even to sing. Although they may retain quite good understanding of speech, they have great difficulty in reading and virtually no ability to write.[36] Eran Zaidel remarks that these patients show the same profile for language skills as does the right hemisphere of the split-brained patient.[37]

The effect of left hemispherectomy therefore depends critically on the age at which the hemisphere was incapacitated (not necessarily the age of the hemispherectomy itself). This ties in with the idea, documented in Chapter 5, of a *critical period* for language that lasts until the age of puberty. That is, the right hemisphere's ability to acquire language skills seems to match that of the left, and depends on roughly the same critical period—although I shall later argue that the right hemisphere's schedule may lag behind that of the left.

The right hemisphere's unexpected ability to acquire language once its domineering partner is removed raises the question of why it does not normally do so. It must be supposed, I think, that the left hemisphere normally exerts an inhibitory influence that prevents language development but not the development of other, nonverbal skills.[38] But if the left hemisphere is removed or otherwise incapacitated, this inhibitory effect is itself removed, and the right hemisphere is then free to develop its potential. There is evidence for a mechanism of this sort in the development of other biological asymmetries. In female birds, for example, the right ovary normally remains undeveloped, but if the left ovary is removed the right one then grows to a size that is larger than normal, suggesting the removal of some inhibitory influence.[39]

Is Equipotentiality Complete?

In his 1967 book *Biological Foundations of language*, Erich Lenneberg argued that language does not start to emerge until around the age of 2 years, which explains why the two hemispheres might be equipotential for language up to that time. We saw above, however, that subsequent research has shown that certain specialized aspects of speech perception emerge much earlier, and indeed are already left-hemispheric. That is, the case for a built-in, biological blueprint appears to be even stronger than Lenneberg supposed. However, if the seeds of language were already planted in the left hemisphere very early in infancy, and probably even before birth, it was not clear how there could still be equipotentiality for language.

Maureen Dennis and her colleagues at the Hospital for Sick Children in Toronto argued that equipotentiality is not in fact complete, and that children whose left hemispheres are incapacitated in the first 2 years of life are impaired in language skills later on. In particular, they have difficulty dealing with grammatical complexity. In one study, for example, those with removals of the left hemisphere were worse than those with removals of the right hemisphere in the understanding of passive negative sentences, but not in their ability to understand active negative or passive affirmative sentences.[40] The implication was that the right hemisphere mediated language in a way that was subtly different from the normal left-hemispheric mode and inferior to it.

In retrospect, though, it seems that the deficiencies of right-hemispheric language in Dennis' patients were greatly exaggerated. Left-hemispherectomized patients were able to *talk* in apparently normal fashion, suggesting no lack of generativity. Dennis' studies have also been questioned on methodological and statistical grounds by Dorothy Bishop of the University of Manchester.[41]

For example, although Dennis and her colleagues claimed that those undergoing left hemispherectomy were poorer in verbal skills than those undergoing right hemispherectomy, both groups were within the normal range of performance. In a further critical review of studies of hemispherectomy and early brain damage, Bishop finds very little evidence that the right hemisphere is in any way deficient in the mediation of language.[42] In a more recent study, Jenni A. Ogden of the University of Auckland has shown that in two left-hemispherectomized patients the remaining right hemisphere mediates language in essentially normal fashion, even with respect to grammatical subtleties.[43]

If there are deficiencies in the way the right hemisphere can take over language, they are slight indeed. Given that left-hemispheric mechanisms for processing speech may emerge very early in infancy, even before birth, we may perhaps expect *some* impediment if the right hemisphere is forced to take over later, even in later infancy. However, any such impediment might be attributed to the delay, and not to the fact that it is the right hemisphere taking over; that is, one might expect a similar impediment if *left*-hemispheric representation were similarly delayed.

In short, the equipotentiality of the two hemispheres with respect to the representation of language does seem to be more or less complete. It should not be inferred from this, though, that we need only half a brain. Although language is largely spared following early hemispherectomy, other abilities may be markedly impaired. Hemispherectomized patients retain at least some degree of weakness or atrophy on the side of the body opposite the removal. More important, they may show impairment of nonverbal capacities. The two left-hemispherectomized patients studied by Jenni A. Ogden, who showed little deficit in language, were severely impaired in memory for nonverbal patterns, in some complex spatial skills, and in spatial orientation.[44]

Patients who undergo removal of the right hemisphere, even in infancy, typically also show impairments in spatial abilities, although their language skills remain largely intact, as one would expect. Summarizing the evidence, Bruno Kohn and Maureen Dennis remark that these patients tend to use verbal ways of approaching spatial problems.[45] This suggests that compensation following hemispherectomy tends to be unidirectional; following early left hemispherectomy the right hemisphere can take over language, although somewhat at the expense of spatial ability, but following early right hemispherectomy the left hemisphere continues to control language and spatial abilities fail to develop. This does not necessarily mean that the left hemisphere has no *potential* for spatial functions. It may, however, be difficult or even impossible to demonstrate that potential.

General Infantile Plasticity?

It is sometimes said that the recovery of language following early incapacitation of the left hemisphere is due simply to a general plasticity of the infant brain and has nothing to do with the equipotentiality of the two hemispheres as such. In the 1930s, Margaret Kennard studied the effects of brain damage on motor skills in monkeys. Her results led her to what has become known as the *Kennard doctrine*, which states that the earlier the brain damage occurs, the less is the behavioral loss.[46] In the years since then, however, this doctrine has been increasingly challenged. The effects of early brain damage may vary quite widely, depending on a number of factors, including the location of the damage and the nature of the functions involved.

Using rats, Bryan Kolb of the University of Lethbridge in Alberta, Canada, compared the effects of hemispherectomy with those of removing the frontal lobes at different stages of development. The rats were given a test of spatial navigation. Kolb found that hemispherectomy had little effect if it was carried out immediately following birth but showed an increasing effect with age. Removal of the frontal lobes, by contrast, had its greatest effect at the earliest stage of development.[47] These results tie in with evidence from humans. We have seen that the effect of hemispherectomy on the development of language increases with age; Kolb also cites evidence that damage to the frontal lobes has a more damaging effect in early than in later years.

There is also evidence that early unilateral brain damage has different effects on different skills. We have seen that language is relatively impervious to early hemispherectomy; it is also relatively immune to *any* early unilateral damage, and is more affected by late than by early damage. By contrast, other functions, including some visuospatial skills, may be quite severely affected, and may be more affected by early damage than by late damage.[48]

It will not do, then, simply to attribute the excellent recovery of language following early hemispherectomy to some global concept of infantile plasticity. Recovery of language does demonstrate plasticity, of course, but that plasticity seems to depend on the fact that one hemisphere remains intact. Indeed, Marie may well have been right in stressing the role of symmetry; that is, the right hemisphere can take over language precisely because it is essentially the mirror image of the left. However, we now need some way to deal with the paradox that lateralization, of language as of other functions, seems already established in early infancy, if not before. How can it be that the two sides of the brain remain equipotential, or nearly so, up to the age of 2 years?

Asymmetry and Growth Gradients

The solution to this problem might well lie in different patterns of growth on the two sides of the brain. As we saw in Chapter 3, a good deal of learning depends on so-called critical periods in development. Language itself is, of course, a prime example, and the very complexity of language may be due to the intricate interplay of growth and environmental influences. The acquisition of language may therefore be orchestrated by genes controlling the rate of development, so that different neural circuits involved in language mature at different times and are modified by appropriate inputs from the linguistic environment. More than any other species, humans have capitalized on this principle, since the period of human development is exceptionally long.

If the rate of growth was programmed differently on the two sides of the brain, then the two sides might acquire different kinds of skill. Indeed, given the importance of growth in the acquisition of skill, and given the overall structural *symmetry* of the brain, this seems the most obvious way in which to achieve functional *asymmetry*. A clever Maker, faced with the problem of constructing functional asymmetry from a fundamentally symmetrical blueprint, might solve the problem not by reformatting the growth plan to build in structural asymmetries, but by prolonging the period of growth and varying the rate of growth between the two sides. Environmental input would do the rest, since children are exposed to different kinds of input at different stages of childhood.

Leading with the Left

In 1865, Paul Broca suggested that both right-handedness and left-cerebral dominance for language might be attributable to earlier growth of the left hemisphere relative to the right. In this he was influenced by the earlier observations of Louis Pierre Gratiolet, who claimed from his observations of fetal brains that the left side developed earlier than the right in the forward parts of the brain, while the reverse was true in the rearward parts.[49] Recall that Broca had located the language area in the frontal cortex and did not know of the involvement of more posterior regions. He was also impressed by the structural symmetry of the brain, arguing that it was only through differential growth that the left side could attain its dominance.

In 1978, Michael J. Morgan (now at the University of Edinburgh) and I made the further point that a left-hemispheric lead might explain how the right hemisphere could take over language if the left was incapacitated.[50] It must also be supposed that the left hemisphere exerts some inhibitory influence that normally prevents the lagging right hemisphere from picking up language. If the left hemisphere is removed or otherwise incapacitated, this inhibitory influence is removed. The right hemisphere is then free to acquire language, provided, of course, that it has not already passed the critical periods for the various aspects of language. If the right hemisphere lags, then its critical periods would occur somewhat later than those of the left hemisphere.

Morgan and I also suggested that this differential growth might be controlled by the RS+ allele postulated by Marian Annett.[51] Those lacking the RS+ allele might be subject to random fluctuations of growth rate between the hemispheres, but this would still have the desired effect of assigning functions laterally, albeit in more haphazard fashion. Again, differential growth might be augmented by a process of inhibition, so that while one hemisphere acquires the circuits that are modified by input, the other is prevented from doing so.

A systematic lead of the left hemisphere may also underlie the asymmetrical control of song in certain passerine birds, such as chaffinches and canaries. We saw in Chapter 6 that singing in these birds is controlled primarily by the left side of the brain through the hypoglossal nerve. Once a bird has developed its song, cutting the nerve on the left virtually destroys the song pattern, while cutting the nerve on the right leaves it more or less intact. However, if the left nerve is cut before the bird begins its spring song, the right side takes over and song develops normally.

Fernando Nottebohm, who made these discoveries, also noted that in the normal development of song, the right hypoglossal nerve seems to take over control of a few high-frequency elements that emerge late in the repertoire. Putting these findings together, he suggested that the lateralized control of song might depend on a growth gradient initially favoring the left, and that a similar gradient might also explain the asymmetrical control of language in humans.[52] Michael Morgan has summarized a number of other biological asymmetries that are consistent with earlier development of the left side.[53] For example, there is evidence that the asymmetry of the heart in vertebrates may be reversed by factors that tend to inhibit embryonic growth on the left. Morgan suggested, in fact, that it may be the *same* gradient, perhaps of molecular origin, that underlies cerebral asymmetry and the asymmetries of the internal organs.[54]

The idea that it is the left hemisphere that leads can also explain the unidirectional character of compensation following hemispherectomy. As we have seen, the right hemisphere can take over the language functions of the left, but at the expense of at least some of its usual spatial abilities. However, the left hemisphere does not seem to take over the spatial abilities of the right, but is limited to a verbal mode that is not always appropriate to spatial tasks.

Morgan and I argued, however, that in later childhood the right hemisphere may develop more rapidly than the left, and so pick up some specialized nonverbal skills that are typically acquired at that time. It is during the latter stages of childhood that children begin more actively to explore the outdoor environment and to indulge in outdoor sporting activities. This might explain why spatial skills tend to be represented in the right cerebral hemisphere. One example of a spatial skill that does seem to be predominantly right-hemispheric is mental rotation, as we saw in Chapter 10, and there is also evidence that mental rotation develops relatively late in childhood.[55]

There is some evidence too that the way in which children identify faces changes in mid- to late childhood. Adults find it very difficult to recognize faces if they are turned upside down, an effect that does not apply with equal force to other familiar shapes, such as houses. However, this disproportionate *inversion effect* does not occur until the age of about 10. It has been suggested that it marks a shift from a piecemeal encoding of faces, which may be left-hemispheric, to a more holistic, configurational code that may be right-hemispheric.[56] There is some evidence that the right-hemispheric advantage in recognizing faces does not emerge until late childhood,[57] although this does seem to conflict with the evidence reviewed earlier that children show a right-hemispheric advantage as early as 4 *months* of age.

The idea of a right-hemispheric lead in the latter part of childhood might also indirectly explain certain differences in ability between men and women. It is sometimes claimed that women tend to be inferior to men in spatial skills but superior in verbal skills, although the differences are tiny and there is considerable overlap.[58] It has also been claimed that language is represented more symmetrically in the brains of women than of men[59]—although this too is controversial.[60] To the extent that these differences are present, they might well follow from the different environmental inputs that girls and boys are exposed to in childhood. During the period of *right*-hemispheric growth in later childhood, boys may be given more freedom and encouragement to indulge in outdoor spatial pursuits, while girls are more likely to be cloistered and to engage in more verbal pursuits. Consequently, boys might be more likely to develop right-hemispheric spatial skills, while girls are more likely to develop right-hemispheric *verbal* skills to supplement the left-hemispheric verbal skills developed earlier. There is in fact some evidence for this. Donna Piazza Gordon found that the right-ear advantage in identifying dichotically presented syllables was clearly present in both boys and girls at age 9; by age 13 it was still present in boys but was no longer so in girls.[61]

According to this account, the difference between males and females in these respects is primarily a function of experience, not of biology. That is, the growth gradient favoring first the left and then the right hemisphere would be common to both sexes, with the differences due essentially to the different kinds of stimulation received in later childhood. This, in turn, may vary among cultures. Ingrid McDougall, a graduate student at the University of Auckland, tried to repeat the finding that sex differences in lateralization did not appear until adolescence but found no sex differences at all. Perhaps New Zealand girls are more tomboyish than their North American sisters.

Chapter 5 discussed the case of Genie, the girl who was deprived of social contact until the age of 13. She subsequently acquired rudimentary language, though with minimal competence in grammar. What is interesting in the present context, however, is that dichotic-listening tests revealed a strong *left*-ear advantage for words, implying that her language skills were represented in the *right* cerebral hemisphere.[62] By the time she was exposed to language her left hemisphere may have passed the critical period, leaving it to the lagging right hemisphere to pick up the remaining accessible crumbs of language.

Replying with the Right

Although this simple picture of a growth gradient favoring first the left and then the right hemisphere can capture much of the evidence, there are some complications. Even in the nineteenth century, Gratiolet's idea of an early left-hemispheric lead had been challenged.[63] Some authors found no evidence between the hemispheres in rate of growth, while others claimed a *right*-sided advantage. For example, M. J. Parrot carried out postmortem examinations of the brains of 96 infants and found that the right side was more developed in four-fifths of them.[64]

History repeated itself when some of the commentators on the theory that Morgan and I proposed, such as Harry Whitaker[65] and George Ettlinger,[66] argued that we were wrong, and that it was actually the *right* hemisphere that developed earlier. For example, there is both anatomical[67] and biochemical[68] evidence that, in the human fetus, the right temporal lobe matures before the left. We also saw earlier that some right-hemispheric skills, such as the discrimination of musical sounds or the recognition of familiar faces, are acquired in early infancy.

Norman Geschwind and Albert M. Galaburda have also noted that the right side of the brain seems to develop earlier than the left in the fetus.[69] They suggest that there is some prenatal influence that actually slows development of parts of the left hemisphere, especially around the Sylvian fissure and Wernicke's area, so that the right hemisphere and other parts of the left hemisphere develop more quickly. This influence may be testosterone or some factor related to it. Testosterone is the male hormone, in that its secretion is essential to the development of male features. However, female fetuses are also exposed to it, although in lesser amounts. Geschwind and Galaburda suggest that the role of testosterone in retarding development of parts of the left hemisphere may explain why males are more likely to be left-handed than females, why they tend to have inferior verbal abilities but superior spatial abilities, and why they are more prone to language disabilities, including stuttering and reading disability. They also suggest that testosterone suppresses a gland known as the *thymus,* increasing vulnerability to automimmune disorders.

As we saw in Chapter 8, the evidence on at least some aspects of this theory is mixed. The difference in the incidence of left-handedness between males and females seems too small to be attributable to testosterone, and seems more plausibly attributed to males being slightly more vulnerable to birth stress. Differences between males and females in verbal and spatial abilities are notoriously difficult to interpret, since even in these liberated times boys and girls may be exposed to rather different experiences, as pointed out earlier.

Nevertheless, Geschwind and Galaburda may well be correct in suggesting that early development, especially prior to birth, may favor the right rather than the left hemisphere. They suggest further that this is not unique to humans but may have a long evolutionary history, and may explain the right-hemispheric advantage in spatial processing and in emotional expression. As we saw in Chapter 10, these asymmetries may be present in species other than humans.

A more specific case for an early right-sided lead has been put by Scania de Schonen and Eric Mathivet.[70] As we saw above, de Schonen and her colleagues have found that infants as young as 4 months of age show a right-hemispheric advantage in recognizing particular faces. At this age the visual system is not properly developed, so that the child receives a rather blurred image in which fine details may be missing. Technically, these characteristics may be expressed in terms of *spatial frequencies*: High spatial frequencies carry details, while lower ones carry more global information about shapes. Consequently, the infant acquires representations of particular faces that are composed of low-frequency (blurred) information, and these representations are biased toward the right side of the brain. There is evidence that this right-hemispheric advantage for the low-frequency components of faces persists into adulthood.[71] De Schonen and Mathivet suggest that this low-frequency representation of faces is picked up by the right hemisphere, since it is more advanced than the left at this stage of development.

More Complex Patterns of Development

The right-lead theory proposed by de Schonen and Mathivet, Geschwind and Galaburda, and others is not necessarily at odds with the left-lead theory suggested by Morgan and I, since relative rates of growth on the two sides might vary from one location to another. Recall, for example, that Gratiolet had actually argued that the left lead applied to the forward parts of the brain, while in the rearward parts the right hemisphere developed earlier. This might explain some of the conflicting evidence, since speech is generally associated with more frontal portions of the brain and recognition of faces with more posterior ones.

A more complex pattern of development has been suggested by Catherine T. Best and is based on observations of anatomical asymmetrics at various stages of growth.[72] It has often been observed that the human brain shows what is sometimes described as a *counterclockwise torque* (or *twist*); as illustrated in Figure 11.2, the frontal lobe is wider and protrudes further forward on the right than on the left, while the reverse is true of the posterior part of the brain. According to Best, this is the result of a growth gradient that proceeds from right to left and from front to back, as shown in the figure.

This gradient seems to be the opposite of that proposed by Gratiolet. However, Best argues that there are two further dimensions to the gradient that must be added. One is that growth also proceeds from the bottom of the brain toward the upper regions, as well as from so-called primary to secondary to tertiary areas. The tertiary areas include both Broca's area in the prefrontal lobes and Wernicke's area in the temporoparietal region, and since these emerge late, they would emerge earlier on the *left*. Best neatly fits many of the facts of the development of asymmetries into this complex growth vector that winds its way through the developing brain.

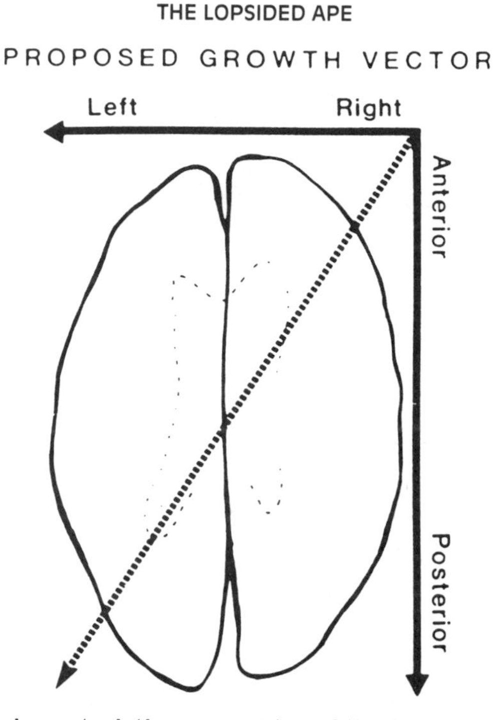

Figure 11.2. A characteristic asymmetry of the human brain, as viewed from the top. The diagram also shows two growth gradients proposed by Best; two others, not shown, are a bottom-to-top gradient and a primary-to-secondary-to-tertiary gradient (from Best, 1988).

There are still many uncertainties, however. Some asymmetries are difficult to interpret. For example, we saw above that most infants show more expressiveness on the right than on the left side of the face, which is the reverse of the asymmetry shown by most adults. This seems most naturally interpreted as an early advantage to the left side of the brain on the grounds that the left side of the brain controls the right side of the face. Still, Best herself has preferred to interpret the asymmetry as reflecting an early advantage to the *right* side, with the right better able to *inhibit* emotional expression.[73] She favors this intepretation because it fits in with other evidence for a right-sided advantage at this age. We are in an odd situation, however, when the same data can be used to support opposing theories.

Some further twists are provided by Gerald Turkewitz of the Albert Einstein College of Medicine in New York.[74] He argues that a right-hemispheric lead gives way to a left-hemispheric lead even before birth. He also cites evidence that the fetus is responsive to sound as early as 24 weeks after conception. At this time, most of the sound reaching the infant is noise produced by the gastrointestinal and cardiovascular systems of the mother and the fetus. The right hemisphere, which is the more advanced, therefore comes to be more specialized in processing noisy input.

Later on, however, the uterine wall becomes thinner and tauter, serving as an amplifier for external sounds, including speech. The speech of the mother is then especially salient in the acoustic environment of the fetus for two reasons: first, because it reaches the uterus via bone conduction as well as by airborne transmission, and second, because it is accompanied by movements that are synchronized with the mother's speech. When the left hemisphere advances in the later part of pregnancy, then, it develops a specialization for the perception of speech sounds. All this happens, according to Turkewitz, before the infant is even born!

After birth, the infant is exposed for the first time to faces, usually in association with speech. But because the left hemisphere is already engaged in the processing of speech, it is left to the right to develop a specialization for processing the faces. So where does this leave my advice to new parents? They might well consider learning the art of ventriloquism, so that their voices can be heard from the right side while their faces beam down from the left.

From 2 to 4—the Really Critical Period?

It is probably premature to draw any clear conclusions from this confusing welter of speculation about the nature of the growth gradients in early infancy and fetal life. I suspect that too much emphasis may have been placed on this early period of development, and in particular on the infant's capacity to discriminate speech sounds and its early left-hemispheric bias. This emphasis derives in part from the view, attributed largely to Alvin M. Liberman and his colleagues, that the perception of speech is special, and involves mechanisms that are uniquely human and in most people left-hemispheric.[75] We have also seen, however, that this view has been questioned. Dominic Massaro has argued that speech perception is not the special modular process that Liberman takes it to be, but obeys quite general perceptual principles.[76] Chapter 6, moreover, reviewed evidence that other species may perceive human speech sounds in much the same way that humans do.

The more critical time in the development of speech as a uniquely human skill is between the ages of 2 and 4. This is the period during which the child proceeds from two-word utterances to the production of sentences—the period during which the child acquires *grammar*. As any parent will know, it is that magical period of development when language skills seem to unfold almost daily. It is also a critical time in the growth of the brain. Noam Chomsky once remarked:

> Lots of strange things are happening in the brain between the ages of 2 and 4; there is a very rich growth of dendritic structures, and this may have something to do with the fact that language is developing. Lots of other things are happening about which very little is known, but certainly it would not be at all surprising if the regular progression we observe relates to biological phenomena having to do with this extraordinarily complex organ [of language] about which remarkably little is known, perhaps because of its complexity.[77]

In view of these observations, it is worth mentioning one study in which there was an attempt to measure the relative rates of growth on the two sides of the brain, since it suggests a pronounced *left*-hemispheric growth spurt precisely between the ages of 2 and 4. R. W. Thatcher and his colleagues took recordings from various points on the scalps of 723 children ranging in age from 2 years to early adulthood.[78] They examined the *coherence* (or, roughly speaking, the relatedness) of electrical rhythms between different areas of the brain, and also the phase differences between different areas—the lead and lag times. These measures give an indication of the development of neural connections.

With electrodes at the frontal and occipital (posterior) parts of the brain, the data showed that the left hemisphere developed much more rapidly than the right between the ages of 2 and 4. By comparison with the left, the right hemisphere developed more smoothly, although there was something of a spurt between the ages of 8 and 10. These findings lend some support to the theory that Morgan and I proposed, on the assumption that the critical period for establishing language as a uniquely human skill is between the end of infancy and the beginning of formal schooling. The later right-hemispheric spurt, although less pronounced, is also consistent with our idea that some right-hemispheric specializations for spatial ability are acquired at about this time.

This view that the period from 2 to 4 years may be the truly critical one in the establishment both of language itself and of its left-hemispheric representation takes us back too to the ideas expressed by Erich Lenneberg in *Biological Foundations of Language*. Lenneberg thought that hemispheric lateralization for language did not begin until the age of about 2, since language itself did not really begin to emerge until then. It is now known, of course, that there is a left-hemispheric advantage for the perception of speech from early infancy, but (*pace* Liberman) it may be questioned whether this has much to do with true generative language. Up to the age of 2, the two hemispheres may indeed be equally ready to pick up generative language, but it is normally the left hemisphere that does so because of its growth spurt. If the left hemisphere has been incapacitated, however, the growth spurt may be transferred to the right side; alternatively, the right hemisphere may simply acquire language in the course of its normal but later growth.

The growth spurt of the left hemisphere between the ages of 2 and 4 is not necessarily incompatible with the ideas of Geschwind and Galaburda. It may be due, in fact, to the release of the factor that inhibits the early growth of the left hemisphere, allowing it to develop rapidly at precisely the time when the developing infant is most receptive to language and is moving about, discovering objects. And it may be the process of development itself, rather than the level of maturity, that is critical in the acquiring of new skills.

Thatcher and his colleagues warn, however, that the data they present are limited, and that different areas of the two sides of the brain develop at different times and different rates. The studies I have mentioned are just a beginning. But there can now be little doubt, I think, that the two sides of the brain *do* have different schedules of growth, and that these play at least some role in the allocation of functions to one side or the other. Whether this is the whole story of cerebral asymmetry remains to be seen.

Conclusions

In this chapter, we link again with themes developed in the early part of the book. One of the characteristics that distinguishes humans from other primates is our prolonged childhood, which begins early and ends late. This allows environmental input to influence growth for an unusually long period of time. This "gene-saving" device may be especially important in the development of language; as we saw in Chapter 5, the native tongue (or tongues) that we speak seem to depend on the setting of language parameters. Their sequence and timing are biologically programmed, but the actual settings are dependent on the linguistic environment.

In this chapter, I have suggested that growth may be guided by systematic gradients that include a left-right component, perhaps controlled by the RS+ allele postulated by Marian Annett. Although this theory remains speculative, it can account, at least in principle, for lateralization of function. In particular, the period between ages

2 and 4 may be especially critical for the development of generative language, and we saw that there is some evidence for a pronounced spurt of left-hemispheric growth at this time.

Although I have focused on language, this is also the age range during which children begin to play with blocks and construction toys. More generally, then, this same period may therefore be critical for the development of generativity, in the representation of objects as well as in language. That is, GAD may be represented in the left hemisphere, not because of any innate *structural* properties of that hemisphere, but because a rapid period of growth in the left hemisphere normally coincides with the period during which generative skills emerge.

Notes

1. Jones (1919, p. 289).
2. Gesell and Ames (1947); Michel (1981).
3. Lewkowicz and Turkewitz (1983).
4. Liederman (1987).
5. Melekian (1981). Not all investigators have found this asymmetry—see Peters (1988).
6. Moss (1929); Taylor (1976). There is a Chinese belief, mentioned by Delaunay (1874), that the fetus moves its right hand at 7 months and its left hand at 8 months (cited by Harris, 1984).
7. Hughey (1985).
8. Overstreet (1938).
9. See Michel (1983) for a review.
10. Vles, Grubben, and Hoogland (1989).
11. Caplan and Kinsbourne (1976).
12. Petrie and Peters (1980).
13. Gesell and Ames (1947).
14. See, for example, Hall (1891), Meyer (1913), Seth (1973), and Woolley (1910).
15. For example, Young (1977).
16. Annett (1985, p. 66).
17. Op. cit.
18. See Maurer (1985) for a review.
19. Entus (1977).
20. Glanville, Best, and Levenson (1977).
21. Best, Hoffman, and Glanville (1982). These authors suggest that their result might explain why Vargha-Khadem and Corballis (1979) were unable to repeat Entus' findings with speech sounds. Most of the infants studies by Entus were over 2 months old, while Vargha-Khadem and Corballis used a younger sample, averaging 2 months of age.
22. Best (1988).

23. Bertocini, Morais, Bijeljac-Babic, McAdams, Peretz, Morais, and Mehler (1989).
24. MacKain, Studdert-Kennedy, Spieker, and Stern (1983).
25. De Schonen, Gil de Diaz, and Mathivet (1986); de Schonen and Mathivet (1989). Best (1988) cites Witelson and Barrera as having obtained a similar result.
26. Molfese, Freeman, and Palermo (1975).
27. Gardiner and Walter (1977).
28. Geschwind and Levitsky (1968).
29. Wada, Clarke, and Hamm (1975).
30. Chi, Dooling, and Gilles (1977).
31. Best and Queen (1989); Rothbart, Taylor, and Tucker (1989).
32. For a critical review of the development of lateralization, see Hiscock (1988).
33. Marie (1922, p. 180), cited in translation by Dennis and Whitaker (1977, p. 93).
34. This is the usual term, although the term *hemidecortication* is technically more correct since subcortical structures are typically not removed.
35. Basser (1962).
36. Burklund and Smith (1977).
37. Zaidel (1985).
38. Corballis and Morgan (1978).
39. Romer (1962).
40. Dennis and Kohn (1975). See also Dennis and Whitaker (1976).
41. Bishop (1983).
42. Bishop (1988).
43. Ogden (1988).
44. Ogden (1989).
45. Kohn and Dennis (1974).
46. Kennard (1936).
47. See Kolb (1989) for review and discussion.
48. See Banich, Levine, Kim, and Huttenlocher (1990) for review and discussion.
49. Leurat and Gratiolet (1857).
50. Corballis and Morgan (1978). We wrote in ignorance of the nineteenth-century ideas of Gratiolet, Broca, and others, which were later pointed out by Harris (1984).
51. Morgan and Corballis (1978).
52. Nottebohm (1972).
53. Morgan (1977).
54. This theme was pursued by Corballis and Morgan (1978).
55. See Corballis (1982) for a review.
56. For review and discussion of these issues, see Levine (1985).
57. Leehey (1976); Phippard (1977); Reynolds and Jeeves (1978). However, not all studies have shown this—see Hiscock (1988) for a review.
58. Sherman (1978).
59. McGlone (1980).

60. Shallice (1988).
61. Gordon (1983).
62. Fromkin, Krashen, Curtiss, Rigler, and Rigler (1974).
63. Again, this was unknown to us, and apparently to 60 or so original commentators on our articles. We are again indebted to Harris' (1984) later commentary for resurrecting the past.
64. Parrot (1879).
65. Whitaker (1978).
66. Ettlinger (1978).
67. Chi, Dooling, and Gilles (1977).
68. Bracco, Tiezzi, Ginanneschi, Campanella, and Amaducci (1984).
69. Geschwind and Galaburda (1987).
70. De Schonen and Mathivet (1989).
71. Sergent (1986, 1988); Sergent and Hellige (1986).
72. Best (1988).
73. Best and Queen (1989); Rothbart et al. (1989) interpret their results similarly.
74. Turkewitz (1988).
75. For example, Liberman (1982); Liberman and Mattingly (1989).
76. Massaro (1987).
77. Chomsky, in Piattelli-Palmarini (1980, pp. 169–170).
78. Thatcher, Walker, and Giudice (1987).

—12—

Conclusions

O most lame and impotent conclusion!
Othello

THE HOMINID LINE diverged from the apes somewhere between 4 and 8 million years ago. Genetically speaking, we remain very close to our nearest ape relative, the chimpanzee. Yet on the face of it we are very different. Even in terms of gross physical characteristics, chimpanzees resemble gorillas more than they resemble humans, though the molecular evidence suggests that they are closer to humans. And where humans are expansive creatures who have populated the globe—overpopulated it, perhaps—chimpanzees are an endangered species drawn to ever-shrinking habitats. Humans also display remarkable control, not only over the environment but also over other species. We may have originally scavenged for meat, but we have evolved into large-scale predators, holding a great variety of animals captive for the kill. We also hold animals captive for labor and amusement. What was it that shaped humans so differently in the course of a few million years of evolution, and with so little genetic change?

It all began when our forebears stood upright and walked about on their hind legs. Bipedal walking seems to have been the main defining characteristic of the hominids as far back as the prehistoric record takes us, to *A. afarensis* some 3 to 4 million years ago. The reasons for it are obscure; perhaps it was a result of an aquatic phase, or perhaps just an adaptation to the changing terrain of East Africa in the late Miocene and early Pliocene. But whatever the reason, bipedalism had a profound effect on human destiny.

Bipedalism completed a trend, already begun in the primates, toward freeing the arms and hands from involvement in weight bearing and locomotion. This allowed the hominids to carry things as they walked, something other apes can do only with difficulty since they cannot walk easily on two legs. The ability to carry no doubt suited a scavenging lifestyle in which the adults gathered food for themselves and their young. Again, there is dispute as to whether food was merely scavenged from carcasses or whether the early hominids were hunters who killed for their supper, and dispute as to whether it was the males or the females who collected food. There may be no single answers to these questions, because the answers probably changed over time and because there were several species of early hominid who probaby adapted in different ways. One critical distinction is between the *robust* form that relied mainly on a coarse vegetarian diet and the *gracile* form that included meat as part of the diet. The robust australopithecines eventually became extinct, while the gracile ones are presumed to have evolved into the species *Homo*.

The upright stance freed the hands not only for carrying but also for using and manufacturing tools. The use of sticks and other objects as implements is common to all primates, and indeed to many other species, and the early australopithecines may not have differed much from the apes in this respect. But with *H. habilis*, beginning something over 2 million years ago, we find the first evidence of stone tools, apparently systematically designed and shaped for the cutting and scraping of carcasses. Tools and their products were to grow in sophistication during the evolution of *Homo*, and manufacture is now perhaps the most dominant feature of human society. Yet it should be remembered that the development of stone tools was at first very slow. It was not really until about 30,000 to 40,000 years ago, with the so-called evolutionary explosion, that manufacture really took off.

In the early evolution of *Homo*, the dramatic increase in brain size was more impressive than the development of tools. *Habilis* already had a larger brain than the australopithecines, but in a period of only a few hundred thousand years, with the emergence of *H. erectus*, brain size showed a further significant increase. It reached its present size with the arrival of *H. sapiens* some 300,000 years ago. Yet during this period the variety and sophistication of stone tools seem to have altered very little, at least compared with the dramatic changes that were to follow—*after* the brain reached its maximum size. This suggests that selection for larger brains may have had little to do with the development of tools and manufacture.

Perhaps it was just the more complex social life that favored larger brains. By the time of *H. erectus* some 1.6 million years ago, the scavenging lifestyle seems to have intensified, and the discovery, disposal, transport, and sharing of meat may have favored a cooperative spirit among groups of adults. At the same time, the nuclear family may have emerged within the larger social units. In this hierarchically organized society, different members would have played different roles, requiring flexibility, learning ability, and the heightened awareness of others (which may have led, reciprocally, to an awareness of self).

By the time of *H. erectus* there is also evidence for the prolonged infancy that characterizes our species. This means that much of the growth of the brain occurs outside of the womb, while the infant is exposed to the vagaries of the environment. Moreover, the physical helplessness of the infant means that the parents are a continuing and salient part of that environment. Since the brain is most plastic during growth, prolonged infancy and childhood would have optimized the molding of people to different roles, largely determined by their parents or caregivers. But early birth may not have been selected simply because of the advantages associated with increased plasticity. The upright stance and bipedal walking may have imposed mechanical restrictions on birth, and infants may have had to be born before their heads grew too large. As in so much of evolution, there was probably not a simple causal chain. Large brains, early birth, bipedal walking, complex social structure—these facets of humanity may have had a synergistic relation to each other that led to rapid evolutionary change.

Perhaps the most critical aspect of the complex social structure that emerged in the species *Homo* was communication, ultimately to become what we now know as language. Despite the argument of some to the contrary, it is my guess that language did evolve gradually, perhaps from *H. habilis* on. The evidence for specific language areas in the left hemisphere of the brain seems to go back to *habilis*. Language also depends on critical periods of growth, so that complex language requires a long childhood. Language may therefore have been quite well developed by the time of *H. erectus* at least.

I have suggested, however, that language was primarily accomplished by manual gestures, at least from *habilis* through archaic *H. sapiens*. No doubt these creatures did vocalize, but at first their vocalizations were probably restricted to the stereotyped calls of other primates. The arms and hands, freed from involvement in locomotion, would have provided the most obvious and flexible means of communication, especially about newly evolved aspects of the lifestyle. The hands could be used to point to objects or other individuals, to indicate the way to places where food might be found, and to demonstrate the use of stone tools. Surely there were many more different gestures that could be made than sounds that could be uttered; gestures also had the advantage that they were not already assigned to built-in functions, as most animal calls are. In short, gesture may have provided the basis for a new form of communication that we may characterize as *propositional.*

However, vocalizations may have accompanied manual gestures, perhaps initially in the form of reflexive grunts induced by physical action or perhaps to attract attention. At some point in evolution vocalization became the dominant mode, relegating manual communication to a minor, supportive role (except in the case of sign languages used by the deaf). I suggest that the switch occurred with the emergence of *H. sapiens sapiens*, or African Eve—the first anatomically modern human. Indeed, vocal language may have been the most important characteristic distinguishing *H. sapiens sapiens* from the more archaic forms and the one that led to their gradual domination.

Vocal speech has several advantages over manual gesturing. It can be carried on at night or when obstacles prevent communicating parties from seeing one another. By shouting, one can communicate over larger distances than by gesturing, and one can more easily address large groups of people. But most important, vocal speech would have freed the hands again, this time from any primary role in language. It would have also freed the eyes from this role. One can talk *about* some tool-making procedure while demonstrating it, or listen while watching it.

I suggest that this final freeing of the hands might have led to a renewed surge of manufacture. Manual skills developed initially in the context of gesture might now be exapted for other purposes, such as sewing, pottery, carving, drawing, or the playing of musical instruments. And the *generativity* that may have evolved in gestural language could now be applied to manufacture. Our clever, manipulative, mischievous hands have simply cut loose and fashioned new environments for humans to inhabit.

The artificial environment has also altered the way we *think*. The proliferation of objects means that we need different and more economical ways to represent them in our minds. Representations need to be flexible since the environment itself is always changing, and a vocabulary-based system provides just this flexibility. Instead of representing each object with a unique template of some sort, a vocabulary-based system allows one to represent virtually anything in terms of different combinations of the basic units of the vocabulary, and indeed, to construct rapidly representations for new objects that we may encounter. Our vocabulary of some 44 phonemes allows us to generate an unlimited number of meaningful utterances, just as the vocabulary of 36 geons proposed by Biederman provides for representation of an unlimited number of objects, real and potential.

This generative mode of representation, which I have attributed to a generative assembling device, or GAD, may be uniquely human, and in most of us it depends primarily on the left cerebral hemisphere. It may, however, have emerged from a left-hemispheric specialization for praxis that goes back much further in evolution. The vocabulary-based, generative mode may nevertheless be contrasted with an earlier, holistic mode of representation that is primarily right-hemispheric. This holistic mode is still valuable, since it is able to capture subtleties of shape that cannot be represented by standardized geons. This applies particularly to shapes in the natural world as distinct from the manufactured one. In the manufactured world we are apt to reproduce the idealized shapes of the geons themselves—buildings, furniture, utensils, vehicles, and so on all tend to have clean, simple, regular shapes that resemble idealized geons or assemblages of geons. The natural world, by contrast, has a more random, textured, *fractal* quality. Neither words nor geons can easily capture the variety of human faces, say, or the texture of a forest.

Generative, vocabulary-based representations are computational in that they follow rules and are organized hierarchically. It is likely that they are acquired during growth as a planned sequence of what has been termed *parameter fixing*. In the development of language, for example, the first year of life may be dedicated primarily to the tuning of the phonological systems, so that the infant learns the sounds of his or her native language. In the second year the infant picks up a basic vocabulary of single words. But the most critical period in the development of generative language occurs between the ages of 2 and 4, when the child develops the ability to construct grammatical sentences. The left hemisphere appears to undergo a growth spurt at this time, and this may be the primary mechanism that normally ensnares language in the left brain. This is also the time at which children begin to play constructively with building blocks and toys, perhaps indicating that the left hemisphere is establishing its vocabulary of geons.

The manner in which language is acquired in childhood is at once "gene-saving" and genetically controlled. The genes probably control the sequence in which different brain areas grow and constrain the nature of the information that is extracted from the linguistic environment at each stage. However, the linguistic environment itself specifies the actual language or languages that the individual learns, so that this information does not have to be coded genetically. This mechanism is also at once restrictive and flexible. Any normal child *could* learn any language but in fact learns perhaps one or two. Other skills no doubt follow similar developmental principles. This critical role of the environment during growth is probably the main factor underlying the extraordinary differences among people in culture, language, skill, and perhaps even perception. In large measure, these aspects of the human condition are set by puberty, and thereafter people are largely set in their ways.

This summary has glossed over many of the issues discussed in earlier chapters and will no doubt turn out to be wrong in some of the details. However, I have tried to present an overall picture that most plausibly portrays the evidence that is currently available—or at least as it seems to me. The next and critical question is whether the evolution of humans from the common ancestor we share with the chimpanzees represents a *discontinuity*.

A Discontinuity?

As we saw in Chapter 1 the question is a loaded one, and we humans have a strong vested interest in believing that we are fundamentally different, not only from the apes, but from all other species. Exploitation of other species is part of our human heritage and often coexists uneasily with another human disposition, that of cooperation and empathy. Religions may owe their existence in part to the need to resolve this conflict (and in part, perhaps, to the need to come to terms with the knowledge of death). But with the notable exception of such figures as Saint Francis of Assisi, the resolution has generally been to the disadvantage of animals, which are deemed to be lacking in grace or in any of the higher virtues.

Even without the blessing of religions, we are inclined to regard ourselves as "higher" than other species. Charles Darwin exhorted himself not to use the terms *higher* and *lower* in discussing evolution, but in fact he did so constantly, especially when discussing human evolution. I too have no doubt failed in this respect. The very notion of ourselves as the highest organism is, of course, dependent on our own definitions as to what is high, and we naturally tend to give weight to activities such as speaking, making things, or displaying drawing-room manners and neglect such superbly skilled activities as flying, navigating by echolocation, or spinning webs.

Our superiority is often asserted in terms of consciousness, free will, or the concept of self. These notions are notoriously difficult to define, to the point where one suspects that they may have been invented for the express purpose of arguing for human superiority. But to the extent that they are meaningful concepts, there seems no sure proof that they are not possessed by other animals, particularly such near relatives as the higher apes. Since the pioneering experiments of Wolfgang Kohler, described in his classic work *The Mentality of Apes*, there seems little doubt that apes can solve problems in purposeful fashion. Modern experiments on apes and dolphins may have failed to demonstrate that these creatures can learn generative grammar, but they leave little doubt as to their intelligence and resourcefulness.

There is also no good reason to link consciousness to our lopsided brains. Although some have argued that consciousness is a property of the left hemisphere, others have argued that it is the preserve of the right! In fact, it seems that the two hemispheres simply represent rather different facets of the sorts of mental processes we regard as conscious. The right hemisphere seems to play the more prominent role in our awareness of space and in the control of emotion, while the left hemisphere is dominant in the more abstract, propositional processes of language and reasoning.

Even the awareness of self does not seem to be confined to one or the other hemisphere, at least from the evidence of split-brain studies and of testing under the effects of sodium amytal injection. However, the two hemispheres may differ in the *ways* in which the self is represented. For example, there is a remarkable form of agnosia, known as *autotopagnosia*, that is characterized by the inability to identify one's own body parts, even though the body parts of others can be identified. There is evidence that this occurs as a result of left-hemispheric damage.[1] It may be simply a further illustration of the specialization of the left hemisphere for partwise representation.

If human uniqueness is not to be found in consciousness or the awareness of self, I have argued nevertheless that there is a fundamental discontinuity between ourselves and other species. It has to do primarily with GAD—the vocabulary based, generative assembling device. It emerged, of course, from a number of other distinctively hominid characteristics, including bipedalism, prolonged infancy and childhood, large brain, and complex social organization. But it was GAD that truly set us apart. It is the basis of a discontinuity that is at least of the order of that which distinguishes animals that fly from those that do not. The proof of this is that GAD *did* eventually permit us to fly, in airplanes and helicopters and space capsules, and only as a particular case of a very general power that it bestows upon us.

That power is an adaptability that goes beyond the demands of the natural environment. Through language we can generate accounts not only of events that do occur but also of those that *might* occur, and even of those that *cannot* occur. Similarly, we can imagine real, potential, and impossible objects—and at a higher level real, potential, and impossible scenes. The flexibility and recursiveness of GAD were not, I think, the result of some fortuitous genetic reshuffle, but were rather an adaptation to a lifestyle in which the hominids gradually assumed control over their own environment. The evolution of characteristics that were adapted to *specific* aspects of the environment would have been too slow to cope with the rapid change that we conferred on ourselves. What we evolved, then, was a more general capacity for adaptation itself.

Manufacture, then, has allowed us to adapt to any environment on earth, and even to extraterrestrial environments, not by bodily change but by changing the environment. We make clothes, construct buildings, install heating and cooling systems, arrange the transportation of food and other necessities, remove or control hostile elements, and so on. Other animals also exert some control, as in such activities as building nests, spinning webs, damming rivers, and the like, but in comparison with human construction these examples are highly stereotyped and inflexible.

GAD has also altered our processes of thought. Although I have argued that consciousness itself is probably not uniquely human, GAD has no doubt expanded the nature of human consciousness, endowing it with a flexibility that is not possible in fixed representational systems. One property of the rules underlying the operation of GAD that may be especially important is recursion. As we saw in Chapter 5, recursion allows us to embed phrases within phrases or even sentences within sentences. Recursion may be a more general property of generative thought, and indeed may allow us to think about thought itself—which is precisely one of the things I have tried to do in this book. My guess is that this recursive quality of thought is uniquely human and is predominantly left-hemispheric.

This recursive quality may also bestow a special kind of self-consciousness. Although humans may not be unique in their capacity for self awareness, they may be unique in being able to use the concept of self recursively. We may therefore know that we know, decide whether to decide, and so on. In such ways, recursion effectively permits introspection, so that we can inspect the act of thought itself, and not just the objects of that thought. The highest executive level of thought is presumably not accessible to inspection, however, for if it were, there would have to be a still higher level that inspected it!

It may be important to distinguish GAD from *creativity*, which is another of those diffusely defined qualities that we humans are apt to bestow uniquely upon ourselves. When we generate a new sentence (such as this one), we are being creative in the literal sense that the sentence has never been generated before, but the process is more or less effortless and rule-governed. However, by creativity we normally mean something rather different from this; for example, creativity in art or poetry takes us beyond the rules of grammar or construction, perhaps to the discovery of new rules. As we saw in Chapter 10, it is sometimes claimed that creativity is a function of the right hemisphere rather than of the left, although it has also been claimed that both are involved.[2] My guess is that, in creativity as in other aspects of thought, each hemisphere makes its own contribution. The left hemisphere may be critically involved in verbal creativity and perhaps in the invention of functional objects, while the right plays the major role in the spatial arts.

For all that, GAD may greatly raise the stakes of creative effort. GAD is a powerful tool of creativity, but it is not creativity itself. The mechanics of generating a sentence or a visual scene are rule-governed and for the most part automatic, but they allow creativity to operate at a high level. In writing a book, for example, it may take creative effort to choose the topics and themes, but the sentences are governed by rules. Similarly a skilled architect can rise above the materials and the mechanics of construction. The processes of creativity and executive control may be common to all higher animals; what sets us apart is the complexity of operations that lie just beneath the executive surface.

My argument for a discontinuity between ourselves and other animals should not be taken as an argument for the exploitation of animals or as some kind of rebuttal to the animal rights movement. I have little doubt that animals are conscious, and experience pain, discomfort, and even embarrassment. Animals possess intricate skills that humans do not. And although I have argued that GAD bestows a quality of thought that other animals do not possess, there is still a great deal of overlap in the mental processes of humans and animals. We still can learn much about the nature of human thought from observations of other animals. The question of whether and when research with animals is justified remains an extraordinarily complex one and is beyond the scope of this book. If anything, my arguments concerning GAD reveal humans to be uniquely exploitative, and recognition of this quality might serve to temper our tendency to exploit other creatures, as well as each other.

The final question, then, must be: Is GAD a benevolent influence? As a result of GAD, our own species has thrived where our closest relative, the chimpanzee, has not. But there is, of course, the danger that our very success may now threaten our existence too. The human population has grown to the point where there are no longer the natural resources to sustain us all. GAD has always been seconded for the purposes of war but now has the power to exterminate friends and enemies alike. Through the influence of GAD, the natural environment is giving way to the artificial environment of cities and highways and flight lanes; in the modern world, nature is not so much harnessed as dominated, and we are beginning to discern the harmful effects of air pollution, acid rain, and distortion of climate. Our lopsidedness may be out of hand, as it were, and we need to find ways to restore the balance with nature.

Notes

1. Ogden (1986).
2. Bogen and Bogen (1988).

References

Adcock, F. (1986). *The incident book*. Oxford: Oxford University Press.

Aimard, G., Devick, M., Lebel, M., Trouillas, P., and Boisson, D. (1975). Agraphie pure (dynamique?) d'origine frontale. *Revue Neurologique, 131*, 505–512.

Akelaitis, A. J. (1941). Studies on the corpus callosum. II: The higher visual functions in each homonymous field following complete section of the corpus callosum. *Archives of Neurology and Psychiatry, 45*, 789–796.

Akelaitis, A. J. (1944). The study of gnosis, praxis, and language following section of the corpus callosum and anterior commissure. *Journal of Neurosurgery, 1*, 94–102.

Alajouanine, J. (1948). Aphasia and artistic realization. *Brain, 71*, 229–241.

Alajouanine, T. (1956). Verbal realization in aphasia. *Brain, 79*, 1–128.

Alvarez, L.W., Alvarez, W., Asaro, F., and Michel, H.V. (1980). Extraterrestrial cause for the Cretaceous-Tertiary extinction. *Science, 208*, 1095–1108.

Anderson, N.G. (1970). Evolutionary significance of viral infection. *Nature, 227*, 1346–1347.

Andral, G. (1840). *Clinique médicale*. Paris: Fortin, Masson.

Andrews, G., Quinn, P.T., and Sorby, W.A. (1972). Stuttering: An investigation into cerebral dominance for speech. *Journal of Neurology, Neurosurgery, and Psychiatry, 35*, 414–418.

Annett, M. (1985). *Left, right, hand and brain: The right shift theory*. London: Erlbaum.

Annett, M., and Manning, M. (1989). *Reading and laterality in primary school children*. Paper presented at the International Conference on Cognitive Neuropsychology, Harrogate, England.

Annett, M., and Manning, M. (1990). Arithmetic and laterality. *Neuropsychologia, 28*, 61–70.

Ardrey, R. (1976). *The hunting hypothesis*. New York: Atheneum.

Arensburg, B., Tillier, A.M., Vandermeersch, B., Duday, H., Schepartz, L.A., and Rak, Y. (1989). A middle Palaeolithic hyoid bone. *Nature, 338*, 758–760.

Argyle, M. (1975). *Bodily communication*. London: Methuen.

Asfaw, B. (1987). The Belohdelie frontal: New evidence of early cranial morpholgy from the Afar of Ethiopia. *Journal of Human Evolution, 16*, 611–624.

Babinski, M.J. (1914). Contribution à l'étude des troubles mentaux dans l'hémiplegie organique cérébrale (Anosognosie). *Revue Neurologique, 12*, 845–848.

Bakan, P. (1971). Birth order and handedness. *Nature, 229*, 195.

Bakan, P., Dibb, G., and Reed, P. (1973). Handedness and birth stress. *Neuropsychologia, 11*, 363–366.

Banich, M.T., Levine, S.C., Kim, H., and Huttenlocher, P. (1990). The effects of developmental factors on IQ in hemiplegic children. *Neuropsychologia, 28*, 35–47.

Barsley, M. (1970). *Left-handed man in a right-handed world.* London: Pitman.

Barton, M., Goodglass, H., and Shai, A. (1965). Differential recognition of tachistoscopically presented English and Hebrew words in right and left visual fields. *Perceptual and Motor Skills, 21*, 431–437.

Basser, L.S. (1962). Hemiplegia with early onset and the faculty of speech with special reference to the effects of hemispherectomy. *Brain, 85*, 427–460.

Beaton, A. (1985). *Left side, right side; A review of laterality research*. London: Batsford.

Beck, B.B. (1980). *Animal tool behavior: The use and manufacture of tools by animals*. New York: Garland STPM Press.

Beck, C.H.M., and Barton, R.L. (1972). Deviation and laterality of hand preference in monkeys. *Cortex, 8*, 339–363.

Behrens, S.J. (1989). Characterizing sentence intonation in a right hemisphere-damaged population. *Brain and Language, 37*, 181–200.

Bellugi, U., and Klima, E.S. (1982). From gesture to sign: Deixis in a visual gestural language. In R.J. Jarvella and W. Klein (Eds.), *Speech, place, and action: Studies of language in context* (pp. 297–313). New York: Wiley.

Benton, A.L. (1975). Developmental dyslexia: Neurological aspects. In W.J. Friedlander (Ed.), *Advances in neurology* (Vol. 7, pp. 1–47). New York: Raven.

Benton, A.L. (1977). The amusias. In M. Critchley and R.A. Henson (Eds.), *Music and the brain*. Springfield, IL: Charles C. Thomas.

Benton, A.L. (1984). Hemispheric dominance before Broca. *Neuropsychologia, 22*, 807–811.

Benton, A.L., Hannay, H.J., and Varney, N.R. (1975). Visual perception of line direction in patients with unilateral brain disease. *Neurology, 25*, 907–910.

Bermudez de Castro, J.M., Bromage, T.G., and Jalvo, Y.F. (1988). Buccal striations on fossil human anterior teeth: Evidence of handedness in the middle and early Upper Pleistocene. *Journal of Human Evolution, 17*, 403–412.

Bertoncini, J., Morais, J., Bijeljac-Babic, R., McAdams, S., Peretz, I., and Mehler, J. (1989). Dichotic perception and laterality in neonates. *Brain and Language, 37*, 591–605.

Best, C.T. (1988). The emergence of cerebral asymmetries in early human development: A review and a neuroembryological model. In D.L. Molfese and S.J. Segalowitz (Eds.), *Brain lateralization in children: Developmental implications* (pp. 5–34). New York: Guilford.

Best, C.T., Hoffman, H., and Glanville, B.B. (1982). Development of ear asymmetries for speech and music. *Perception and Psychophysics, 31*, 75–85.

Best, C.T., and Queen, H.F. (1989). Baby, it's in your smile: Right hemiface bias in infant emotional expressions. *Developmental Psychology, 25*, 264–276.

Bever, T., and Chiarello, R.J. (1974). Cerebral dominance in musicians. *Science, 185*, 537–537.

Bhatt, R.S., Wasserman, E.A., Reynolds, W.F., Jr., and Knauss, K.S. (1988). Conceptual behavior in pigeons: Categorization of both familiar and novel examples from four classes of natural and artificial stimuli. *Journal of Experimental Psychology: Animal Behavior Processes, 14*, 219–234.

Bianki, V.L. (1988). *The right and left hemispheres of the animal brain*. New York: Gordon and Breach.

Bickerton, D. (1984). The language bioprogram hypothesis. *Behavioral and Brain Sciences, 7*, 173–221.

Bickerton, D. (1986). More than nature needs? A reply to Premack. *Cognition, 23*, 73–79.

Biederman, I. (1987). Recognition-by-components: A theory of human image understanding. *Psychological Review, 94*, 115–147.

Bishop, D.V.M. (1983). Linguistic impairment after left hemidecortication for infantile hemiplegia? A reappraisal. *Quarterly Journal of Experimental Psychology, 35A*, 199–207.

Bishop, D.V.M. (1987). Is there a link between handedness and hypersensitivity? *Cortex, 22*, 289–296.

Bishop, D.V.M. (1988). Can the right hemisphere mediate language as well as the left? A critical review of recent research. *Cognitive Neuropsychology, 5*, 353–367.

Bishop, D.V.M. (1989). Does hand proficiency determine hand preference? *British Journal of Psychology, 80*, 191–199.

Bisiach, E., Capitani, E., Luzzatti, C., and Perani, D. (1981). Brain and conscious representation of outside reality. *Neuropsychologia, 19*, 543–522.

Bisiach, E., and Luzzatti, C. (1978). Unilateral neglect of representational space. *Cortex, 14*, 129–133.

Bisiach, E., Luzzatti, C., and Perani, D. (1979). Unilateral neglect, representational schema, and consciousness. *Brain, 102*, 609–618.

Bisiach, E., Mini, M., Sterzi, R., and Vallar, G. (1982). Hemispheric lateralization of the decisional stage in choice reaction times to visual unstructured stimuli. *Cortex, 18*, 191–198.

Bisiach, E., and Vallar, G. (1988). Hemineglect in humans. In F. Boller and J. Grafman (Eds.), *Handbook of neuropsychology* (Vol. 1, pp. 195–222). Amsterdam: Elsevier Science Publishers.

Blakemore, C., Iversen, S.D., and Zangwill, O.L. (1972). Brain functions. *Annual Review of Psychology, 23,* 413–456.

Blau, A. (1946). *The master hand.* New York: American Orthopsychiatric Association.

Blumenberg, B. (1983). The evolution of the advanced hominid brain. *Current Anthropology, 24,* 589–623.

Blumstein, S., and Cooper, W.E. (1974). Hemisphere processing of intonation contours. *Cortex, 10,* 146–158.

Boesch, C., and Boesch, H. (1983). Optimization of nut-cracking with natural hammers by wild chimpanzees. *Behavior, 83,* 265–285.

Boesch, C., and Boesch, H. (1984). Mental map in chimpanzees: An analysis of hammer transport for nutcracking. *Primates, 25,* 160–170.

Bogen, J.E. (1969a). The other side of the brain I: Dysgraphia and dyscopia following cerebral commissurotomy. *Bulletin of the Los Angeles Neurological Society, 34,* 73–105.

Bogen, J.E. (1969b). The other side of the brain II: An appositional mind. *Bulletin of the Los Angeles Neurological Society, 34,* 135–162.

Bogen, J.E. (1985). The callosal syndromes. In K.M. Heilman and E. Valenstein (Eds.), *Clinical neuropsychology* (2nd edition, pp. 295–338). Oxford: Oxford University Press.

Bogen, J.E., and Bogen, G.M. (1969). The other side of the brain III: The corpus callosum and creativity. *Bulletin of the Los Angeles Neurological Society, 34,* 191–220.

Bogen, J.E., and Bogen, G.M. (1988). Creativity and the corpus callosum. *Psychiatric Clinics of North America, 11,* 293–301.

Bolk, L. (1926). *Das Problem der Menschwerdung.* Jena: Gustav Fischer.

Boller, F., and Grafman, J. (1983). Acalculia: Historical development and current significance. *Brain and Cognition, 2,* 205–223.

Bordon, G.J. (1983). Initiation versus execution time during manual and oral counting by stutterers. *Journal of Speech and Hearing Research, 26,* 389–396.

Borod, J.C., and Caron, H.S. (1980). Facedness and emotion related to lateral dominance, sex, and expression type. *Neuropsychologia, 18,* 237–242.

Boswall, J. (1977). Tool using by birds and related behavior. *Aviculture Magazine, 83,* 88–97, 146–159.

Bouillaud, J.B. (1825). *Traité clinique et physiologique de l'encéphale.* Paris: J.B. Baillière.

Boyd, W. (1956). *Emile for today: The Emile of Jean Jacques Rousseau.* London: Heinemann.

Bracco, L., Tiezzi, A., Ginanneschi, A., Campanella, C., and Amaducci, L. (1984). Lateralization of choline acetyltransferase (ChAT) activity in fetus and adult human brain. *Neuroscience Letter, 50,* 301–305.

Brackenridge, C.J. (1981). Secular variations in handedness over ninety years. *Neuropsychologia, 19,* 459–462.

Bradshaw, J.L., Burden, V., and Nettleton, N.C. (1986). Dichoptic and dichaptic techniques. *Neuropsychologia, 24,* 74–90.

Bradshaw, J.L., and Nettleton, N.C. (1983). *Human cerebral asymmetry.* Englewood Cliffs, NJ: Prentice-Hall.

Bradshaw, J.L., Nettleton, N.C., and Taylor, M.J. (1981). Right hemisphere language and cognitive deficit in sinistrals? *Neuropsychologia, 19*, 113–132.

Brain, C.K. (1981). *The hunters or the hunted? An introduction to African cave taphonomy*. Chicago: University of Chicago Press.

Brain, C.K. (1986). Interpreting early hominid death assemblies: The rise of taphonomy since 1925. In P.V. Tobias (Ed.), *Hominid evolution, past, present, and future* (pp. 41–46). New York: Alan R. Liss.

Bresard, B., and Bresson, F. (1987). Reaching or manipulation: Left or right? *Behavioral and Brain Sciences, 10*, 265–266.

Brinton, D.G. (1896). Left handedness in North American aboriginal art. *American Anthropologist, 9*, 175.

Briquet, P. (1859). *Traité clinique et thérapeutique de l'hystérie*. Cited by Harrington (1987).

Broca, P. (1861). Remarques sur la siege de la faculte du langage articule, suivies d'une observation d'aphemie. *Bulletin de la Societe anatomique de Paris, 2*, 330–357.

Broca, P. (1864). Deux cas d'aphemie traumatique, produite par des lesions de la troisieme circonvolution frontale gauche. *Bulletin de la Societe de Chirurgie, 5*, 51–54.

Broca, P. (1865). Sur la siege de la faculte du langage articule. *Bulletins de la Societe d'Anthropologie de Paris, 6*, 377–393.

Bronowski, J., and Bellugi, U. (1970). Language, name, and concept. *Science, 168*, 669–673.

Brown, F., Harris, J., Leakey, R., and Walker, A. (1985). Early *Homo erectus* skeleton from west Lake Turkana, Kenya. *Nature, 316*, 788–792.

Brown, K. (1984). *Linguistics today*. Bungay, Suffolk, England: Fontana Paperbacks.

Brown, R.W. (1957). *Words and things*. Glencoe, IL: Free Press.

Bruner, J.S. (1968). *Processes of cognitive growth: Infancy*. Worcester, MA: Clark University Press.

Bruner, J.S. (1983). *Child's talk: Learning to use the language*. Oxford: Oxford University Press.

Bryden, M.P. (1976). Response bias and hemispheric differences in dot localization. *Perception and Psychophysics, 19*, 23–28.

Bryden, M.P. (1982). *Laterality: Functional asymmetry in the intact brain*. New York: Academic Press.

Buchanan, A. (1862). Mechanical theory of the preponderance of the right hand over the left; or, more generally, of the limbs of the right side over the left side of the body. *Proceedings of the Philosophical Society of Glasgow, 5*, 142–167.

Buckingham, H.W., Jr., and Kertesz, A. (1974). A linguistic analysis of fluent aphasia. *Brain and Language, 1*, 29–42.

Buettner-Janusch, J. (1963). An introduction to the primates. In J. Buettner-Janusch (Ed.), *Evolutionary and genetic biology of primates*. Vol. 1 (pp. 1–64). New York: Academic Press.

Buffon, G.L.L., comte de. (1792). *Buffon's natural history*. London: J.S. Barr.

Bures, J., Buresova. O., and Krinavek, J. (1981). An asymmetric view of brain laterality. *Behavioral and Brain Sciences, 4,* 22–23.
Burklund, C.W., and Smith, A. (1977). Language and the cerebral hemispheres: Observations of verbal and nonverbal responses during 18 months following left hemispherectomy. *Neurology, 27,* 627–633.
Burt, C. (1937). *The backward child.* New York: Appleton.
Butterworth, G., and Grover, L. (1988). The origins of referential communication in human infancy. In L. Weiskrantz (Ed.), *Thought without language* (pp. 5–24). Oxford: Clarendon Press.
Calvert, D.R. (1980). *Descriptive phonetics.* New York: Brian C. Decker, Division of Thieme Stratton.
Calvin, W.H. (1983). *The throwing madonna: Essays on the brain.* New York: McGraw-Hill.
Calvin, W.H. (1987). On evolutionary expectations of symmetry and toolmaking. *Behavioral and Brain Sciences, 10,* 267–268.
Campbell, R. (1978). Asymmetries in interpreting and expressing a posed facial expression. *Cortex, 14,* 327–342.
Cann, R.L. (1987). In search of Eve. *The Sciences, 27,* 30–37.
Caplan, P.J., and Kinsbourne, M. (1976). Baby drops the rattle: Asymmetry of duration of grasp by infants. *Child Development, 47,* 532–534.
Carmon, A., and Bechtold, H.P. (1973). Dominance of the right cerebral hemisphere for stereopsis. *Acta Psychologica, 37,* 351–357.
Carmon, A., and Nachshon, I. (1973). Ear asymmetry in perception of emotional nonverbal stimuli. *Acta Psychologica, 37,* 351–357.
Cartmill, M., Pilbeam, D., and Isaac, G. (1986). One hundred years of paleoanthropology. *American Scientist, 74,* 410–420.
Cerella, J. (1979). Visual classes and natural categories in the pigeon. *Journal of Experimental Psychology: Human Perception and Performance, 5,* 68–77.
Chaney, R.B., and Webster, J.C. (1966). Information in certain multidimensional sounds. *Journal of the Acoustical Society of America, 40,* 449–455.
Changeux, J.-P. (1980). Genetic determinism and epigenesis of the neuronal networks: Is there a biological compromise between Chomsky and Piaget? In M. Piattelli-Palmarini (Ed.), *Language and learning: The debate between Jean Piaget and Noam Chomsky* (pp. 185–197). Cambridge, MA: Harvard University Press.
Charcot, J.-M. (1878). Phénomène divers de l'hystéro—epilepsie.—Catalepsie provoqué artificiellement. *La Lancette française: Gazette des hôpitaux, 51,* 217–243.
Chaurasia, B.D., and Goswami, H.K. (1975). Functional asymmetry in the face. *Acta Anatomica, 91,* 154–160.
Chevalier-Skolnikoff, S. (1989). Spontaneous tool use and sensorimotor intelligence in *Cebus* compared with other monkeys and apes. *Behavioral and Brain Sciences, 12,* 561–627.
Chi, J.E., Dooling, E.C., and Gilles, F.H. (1977). Left-right asymmetries of the temporal speech areas of the human brain. *Archives of Neurology, 34,* 346–348.

Chomsky, N. (1957). *Syntactic structures*. The Hague: Mouton.
Chomsky, N. (1959). A review of B.F. Skinner's "Verbal behavior." *Language, 35*, 26–58.
Chomsky, N. (1966a). *Cartesian linguistics: A chapter in the history of rationalist thought*. New York: Harper and Row.
Chomsky, N. (1966b). *Language and mind*. New York: Harcourt Brace Jovanovich.
Chomsky, N. (1980). On cognitive structures and their development: A reply to Piaget. In M. Piattelli-Palmarini (Ed.), *Language and learning: The debate between Jean Piaget and Noam Chomsky* (pp. 35–52). Cambridge, MA: Harvard University Press.
Churchland, P.A. (1986). *Neurophilosophy*. Cambridge, MA: The MIT Press/A Bradford Book.
Ciochon, R.L., and Corrucini, R.S. (Eds.). (1983). *New interpretations of ape and human ancestry*. New York: Plenum Press.
Claiborne, R. (1983). *Our marvellous native tongue: The life and times of the English language*. New York: Times Books.
Clark, G. (1969). *World prehistory: A new outline* (2nd edition). Cambridge: Cambridge University Press.
Clarke, B. (1987). Arthur Wigan and the Duality of Mind. *Psychological Medicine, Monograph Supplement II*.
Cohen, G. (1982). Theoretical interpretations of lateral asymmetries. In J.G. Beaumont (Ed.), *Divided visual-field studies of cerebral organization* (pp. 87–111). London: Academic Press.
Cohen, H., and Levy, J.J. (1986). Cerebral and sex differences in the categorization of haptic information. *Cortex, 22*, 253–259.
Cole, J. (1955). Paw preference in cats related to hand preference in animals and man. *Journal of Comparative and Physiological Psychology, 48*, 337–345.
Collins, R.L. (1970). The sound of one paw clapping: An enquiry into the origins of left handedness. In G. Lindzey and D.D. Thiessen (Eds.), *Contributions to behavior-genetic analysis—the mouse as a prototype*. New York: Meredith.
Coltheart, M. (1980). Deep dyslexia: A right-hemisphere hypothesis. In M. Coltheart, K.E. Patterson, and J.C. Marshall (Eds.), *Deep dyslexia*. London: Routledge & Kegan Paul.
Condillac, Etienne Bonnot de. (1947). Essai sur l'origine des connaissances humaines, ouvrage ou l'on reduit à un seul principe tout ce concerne l'entendement. *Oeuvres philosophiques de Condillac*. Paris: Georges Leroy (originally published 1746).
Confer, W.N., and Ables, B.S. (1983). *Multiple personality: Etiology, diagnosis, and treatment*. New York: Human Sciences Press.
Conroy, G.C., Vannier, M.W., and Tobias, P.V. (1990). Endocranial features of *Australopithecus africanus* revealed by 2- and 3-D computed tomography. *Science, 247*, 838–841.
Cooper, L.A., and Shepard, R.N. (1973). Chronometric studies of the rotation of mental images. In W.G. Chase (Ed.), *Visual information processing* (pp. 75–176). New York: Academic Press.

Corballis, M.C. (1980). Laterality and myth. *American Psychologist, 35*, 284–295.

Corballis, M.C. (1982). Mental rotation: Anatomy of a paradigm. In M. Potegal (Ed.), *Spatial abilities: Developmental and physiological foundations* (pp. 173–198). New York: Academic Press.

Corballis, M.C. (1983). *Human laterality*. New York: Academic Press.

Corballis, M.C. (1988). Recognition of disoriented shapes. *Psychological Review, 95*, 115–123.

Corballis, M.C. (1989). Laterality and human evolution. *Psychological Review, 96*, 492–505.

Corballis, M.C., and Beale, I.L. (1970). Bilateral symmetry and behavior. *Psychological Review, 77*, 451–464.

Corballis, M.C., and Morgan, M.J. (1978). On the biological basis of human laterality: I. Evidence for a maturational left-right gradient. *Behavioral and Brain Sciences, 1*, 261–269.

Corballis, M.C., and Sergent, J. (1988). Imagery in a commissurotomized patient. *Neuropsychologia, 26*, 13–26.

Corballis, M.C., and Sergent, J. (1989a). Hemispheric specialization for mental rotation. *Cortex, 15*, 15–26.

Corballis, M.C., and Sergent, J. (1989b). Mental rotation in a commissurotomized subject. *Neuropsychologia, 27*, 585–597.

Coren, S., and Porac, C. (1977). Fifty centuries of right-handedness: The historical record. *Science, 198*, 631–632.

Cornford, J.M. (1986). Specialized resharpening techniques and evidence of handedness. In P. Callow and J.M. Cornford (Eds.), *La Cotte de St. Brelade 1961–1978: Excavations by C.B.M. McBurney* (pp. 337–362). Norwich, England: Geo Books.

Costello, A., and Warrington, E.K. (1987). The dissociation of visual neglect and neglect dyslexia. *Journal of Neurology, Neurosurgery, and Psychiatry, 50*, 1110–1116.

Critchley, M. (1964). *Developmental dyslexia*. London: Heinemann.

Critchley, M. (1975). Developmental dyslexia: Its history, nature, and prospects. In D.D. Duane and M.B. Rawson (Eds.), *Reading, perception, and language*. Baltimore: York Press.

Curry, F.K.W. (1967). A comparison of left-handed and right-handed subjects on verbal and nonverbal listening tasks. *Cortex, 3*, 343–352.

Curtiss, S. (1977). *Genie: A psycholinguistic study of a modern day "wild child."* New York: Academic Press.

Damasio, A.R., Damasio, H., and Van Hoesen, G.W. (1982). Prosopagnosia: Anatomic and behavioral mechanisms. *Neurology, 32*, 331–341.

Dandy, W.E. (1936). Operative experience in cases of pineal tumor. *Archives of Surgery, 33*, 19–46.

Dart, R.A. (1925). *Australopithecus africanus*: The man-ape of South Africa. *Nature, 115*, 195–199.

Dart, R.A. (1949). The predatory implemental technique of *Australopithecus*. *American Journal of Physical Anthropology, 7*, 1–38.

Dart, R.A. (with Dennis Craig) (1959). *Adventures with the missing link*. London: Hamish Hamilton.

Darwin, C. (1859). *The origin of species by means of natural selection*. London: John Murray.

Darwin, C. (1872). *The expression of the emotions in man and animals*. London: John Murray.

Darwin, C. (1901). *The descent of man and selection in relation to sex*. (Originally published in 1871). New York: Appleton.

Davenport, R.K., and Rogers, C.M. (1971). Perception of photographs by apes. *Behavior, 39*, 318–320.

Davidoff, J.B. (1975). Hemispheric differences in the perception of lightness. *Neuropsychologia, 13*, 121–124.

Davidoff, J.B. (1976). Hemispheric sensitivity differences in the perception of color. *Quarterly Journal of Experimental Psychology, 28*, 387–394.

Davidson, I., and Noble, W. (1989). The archeology of perception: Traces of depiction and language. *Current Anthropology, 30*, 125–155.

Davies, R. (1989). *The lyre of Orpheus*. Harmondsworth, England: Penguin Books.

Dax, M. (1865). Lésions de la moitié gauche de l'encéphale coincident avec l'oubli des signes de la pensée. *Gazette Hebdomadaire de Médicine et de Chirurgie (Paris), 2*, 259–260.

Day, M.E. (1964). An eye-movement phenomenon relating to attention, thought, and anxiety. *Perceptual and Motor Skills, 19*, 443–446.

Day, M.H., Leakey, R.E., Walker, A.C., and Wood, B.A. (1975). New hominids from East Rudolf, Kenya, I. *American Journal of Physical Anthropology, 42*, 461–476.

de Fleury, A. (1872). Du dynamisme comparé des hémisphères cerebraux dans l'homme. *Association Française pour l'Avancement des Sciences, 1*, 834–845.

de Mortillet, G. (1890). Formation des variétés, albinisme et gauchissement. *Bulletin de la Société d'Anthropologie de Paris, Séance de 3 Juillet*.

De Renzi, E. (1982). *Disorders of space exploration and cognition*. New York: Wiley.

De Renzi, E. (1986). Prosopagnosia in two patients with CT scan evidence of damage confined to the right hemisphere. *Neuropsychologia, 24*, 385–389.

De Renzi, E., Scotti, G., and Spinnler, H. (1969). Perceptual and associative disorders of visual recognition. Relationship to the side of the cerebral lesion. *Neurology, 19*, 634–642.

De Schonen, S., Gil de Diaz, M., and Mathivet, E. (1986). Hemispheric asymmetry for face processing in infancy. In H.D. Ellis, M.A. Jeeves, F. Newcombe, and A. Young (Eds.), *Aspects of face processing* (pp. 199–208). Dordrecht: Martinus Nijhof.

De Schonen, S., and Mathivet, E. (1989). First come, first served: A scenario about the development of hemispheric specialization for face recognition during infancy. *European Bulletin of Cognitive Psychology, 9*, 3–44.

Deevey, E.S. (1960). The human population. *Scientific American, 203*, 194–205.

Déjérine, J. (1892). Contributions à l'étude anatomoclinique et clinique des différentes variétés de cécité verbale. *Mémoires de la Société de Biologie, 4*, 61–90.

Delaunay, C.-G. (1874). *Biologie comparée du côté droit et du côté gauche chez l'homme et chez les êtres vivants*. Paris: N. Blanpain.

Deleval, J., De Mol, J., and Noterman, J. (1983). La perte des images souvenirs. *Acta Neurologica Belgique, 83*, 61–79.

Delis, D.C., Knight, R.T., and Simpson, G. (1983). Reversed hemispheric organization in a left-hander. *Neuropsychologia, 21*, 13–24.

Denckla, M.B. (1979). Childhood learning disabilities. In K.M. Heilman and E. Valenstein (Eds.), *Clinical neuropsychology*. New York: Oxford University Press.

Denenberg, V.H. (1981). Hemispheric laterality in animals and the effects of early experience. *Behavioral and Brain Sciences, 4*, 1–49.

Denenberg, V.H. (1988). Laterality in animals: Brain and behavioral asymmetries and the role of early experience. In D.L. Molfese and S.J. Segalowitz (Eds.), *Brain lateralization in children: Developmental implications* (pp. 59–72). New York: Guilford Press.

Denenberg, V.H., and Yutzey, D.A. (1985). Hemispheric laterality, behavioral asymmetry, and the effects of early experience in rats. In S.D. Glick (Ed.), *Cerebral lateralization in nonhuman species* (pp. 110–135). New York: Academic Press.

Dennis, M. (1976). Dissociated naming and locating of body parts after left anterior temporal lobe resection: An experimental case study. *Brain and Language, 3*, 147–163.

Dennis, M., and Kohn, B. (1975). Comprehension of syntax in infantile hemiplegics after cerebral hemidecortication. *Brain and Language, 2*, 472–482.

Dennis, M., and Whitaker, H.A. (1976). Language acquisition following hemidecortication: Linguistic superiority of the left over the right hemisphere. *Brain and Language, 3*, 404–433.

Dennis, M., and Whitaker, H.A. (1977). Hemispheric equipotentiality and language acquisition. In S.J. Segalowitz and F. Gruber (Eds.), *Language development and neurological theory*. New York: Academic Press.

Dennis, W. (1958). Early graphic evidence of dextrality in man. *Perceptual and Motor Skills, 8*, 147–149.

Descartes, R. (1985). The philosophical writings of Descartes. J. Cottingham, R. Stoothoff, and D. Murdock (Ed. and trans.) Cambridge: Cambridge University Press.

Deutsch, G., Bourbon, T., Papanicolaou, A.C., and Eisenberg, H.M. (1988). Visuospatial tasks compared via activation of regional cerebral blood flow. *Neuropsychologia, 26*, 445–452.

Diamond, A.S. (1959). *The history and origin of language*. London: Methuen.

Dimond, S., and Harries, R. (1984). Face touching in monkeys, apes, and man: Evolutionary origins and cerebral asymmetry. *Neuropsychologia, 22*, 227–233.

Duputte, B.L. (1982). Duetting in male and female songs of the white cheeked gibbon (*Hylobytes concolor leucogenys*). In C.T. Snowdon, C.H. Brown, and M.R. Petersen (Eds.), *Primate communication* (pp. 67–93). Cambridge: Cambridge University Press.

Durnford, M., and Kimura, D. (1971). Right hemisphere specialization for depth perception reflected in visual field differences. *Nature, 231*, 394–395.

Eccles, J.C. (1965). *The brain and the unity of conscious experience*. Cambridge: Cambridge University Press.

Eccles, J.C. (1981). Mental dualism and commissurotomy. *Behavioral and Brain Sciences, 4*, 105.

Edelman, G.M., and Gall, W.E. (1969). The antibody problem. *Annual Review of Biochemistry, 38*, 699–766.

Edwards, B. (1979). *Drawing on the right side of the brain: A course in enhancing creativity and artistic confidence*. Los Angeles: Tarcher.

Ehret, G. (1987). Left hemisphere advantage in the mouse brain for recognizing ultrasonic communication calls. *Nature, 325*, 249–251.

Ehrlichman, H., and Barrett, J. (1983). Right hemisphere specialization for mental imagery: A review of the evidence. *Brain and Cognition, 2*, 55–76.

Ehrlichman, H., and Weinberger, A. (1978). Lateral eye movments and hemispheric asymmetry: A critical review. *Psychological Bulletin, 85*, 1080–1101.

Eiseley, L.C. (1959). Alfred Russel Wallace. *Scientific American, 200*, 70–84.

Ellis, S.J., Ellis, P.J., and Marshall, E. (1988). Hand preference in a normal population. *Cortex, 24*, 157–163.

Entus, A. (1977). Hemispheric asymmetry in processing of dichotically presented speech and nonspeech by infants. In S.J. Segalowitz and G. Gruber (Eds.), *Language development and neurological theory*. New York: Academic Press.

Ettlinger, G. (1978). Have we forgotten the infant? *Behavioral and Brain Sciences, 2*, 294–295.

Exner, S. (1881). *Untersuchungen uber die localisation der Functionen in der Grosshirnrinde des Menschen*. Vienna: Wilhelm Braumuller.

Falk, D. (1975). Comparative anatomy of the larynx in man and the chimpanzee: Implications for language in Neanderthal. *American Journal of Physical Anthropology, 43*, 123–132.

Falk, D. (1980). Language, handedness, and primate brains: Did the australopithecines sign? *American Anthropologist, 82*, 72–78.

Falk, D. (1982). Mapping fossil endocasts. In E. Armstrong and D. Falk (Eds.), *Primate brain evolution: Methods and concepts* (pp. 217–226). New York: Plenum.

Falk, D. (1983a). Cerebral cortices of East African early hominids. *Science, 222*, 1072–1074.

Falk, D. (1983b). The Taung endocast: A reply to Holloway. *American Journal of Physical Anthropology, 60*, 17–45.

Falk, D. (1987a). Brain lateralization in primates and its evolution in hominids. *Yearbook of Physical Anthropology, 30*, 107–125.

Falk, D. (1987b). Hominid paleoneurology. *Annual Review of Anthropology, 16*, 13–30.

Farah, M.J. (1984). The neurological basis of mental imagery: A componential analysis. *Cognition, 18*, 245–272.

Farah, M.J. (1988). Is visual imagery really visual? Overlooked evidence from neuropsychology. *Psychological Review, 95*, 307–317.
Farah, M.J. (in press). Patterns of co-occurrence among the associative agnosias. *Cognitive Neuropsychology.*
Farah, M.J., Gazzaniga, M.S., Holtzman, J.D., and Kosslyn, S.M. (1985). A left hemisphere basis for mental imagery? *Neuropsychologia, 23*, 115–118.
Farah, M.J., Levine, D.N., and Calvanio, R. (1988). A case study of mental imagery deficit. *Brain and Cognition, 8*, 147–164.
Farah, M.J., Peronnet, F., Weisberg, L.L., and Monheit, M.A. (1989). Brain activity underlying mental imagery: Event-related potentials during mental image generation. *Journal of Cognitive Neuroscience, 1*, 302–316.
Fechner, G.T. (1860). *Elemente der Psychophysik,* Vol. 2. Leipzig: Breitkopf and Hartel.
Flor-Henry, P. (1983). Functional hemispheric asymmetry and psychopathology. *Integrative Psychiatry, 1*, 46–52.
Flor-Henry, P. (1985). Psychiatric aspects of cerebral lateralization. *Psychiatric Annals, 15*, 429–434.
Fodor, J.A. (1983). *The modularity of mind.* Cambridge, MA: MIT Press.
Fodor, J.A., and Pylyshyn, Z.W. (1988). Connectionism and cognitive architecture. *Cognition, 28*, 3–71.
Foley, R. (1987). Hominid species and stone tool assemblages. *Antiquity, 61*, 380–392.
Foley, R.A., and Lee, P.C. (1989). Finite social space, evolutionary pathways, and reconstructing hominid behavior. *Science, 243*, 901–906.
Franco, L., and Sperry, R.W. (1977). Hemispheric lateralization for cognitive processing of geometry. *Neuropsychologia, 15*, 107–114.
Friedman, H., and Davis, M. (1938). "Left handedness" in parrots. *Auk, 55*, 478–480.
Fromkin, V.A., Krashen, S., Curtiss, S., Rigler, D., and Rigler, M. (1974). The development of language in Genie: A case of language acquisition beyond the critical period. *Brain and Language, 1*, 81–107.
Frost, G.T. (1980). Tool behavior and the origins of laterality. *Journal of Human Evolution, 9*, 447–459.
Gainotti, G. (1972). Emotional behavior and the hemispheric side of the lesion. *Cortex, 8*, 41–54.
Gainotti, G., and Tiacci, C. (1970). Patterns of drawing disability in right and left hemispheric patients. *Neuropsychologia, 8*, 379–384.
Gallup, G.G., Jr. (1977). Self-recognition in primates: A comparative approach to the bidirectional properties of consciousness. *American Psychologist, 32*, 329–338.
Gardiner, M.F., and Walter, D.O. (1977). Evidence of hemispheric specialization from infant EEG. In S. Harnad, R. Doty, L. Goldstein, J. Jaynes, and G Krauthamer (Eds.), *Lateralization in the nervous system* (pp. 481–500). New York: Academic Press.
Gardner, H. (1985). *The mind's new science: A history of the cognitive revolution.* New York: Basic Books.

Gardner, M. (1965). *The annotated Alice*. Harmondsworth, England: Penguin Books.

Gardner, R.A., and Gardner, B.T. (1969). Teaching sign language to a chimpanzee. *Science, 165*, 664–672.

Gardner, R.A., and Gardner, B.T. (1985). A vocabulary test for chimpanzees (*Pan troglodytes*). *Journal of Comparative Psychology, 98*, 381–404.

Garrett, S.V. (1976). Putting our whole brain to use: A fresh look at the creative process. *Journal of Creative Behavior, 10*, 239–249.

Gaur, A. (1987). *A history of writing*. London: British Library.

Gautier, J.-P., and Gautier-Hion, A. (1982). Vocal communication within a group of monkeys: An analysis by telemetry. In C.T. Snowdon, C.H. Brown, and M.R. Petersen (Eds.), *Primate communication* (pp. 5–29). Cambridge: Cambridge University Press.

Gazzaniga, M.S. (1970). *The bisected brain*. New York: Appleton.

Gazzaniga, M.S. (1983). Right hemisphere language following bisection: A 20-year perspective. *American Psychologist, 38*, 525–537.

Gazzaniga, M.S. (1988). The dynamics of cerebral specialization and modular interactions. In L. Weiskrantz (Ed.), *Thought without language* (pp. 430–450). Oxford: Clarendon Press.

Gazzaniga, M.S., Bogen, J.E., and Sperry, R.W. (1965). Observations of visual perception after disconnexion of the cerebral hemispheres in man. *Brain, 88*, 221–230.

Gazzaniga, M.S., Bogen, J.E., and Sperry, R.W. (1967). Dyspraxia following division of the cerebral hemispheres. *Archives of Neurology, 16*, 606–612.

Gazzaniga, M.S., and LeDoux, J.E. (1978). *The integrated mind*. New York: Plenum.

Gazzaniga, M.S., and Smylie, C.S. (1983). Facial recognition and brain asymmetries: Clues to underlying mechanisms. *Annals of Neurology, 13*, 536–540.

Gazzaniga, M.S., and Sperry, R.W. (1967). Language after section of the cerebral commissures. *Brain, 90*, 131–148.

Gelb, I.J. (1952). *A study of writing: The foundations of grammatology*. London: Routledge and Kegan Paul.

Geschwind, N. (1969). Problems in the anatomical understanding of the aphasias. In A.L. Benton (Ed.), *Contributions to clinical neuropsychology*. Chicago: Aldine.

Geschwind, N. (1985). Implications for evolution, genetics, and clinical syndromes. In S.D. Glick (Ed.), *Cerebral lateralization in nonhuman species* (pp. 247–278). Orlando, FL: Academic Press.

Geschwind, N., and Behan, P. (1982). Left-handedness: Association with immune disease, migraine, and developmental learning disorder. *Proceedings of the National Academy of Sciences, 79*, 5097–5100.

Geschwind, N., and Galaburda, A.M. (1987). *Cerebral lateralization: Biological mechanisms, associations, and pathology*. Cambridge, MA: Bradford MIT Press.

Geschwind, N., and Levitsky, W. (1968). Human brain: Right-left asymmetries in temporal speech region. *Science, 161*, 186–187.

Gesell, A., and Ames, L.B. (1947). The development of handedness. *Journal of Genetic Psychology, 70*, 155–175.

Gibson, J.J. (1966). *The senses considered as perceptual systems*. Boston: Houghton Mifflin.

Gibson, J.J. (1979). *The ecological approach to visual perception*. Boston, MA: Houghton-Mifflin.

Gilbert, C., and Bakan, P. (1973). Visual asymmetry in perception of faces. *Neuropsychologia, 11*, 355–362.

Glanville, B.B., Best, C.T., and Levenson, R. (1977). A cardiac measure of cerebral asymmetries in infant auditory perception. *Developmental Psychology, 13*, 54–59.

Glezer, I.I. (1987). The riddle of Carlyle: The unsolved problem of the origin of handedness. *Behavioral and Brain Sciences, 10*, 273–275.

Glick, S.D. (Ed.). (1985). *Cerebral lateralization in nonhuman species*. New York: Academic Press.

Glick, S.D., and Shapiro, R.M. (1985). Functional and neurochemical mechanisms of cerebral lateralization in rats. In S.D. Glick (Ed.), *Cerebral lateralization in nonhuman species* (pp. 158–184). New York: Academic Press.

Godfrey, J.J. (1974). Perceptual difficulty and the right ear advantage for vowels. *Brain and Language, 1*, 323–336.

Godfrey, L., and Jacobs, K.E. (1981). Gradual, autocatalytic and punctuational models of hominid brain evolution: A cautionary tale. *Journal of Human Evolution, 10*, 255–272.

Goldenberg, G., Podreka, I., Steiner, M., Willmes, K., Suess, E., and Deecke, L. (1989). Regional cerebral blood flow patterns in visual imagery. *Neuropsychologia, 27*, 641–664.

Gomori, A.J., and Hawryluk. G.A. (1984). Visual agnosia without alexia. *Neurology, 34*, 947–950.

Goodall, J. (1970). Tool use in primates and other vertebrates. In D.S. Lehrman, R.A. Hinde, and E. Shaw (Eds.), *Advances in the study of behavior* (Vol. 1, pp. 195–249). New York: Academic Press.

Goodglass, H., Blumstein, S.E., Gleason, J.B., Hyde, M.R., Green, E., and Statlender, S. (1979). The effect of syntactic coding on sentence comprehension in aphasia. *Brain and Language, 7*, 201–209.

Gordon, D.P. (1983). The influence of sex on the development of lateralization of speech. *Neuropsychologia, 21*, 139–146.

Gordon, H.W. (1970). Hemispheric asymmetries in the perception of musical chords. *Cortex, 6*, 387–398.

Gould, J.L., and Marler, P. (1987). Learning by instinct. *Scientific American, 256*, 62–73.

Gould, S.J. (1977). *Ontogeny and phylogeny*. Cambridge, MA: Harvard University Press.

Gould, S.J. (1980). *Ever since Darwin*. Harmondsworth, England: Penguin Books.

Gould, S.J. (1983). *The panda's thumb*. Harmondsworth, England: Penguin Books.

Gould, S.J. (1984). *Hen's teeth and horse's toes*. Harmondsworth, England: Penguin Books.

Gould, S.J. (1987). *An urchin in the storm*. New York: Norton.
Gould, S.J., and Eldredge, N. (1977). Punctuated equilibria: The "tempo" and "mode" of evolution reconsidered. *Paleobiology, 3*, 115–151.
Gould, S.J., and Vrba, E.S. (1982). Exaptation—a missing term in the science of form. *Paleobiology, 8*, 4–15.
Govind, C.K. (1989). Asymmetry in lobster claws. *American Scientist, 77*, 468–474.
Gowlett, J. (1984). *Ascent to civilization: The archeology of early man*. London: Collins.
Greenberg, M.S., and Farah, M.J. (1986). The laterality of dreaming. *Brain and Cognition, 5*, 307–321.
Griffin, D.R. (1976). *The question of animal awareness: Evolutionary continuity of mental experience* (2nd edition published 1981). New York: Rockefeller University Press.
Grossman, M. (1988). Drawing deficits in brain-damaged patients' freehand pictures. *Brain and Cognition, 8*, 189–205.
Guthrie-Smith, H. (1936). *Sorrows and joys of a New Zealand naturalist*. Dunedin, New Zealand: Reed.
Haggard, M.P., and Parkinson, A.M. (1971). Stimulus task factors as determinants of ear advantages. *Quarterly Journal of Experimental Psychology, 23*, 168–177.
Hailman, J.P., and Ficken, M.S. (1986). Combinatorial animal communication with computable syntax: Chick-a-dee calling qualifies as "language" by structural linguistics. *Animal Behaviour, 34*, 1899–1901.
Hall, G.S. (1891). Notes on the study of infants. *Pedagogical Seminary, 1*, 128–138.
Hall, R.A., Jr. (1959). Pidgin languages. *Scientific American, 200*, 124–134.
Halle, M. (1988). The immanent form of phonemes. In W. Hirst (Ed.), *The making of cognitive science: Essays in honor of George A. Miller*. Cambridge: Cambridge University Press.
Halpern, D.F., and Coren, S. (1988). Do right handers live longer? *Nature, 333*, 213.
Hames, B.D., and Glover, D.M. (1988). *Molecular immunology*. Eynsham, Oxford: IRL Press.
Hamilton, C.R., and Vermeire, B.A. (1983). Discrimination of monkey faces by split-brain monkeys. *Behavioral Brain Research, 9*, 263–275.
Hamilton, C.R., and Vermeire, B.A. (1988). Complementary hemispheric specialization in monkeys. *Science, 242*, 1691–1694.
Hannay, H.J., Varney, N.R., and Benton, A.L. (1976). Visual localization in patients with unilateral brain disease. *Journal of Neurology, Neurosurgery, and Psychiatry, 39*, 307–313.
Hardyck, C., Petrinovich, L., and Goldman, R. (1976). Left handedness and cognitive deficit. *Cortex, 12*, 266–278.
Harrington, A. (1987). *Medicine, mind, and the double brain*. Princeton, NJ: Princeton University Press.
Harris, L.J. (1980). Left-handedness: Early theories, facts, and fancies. In J. Herron (Ed.), *Neuropsychology of left-handedness*. New York: Academic Press.

Harris, L.J. (1984). Louis Pierre Gratiolet, Paul Broca, et al. on the question of a maturational left-right gradient: Some forerunners of current-day models. *Behavioral and Brain Sciences, 7*, 730–731.

Harris, L.J. (1988). Right-brain training: Some reflections on the application of research on cerebral hemispheric specialization to education. In D.L. Molfese and S.J. Segalowitz (Eds.), *Brain lateralization in children* (pp. 207–235). New York: Guilford Press.

Harris, L.J., and Carlson, D.F. (1988). Pathological left-handedness: An analysis of theories and evidence. In D.L. Molfese and S.J. Segalowitz (Eds.), *Brain lateralization in children*. New York: Guilford Press.

Haslam, R.H., Dalby, J.T., Johns, R.D., and Rademaker, A.W. (1981). Cerebral asymmetry in developmental dyslexia. *Archives of Neurology, 38*, 679–682.

Hay, D.A., and Howie, P.M. (1980). Handedness and differences in birthweight of twins. *Perceptual and Motor Skills, 51*, 666.

Hayes, C. (1952). *The ape in our house.* London: Gollanez.

Hayes, K.J., and Hayes, C. (1953). Picture perception in a home-raised chimpanzee. *Journal of Comparative and Physiological Psychology, 46*, 470–474.

Heilman, K.M., and Rothi, L.J.G. (1985). Apraxia. In K.M. Heilman and E. Valenstein (Eds.), *Clinical neuropsychology* (2nd edition, pp. 131–150). Oxford: Oxford University Press.

Heilman, K.M., Rothi, L.J.G., and Valenstein, E. (1982). Two forms of ideomotor apraxia. *Neurology, 32*, 342–346.

Henderson, L. (1982). *Orthography and word recognition in reading.* New York: Academic Press.

Henschen, S.E. (1919). Uber Sprach-Musik und Rechenmechanisme und ihre Lokalizasion im Gehirn. *Zeitschrift für die Gesante Neurologie und Psychiatrie, 52*, 273–298.

Herman, L.M., Richards, D.G., and Wolz, J.P. (1984). Comprehension of sentences by bottle-nosed dolphins. *Cognition, 16*, 129–219.

Herrnstein, R.J. (1985). Riddles of natural categorization. *Philosophical Transactions of the Royal Society, B, 308*, 129–144.

Herrnstein, R.J., and Loveland, D.H. (1964). Complex visual concept in the pigeon. *Science, 146*, 549–551.

Herrnstein, R.J., Loveland, D.H., and Cable, C. (1976). Natural concepts in pigeons. *Journal of Experimental Psychology: Animal Behavior Processes, 2*, 285–302.

Hertz, R. (1909). La préeminence de la main droite: Etude sur la polarité religieuse. *Revue Philosophique, 68*, 553–580.

Hertz, R. (1960). *Death and the right hand.* Aberdeen: Cohen and West.

Hewes, G. (1949). Lateral dominance, culture, and writing systems. *Human Biology, 21*, 233–245.

Hewes, G.W. (1973). Primate communication and the gestural origins of language. *Current Anthropology, 14*, 5–24.

Hier, D.B., LeMay, M., Rosenberger, P.B., and Perlo, V.P. (1978). Developmental dyslexia: Evidence for a subgroup with a reversal of cerebral asymmetry. *Archives of Neurology, 35*, 90–92.

Hilton, C.E. (1986). *Hands across the old world: The changing hand morphology of the hominids*. Department of Anthropology, University of New Mexico.

Hines, M., and Shipley, C. (1984). Prenatal exposure to diethylstilbestrol (DES) and the development of sexually dimorphic cognitive abilities and cerebral lateralization. *Developmental Psychology, 20*, 81–94.

Hirsch-Pasek, K., Kemler Nelson, D.G., Juszcyk, P.W., Wright Cassidy, K., Druss, B., and Kennedy, L. (1987). Clauses are perceived as perceptual units for young infants. *Cognition, 26*, 269–286.

Hiscock, M. (1988). Behavioral asymmetries in normal children. In D.L. Molfese and S.J. Segalowitz (Eds.), *Brain lateralization in children: Developmental implications* (pp. 85–169). New York: Guilford Press.

Hoffmeister, R., and Wilbur, R.B. (1980). Developmental: The acquisition of sign language. In H. Lane and F. Grosjean (Eds.), *Recent perspectives on American Sign Language* (pp. 61–78). Hillsdale, NJ: Erlbaum.

Hofstadter, D.R., and Dennett, D.C. (1981). *The mind's I: Fantasies and reflections on mind and soul*. New York: Basic Books.

Holloway, R.L. (1969). Culture: A human domain. *Current Anthropology, 4*, 135–168.

Holloway, R.L. (1981a). Culture, symbols, and human brain evolution. *Dialectical Anthropology, 5*, 287–303.

Holloway, R.L. (1981b). Revisiting the South African *Australopithecus* endocast: The position of the lunate sulcus as determined by the stereoplotting technique. *American Journal of Physical Anthropology, 56*, 43–58.

Holloway, R.L. (1983). Human paleontological evidence relevant to language behavior. *Human Neurobiology, 2*, 105–114.

Holloway, R.L. (1985). The past, present, and future significance of the lunate sulcus in early hominid evolution. In P.V. Tobias (Ed.), *Hominid evolution: Past, present, and future* (pp. 47–62). New York: Alan R. Liss.

Holloway, T. (1956). Left-handedness is no handicap. *Psychology, 20*, 27.

Hopkins, W.D. (in press). Priming as a technique in the study of lateralized and generalized cognitive processes in non-human primates: Some recent findings. In H. Roitblat, L. Herman, and P. Natigall (Eds.), *Comparative cognition*. New York: Erlbaum.

Hopkins, W.D., Morris, R.D., Savage-Rumbaugh, S., and Rumbaugh, D.M. (1989). *Hemispheric activation for meaningful and nonmeaningful symbols in language-trained chimpanzees: Evidence for a left hemisphere advantage*. Unpublished manuscript.

Hopkins, W.D., Washburn, D.A., and Rumbaugh, D.M. (1990). Processing of form stimuli presented unilaterally in humans, chimpanzees (*Pan troglodytes*), and monkeys (*Macaca mulatta*). *Behavioral Neuroscience, 104*, 577–582.

Hubel, D.H. (1982). Explorations of the primary visual cortex, 1955–78. *Nature, 299*, 515–524.

Hughey, M.J. (1985). Fetal position during pregnancy. *American Journal of Obstetrics and Gynecology, 153*, 885–886.

Humboldt, W. von (1792). *Ideen zu einem Versuch die Grenzen der Wirksamkeit des Staats zu bestimmen.* (Cited by Chomsky, 1966b.)

Humphreys, G.W., and Riddoch, M.J. (1984). Routes to object constancy: Implications from neurological impairments of object constancy. *Cognitive Neuropsychology, 36A,* 385–415.

Humphreys, G.W., and Riddoch, M.J. (1987). *To see but not to see: A case study of visual agnosia.* London: Erlbaum.

Hung, C.-C., Tu, Y.-K., Chen, S.-H., and Chen, R.-C. (1985). A study of handedness and cerebral speech dominance in right-handed Chinese. *Journal of Neurolinguistics, 1,* 143–163.

Huxley, A. (1957). *Adonis and the alphabet.* London: Chatto and Windus.

Huxley, A.L. (1950). *After many a summer.* London: Chatto and Windus.

Hynd, G.W., and Semrud-Clikeman, M. (1989). Dyslexia and brain morphology. *Psychological Bulletin, 106,* 447–482.

Ibbotson, N.R., and Morton, J. (1981). Rhythm and dominance. *Cognition, 9,* 125–138.

Ifune, C.K., Vermeire, B.A., and Hamilton, C.R. (1984). Hemispheric differences in split-brain monkeys viewing and responding to videotaped recordings. *Behavioral and Neural Biology, 41,* 231–235.

Isaac, G.L. (1972). Chronology and tempo of cultural change during the Pleistocene. In W.W. Bishop and J.A. Miller (Eds.), *Calibration of hominid evolution* (pp. 381–430). Edinburgh: Scottish Academic Press.

Isaac, G.L. (1976). Stages of cultural elaboration in the Pleistocene: Possible archeological indicators of the development of language capabilities. *Annals of the New York Academy of Sciences, 280,* 275–288.

Isserlin, M. (1922). Uber Agrammatismus. *Zeitschrift für die Gesamte Neurologie und Psychiatrie, 75,* 332–416.

Jackendoff, R. (1987). *Consciousness and the computational mind.* Cambridge, MA: Bradford/MIT Press.

Jackson, J.H. (1864). Clinical remarks on cases of defects of expression (by words, writing, signs, etc.) in diseases of the nervous system. *Lancet, 2,* 604.

Jackson, J.H. (1868). Defect of intellectual expression (aphasia) with left hemiplegia. *Lancet, 2,* 457.

Jackson, J.H. (1876). Case of large tumor without optic neuritis and with left hemiplegia and imperception. *Royal London Ophthalmic Hospital Report, 8,* 434.

Jackson, J.H. (1879). On affections of speech from disease of the brain. *Brain, 1,* 304–330.

Janet, P. (1907). *The major symptoms of hysteria.* New York: Macmillan.

Janet, P. (1899). *L'automatisme psychologique.* Paris: Libraire Felix Alcan.

Jansons, K.M. (1988). A personal view of dyslexia and of thought without language. In L. Weiskrantz (Ed.), *Thought without language* (pp. 498–503). Oxford: Clarendon Press.

Jason, G.W., Cowey, A., and Weiskrantz (1984). Hemispheric asymmetry for a visuo-spatial task in monkeys. *Neuropsychologia, 22,* 777–784.

Jaynes, J. (1976). *The origin of consciousness in the breakdown of the bicameral mind.* Boston: Houghton Mifflin.

Jerison, H. (1973). *Evolution of brain and intelligence*. New York: Academic Press.

Jerne, N.K. (1967). Antibodies and learning: Selection versus instruction. In G.C. Quarton, T. Melnechuk, and F.O. Schmitt (Eds.), *The neurosciences: A study program*. New York: Rockefeller University Press.

Johanson, D.C., and Edey, M. (1981). *Lucy: The beginnings of humankind*. London: Granada.

Johnson, J.S., and Newport, E.L. (1989). Critical period effects in second language learning: The influence of maturational state on the acquisition of English as a second language. *Cognitive Psychology, 21*, 60–99.

Johnson, L.E. (1984). Vocal responses to left visual stimuli following forebrain commissurotomy. *Neuropsychologia, 22*, 153–166.

Johnson, R., Bowers, J., Gamble, M., Lyons, F., Presbey, T., and Vetter, R. (1977). Ability to transcribe music and ear superiority for tone sequences. *Cortex, 13*, 295–299.

Johnson, S.C. (1981). Bonobos: Generalized hominid prototypes or specialized insular dwarfs? *Current Anthropology, 22*, 363–375.

Johnson, W.E. (1959). *The onset of stuttering*. Minneapolis: University of Minnesota Press.

Johnson, W.E. (Ed.). (1955). *Stuttering in children and adults*. Minneapolis: University of Minnesota Press.

Johnson-Laird, P.N. (1988). *The computer and the mind: An introduction to cognitive science*. London: Fontana.

Jolly, A. (1972). *The evolution of primates*. New York: Macmillan.

Jones, H.F. (Ed.) (1919). *The note-books of Samuel Butler*. London: A.C. Fifield.

Jones, R.K. (1966). Observations on stammering after localized cerebral injury. *Journal of Neurology, Neurosurgery, and Psychiatry, 29*, 192–195.

Jones, T., and Kamil, A.C. (1973). Tool-making and tool-using in the northern blue jay. *Science, 180*, 1076–1078.

Joynt, R.J. (1964). Paul Pierre Broca: His contribution to the study of aphasia. *Cortex, 1*, 206–213.

Kallman, H.J., and Corballis, M.C. (1975). Ear asymmetry in reaction time to musical sounds. *Perception and Psychophysics, 17*, 368–370.

Kamin, L. (1974). *The science and politics of IQ*. Hillsdale, NJ: Erlbaum.

Kashiwagi, A., Kashiwagi, T., Nishikawa, T., and Okuda, J.-I. (1989). Hemispheric asymmetry of processing temporal aspects of repetitive movement in two patients with infarction involving the corpus callosum. *Neuropsychologia, 27*, 799–809.

Keeley, J.H. (1977). The functions of paleolithic flint tools. *Scientific American, 237*, 108–127.

Kennard, M.A. (1936). Age and other factors in motor recovery from precentral lesions in monkeys. *Journal of Neurophysiology, 1*, 477–496.

Kimura, D. (1961). Cerebral dominance and the perception of verbal stimuli. *Canadian Journal of Psychology, 15*, 166–171.

Kimura, D. (1964). Left-right differences in the perception of melodies. *Quarterly Journal of Experimental Psychology, 16*, 355–358.

Kimura, D. (1967). Functional asymmetry of the brain in dichotic listening. *Cortex, 3*, 163–178.

Kimura, D. (1973). Manual activity during speaking. I. Right-handers. *Neuropsychologia, 11*, 45–50.

Kimura, D., and Archibald, Y. (1972). Motor functions of the left hemisphere. *Brain, 97*, 337–350.

Kimura, D., and Folb, S. (1968). Neural processing of backwards-speech sounds. *Science, 161*, 395–396.

King, F.L., and Kimura, D. (1972). Left-ear superiority in dichotic perception of vocal nonverbal sounds. *Canadian Journal of Psychology, 26*, 111–116.

Kingsley, C. (1886). *The water babies*. London: Macmillan.

Kingsolver, J.G., and Koehl, M.A.R. (1985). Aerodynamics, thermoregulation, and the evolution of insect wings. *Evolution, 39*, 488–504.

Kinsbourne, M. (1970). The cerebral basis of lateral asymmetries in attention. *Acta Psychologica, 33*, 193–201.

Kinsbourne, M. (1972). Eye and head turning indicates cerebral localization. *Science, 176*, 539–541.

Kinsbourne, M. (1975). The mechanism of hemispheric control of the lateral gradient of attention. In P.M.A. Rabbitt and S. Dornic (Eds.), *Attention and performance V*. (pp. 81–97). New York: Academic Press.

Kinsbourne, M. (1987). Mechanisms of unilateral neglect. In M. Jeannerod (Ed.), *Neurophysiological and neuropsychological aspects of spatial neglect* (pp. 69–86). Amsterdam: North-Holland.

Kipling, R. (1927). *Rudyard Kipling's verse*. London: Hodder and Stoughton.

Kirk, A., and Kertesz, A. (1989). Hemispheric contribution to drawing. *Neuropsychologia, 27*, 881–886.

Kohler, W. (1925). *The mentality of apes*. (E. Winter, translator.) New York: Humanities Press.

Kohn, B., and Dennis, M. (1974). Patterns of hemispheric specialization after hemidecortication for infantile hemiplegia. In M. Kinsbourne and W.L. Smith (Eds.), *Hemispheric disconnection and cerebral function* (pp. 34–47). Springfield, IL: Charles C. Thomas.

Kohne, D.E. (1975). DNA evolution data and its relevance to mammalian phylogeny. In W.P. Luckett and F.S. Szalay (Eds.), *Phylogeny of the primates* (pp. 249–261). New York: Plenum.

Kolb, B. (1989). Brain development, plasticity, and behavior. *American Psychologist, 44*, 1203–1212.

Kortlandt, A. (1972). *New perspectives on ape and human evolution*. Amsterdam: Stichting voor Psychobiologie.

Kosslyn, S.M. (1980). *Image and mind*. Cambridge, MA: Harvard University Press.

Kosslyn, S.M. (1987). Seeing and imagining in the cerebral hemispheres. *Psychological Review, 94*, 148–175.

Kosslyn, S.M. (1988). Aspects of a cognitive neuroscience of mental imagery. *Science, 240*, 1621–1626.

Kosslyn, S.M., Holtzman, J.D., Farah, M.J., and Gazzaniga, M.S. (1985). A computational analysis of mental image generation: Evidence from functional dissociations in split-brain patients. *Journal of Experimental Psychology: General, 114*, 311–341.

Krogman, W.M. (1972). *Child growth*. Ann Arbor: University of Michigan Press.

Kuhl, P.K., and Miller, J.D. (1975). Speech perception by the chinchilla: Voice-voiceless distinction in alveolar plosive consonants. *Science, 190*, 69–72.

Kuhl, P.K., and Padden, D.M. (1982). Enhanced discriminability at the phonetic boundaries for the voicing feature in macaques. *Perception and Psychophysics, 32*, 542–558.

Kuhl, P.K., and Padden, D.M. (1983). Enhanced discriminability at the phonetic boundaries for the place feature in macaques. *Journal of the Acoustical Society of America, 73*, 1003–1010.

La Mettrie, J.O. de (1747). *L'homme machine*. 1960 critical edition, edited by A. Vartanian. Princeton, NJ: Princeton University Press. Also published in 1912 in English translation as *Man a machine*. La Salle, IL: Open Court.

Lakoff, G., and Ross, J.R. (1976). Is deep structure necessary? In J. D. McCawley (Ed.), *Syntax and semantics* (Vol. 7, pp. 159–164). New York: Academic Press.

Lancaster, J.B. (1973). On the evolution of tool-using behavior. In C.L. Brace and J. Metress (Eds.), *Man in evolutionary perspective* (pp. 79–90). New York: Wiley.

Landis, T., Cummings, J.L., Christen, L., Bogen, J.E., and Imhof, H.-G. (1986). Are unilateral right posterior lesions sufficient to cause prosopagnosia? Clinical and radiological findings in six additional patients. *Cortex, 22*, 243–252.

Lapointe, S.G. (1985). A theory of verb form use in the speech of agrammatic aphasics. *Brain and Language, 24*, 100–155.

Lashley, K.S. (1949). Persisting problems in the evolution of mind. *Quarterly Review of Biology, 24*, 28–42.

Leakey, M.D. (1976). A summary and discussion of the archeological evidence from Bed I and Bed II, Olduvai Gorge, Tanzania. In G.L. Isaac and E.R. McGowan (Eds.), *Human origins: Louis Leakey and the East African evidence* (pp. 431–459). Menlo Park, CA: Benjamin.

Leakey, M.D. (1979). Footprints in the ashes of time. *National Geographic, 155*, 446–457.

Leakey, R.E., and Lewin, R. (1979). *Origins*. London: Rainbird Publishing Group.

Lecours, A.R., and Lhermitte, F. (1976). The "pure" form of the phonetic disintegration syndrome (pure anarthria): Anatomo-clinical report of a historical case. *Brain and Language, 3*, 88–113.

LeDoux, J.E. (1983). Cerebral asymmetry and the integrated function of the brain. In A.W. Young (Ed.), *Functions of the right cerebral hemisphere* (pp. 203–216). New York: Academic Press.

LeDoux, J.E., Wilson, D.H., and Gazzaniga, M.S. (1977). Manipulo-spatial aspects of cerebral lateralization: Clues to the origin of lateralization. *Neuropsychologia, 15*, 743–750.

Leehey, S.C. (1976). *Development of right-hemispheric specialization in children*. Paper presented at animal meeting of Eastern Psychological Association, New York.

LeMay, M. (1976). Morphological cerebral asymmetries of modern man, fossil man, and nonhuman primates. *Annals of the New York Academy of Sciences, 280*, 349–366.

Lemon, R.E. (1973). Nervous control of the syrinx in white-throated sparrows (*Zonotrichia albicollis*). *Journal of Zoology (London), 71*, 131–140.

Lenneberg, E.H. (1967). *Biological foundations of language*. New York: Wiley.

Lenneberg, E.H. (1969). On explaining language. *Science, 164*, 635–643.

Lennon, F.B. (1945). *Victoria through the looking glass: The life of Lewis Carroll*. New York: Simon and Schuster.

Lerdahl, F., and Jackendoff, R. (1983). *A generative theory of tonal music*. Cambridge, MA: MIT Press.

Leurat, F., and Gratiolet, P. (1857). *Anatomie comparée du système nerveux, considérée dans ses rapports avec l'intelligence*. (Vol. 2, written by Gratiolet.) Paris: J.B. Baillière.

Levine, S.C. (1985). Developmental changes in right hemisphere involvement in face recognition. In C.T. Best (Ed.), *Hemispheric function and collaboration in the child* (pp. 157–191). New York: Academic Press.

Levy, J. (1974). Psychobiological implications of bilateral asymmetry. In S.J. Dimond and J.G. Beaumont (Eds.), *Hemispheric function in the human brain*. London: Paul Elek.

Levy, J., Heller, W., Banich, M.T., and Burton, L.A. (1983). Asymmetry of perception in free viewing of chimeric faces. *Brain and Cognition, 2*, 404–419.

Levy, J., Trevarthen, C., and Sperry, R.W. (1972). Perception of bilateral chimeric figures following hemispheric disconnection. *Brain, 95*, 61–78.

Lewin, R. (1987). *Bones of contention: Controversies in the search for human origins*. New York: Simon and Schuster.

Lewkowicz, D.J., and Turkewitz, G. (1983). Relationships between processing and motor asymmetries in early development. In G. Young, S.J. Segalowitz, C.M. Corter, and S.E. Trehub (Eds.), *Manual specialization and the developing brain* (pp. 375–393). New York: Academic Press.

Ley, R.G. (1983). Cerebral asymmetry and imagery. In A.A. Sheik (Ed.), *Imagery: Current theory, research, and applications* (pp. 252–287). New York: Wiley.

Ley, R.G., and Bryden, M.P. (1979). Hemispheric differences in processing emotions and faces. *Brain and Language, 7*, 127–138.

Liberman, A.M. (1982). On finding that speech is special. *American Psychologist, 37*, 148–167.

Liberman, A.M., Cooper, F.S., Shankweiler, D.P., and Studdert-Kennedy, M. (1967). Perception of the speech code. *Psychological Review, 74*, 431–461.

Liberman, A.M., and Mattingly, I.G. (1985). The motor theory of speech perception revisited. *Cognition, 21*, 1–36.

Liberman, A.M., and Mattingly, I.G. (1989). A specialization for speech perception. *Science, 243*, 489–494.
Lichtheim, L. (1885). On aphasia. *Brain, 7*, 433–484.
Lieberman, P. (1984). *The biology and evolution of language*. Cambridge, MA: Harvard University Press.
Lieberman, P., Crelin, E.S., and Klatt, D.H. (1972). Phonetic ability and related anatomy of the newborn, adult human, Neanderthal man, and the chimpanzee. *American Anthropologist, 74*, 287–307.
Liederman, J. (1987). Neonates show an asymmetric degree of head rotation but lack an asymmetric tonic neck reflex. *Developmental Neuropsychology, 3*, 439–450.
Liepmann, H. (1908). *Drei Aufsätze aus dem Apraxiegebiet*. Berlin: Karger.
Lightfoot, D. (1989). The child's trigger experience: Degree-0 learnability. *Behavioral and Brain Sciences, 12*, 321–375.
Lissauer, H. (1890). Ein Fall von Seelenbleindheit nebsteinen Betrag zur Theorie derselben. *Archiv für Psychiatrie und Nervenkrankheiten, 21*, 222–270.
Lock, A. (1980). *The guided reinvention of language*. London: Academic Press.
Longden, K., Ellis, C., and Iversen, D.S. (1976). Hemispheric differences in the perception of curvature. *Neuropsychologia, 14*, 195–202.
Lovejoy, O.C. (1981). The origin of man. *Science, 221*, 341–350.
Lovejoy, O.C. (1988). The evolution of human walking. *Scientific American, 259*, 118–125.
Luchsinger, R., and Arnold, G.E. (1965). *Voice—speech—language*. London: Constable.
Lumsden, C.L., and Wilson, E.O. (1981). *Genes, mind, and culture: The coevolutionary process*. Cambridge, MA: Harvard University Press.
Lumsden, C.L., and Wilson, E.O. (1983). *Promethean fire: Reflections on the origin of mind*. Cambridge, MA: Harvard University Press.
Luria, A.R., Tsvetkova, L.S., and Futer, D.S. (1965). Aphasia in a composer. *Journal of Neurological Science, 2*, 288–292.
Lussenhop, A.J., Boggs, J.S., LaBorwit, L.J., and Walle, E.L. (1973). Cerebral dominance in stutterers determined by Wada testing. *Neurology, 23*, 1190–1192.
Luys, J.B. (1881). Recherches nouvelles sur les hémiplégies émotives. *Encéphale, 1*, 644–646.
MacBain, K.S., Best, C.T., and Strange, W. (1981). Categorical perception of English /r/ and /l/ by Japanese bilinguals. *Applied Psycholinguistics, 2*, 269–290.
MacKain, K., Studdert-Kennedy, M., Spieker, S., and Stern, D. (1983). Infant intermodal speech perception is a left-hemisphere function. *Science, 219*, 1347–1349.
MacNeilage, P.F., Studdert-Kennedy, M.G., and Lindblom, B. (1987). Primate handedness reconsidered. *Behavioral and Brain Sciences, 10*, 247–303.
Makita, K. (1968). The rarity of reading disability in Japanese children. *American Journal of Orthopsychiatry, 38*, 599–614.

Marie, P. (1922). Existe-t-il dans le cerveau humain des centres innés ou préformés de langage? *La Presse Médicale, 17*, 117–181.

Marks, D.F. (1990). On the relationship between imagery, body and mind. In P.J. Hampson, D.F. Marks, and J.T.E. Richardson (Eds.), *Imagery: Current developments* (pp. 1–38). London: Routledge.

Marler, P. (1977). The structure of animal communication sounds. In T.H. Bullock (Ed.), *Recognition of complex acoustic signals* (pp. 17–35). Berlin: Dahlem Konferenzen.

Marquez, G.G. (1971). *One hundred years of solitude*. New York: Avon Books.

Marr, D. (1982). *Vision*. San Francisco: Freeman.

Marr, D., and Nishihara, H.K. (1978). Representation and recognition of the spatial organization of three-dimensional shapes. *Proceedings of the Royal Society of London B, 200*, 269–294.

Marshall, J.C. (1989). The descent of the larynx? *Nature, 338*, 702–703.

Massaro, D.W. (1987). *Speech perception by ear and eye: A paradigm for psychological inquiry*. Hillsdale, NJ: Erlbaum.

Maurer, D. (1985). Infant's perception of facedness. In T.M. Field and N.A. Fox (Eds.), *Social perception in infants* (pp. 73–100). Norwood, NJ: Ablex.

Maynard Smith, J., and Haigh, J. (1974). The hitch-hiking effect of a favorable gene. *Genetical Research (Cambridge), 23*, 23–35.

Mazars, G., Hécaen, H., Tzavaras, A., and Merreune, L. (1970). Contribution à la chirurgie de certain bégaiements et à la compréhension de leur physiopathologie. *Revue Neurologique, 122*, 213–220.

McCarthy, R.A., and Warrington, E.K. (1986). Visual associative agnosia: A clinico-anatomical study of a single case. *Journal of Neurology, Neurosurgery, and Psychiatry, 49*, 1233–1240.

McCasland, J.S. (1987). Neuronal control of bird song production. *Journal of Neuroscience, 7*, 23–39.

McClelland, J.L., Rumelhart, D.E., and the PDP Research Group (1986). *Parallel distributed processing: Explorations in the microstructure of cognition*. Vol. 2: *Psychological and biological models*. Cambridge, MA: Bradford/MIT Press.

McClintock, B. (1984). The significance of responses to the genome to challenge. *Science, 226*, 792–801.

McGlone, J. (1980). Sex differences in the human brain: A critical survey. *Behavioral and Brain Sciences, 3*, 215–263.

McGlynn, S.M., and Schacter, D.L. (1989). Unawareness of deficits in neuropsychological syndromes. *Journal of Clinical and Experimental Neuropsychology, 11*, 143–205.

McGrew, W.C., Tutin, C.E., and Baldwin, P.J. (1979). Chimpanzees, tools, and termites: Cross-cultural comparisons of Senegal, Tanzania, and Rio Muni. *Man, 14*, 185–214.

McKeever, W.F., and Dixon, M.S. (1981). Right-hemisphere superiority for discriminating memorized from nonmemorized faces: Affective imagery, sex, and perceived emotionality effects. *Brain and Language, 12*, 246–260.

McKenna, P., and Warrington, E.K. (1978). Category-specific naming preservation: A single-case study. *Journal of Neurology, Neurosurgery, and Psychiatry, 41*, 571–574.

McManus, I.C. (1980). Handedness in twins: A critical review. *Neuropsychologia, 18*, 347–355.
McNeill, D. (1985). So you think gestures are nonverbal? *Psychological Review, 92*, 350–371.
Meadows, J.C. (1974). The anatomical basis of prosopagnosia. *Journal of Neurology, Neurosurgery, and Psychiatry, 37*, 489–501.
Mehler, J., Jusczyk, P., Lambertz, G., Halsted, N., Bertoncini, J., and Amiel-Tison, C. (1988). A precursor of language acquisition in young infacts. *Cognition, 29*, 143–178.
Melekian, B. (1981). Lateralization in the human newborn at birth: Asymmetry of the stepping reflex. *Neuropsychologia, 19*, 707–711.
Mellars, P. (1989). Major issues in the emergence of modern humans. *Current Anthropology, 30*, 349–385.
Meyer, M. (1913). Left-handedness and right-handedness in infancy. *Psychological Bulletin, 10*, 52–53.
Meyer-Bahlburg, F.L. (1983). Functional hemispheric asymmetry and psychopathology: Commentary. *Integrative Psychiatry, 1*, 57–58.
Michel, G. (1981). Right-handedness: A consequence of infant supine head-orientation preference? *Science, 212*, 685–687.
Michel, G. (1983). Development of hand use preference in infancy. In G. Young, S.J. Segalowitz, C.M. Carter, and S.E. Trehub (Eds.), *Manual specialization and the developing brain* (pp. 33–70). New York: Academic Press.
Miller, G.A. (1968). *The psychology of communication: Seven essays*. Harmondsworth, England: Penguin Books.
Milner, B. (1958). Psychological effects produced by temporal lobe excision. *Research Publications of the Association for Nervous and Mental Diseases, 36*, 244–257.
Milner, B. (1962). Laterality effects in audition. In V.B. Mountcastle (Ed.), *Interhemispheric relations and cerebral dominance* (pp. 177–195). Baltimore: Johns Hopkins University Press.
Milner, B. (1975). Psychological aspects of focal epilepsy and its neurosurgical management. In D.P. Purpura, J.K. Penry, and R.D. Walters (Eds.), *Advances in Neurology* (Vol. 8, pp. 299–321). New York: Raven.
Milner, B., Taylor, L., and Sperry, R.W. (1968). Lateralized suppression of dichotically presented digits after commissural section in man. *Science, 161*, 184–186.
Mishkin, M., and Forgays, D.G. (1952). Word recognition as a function of retinal locus. *Journal of Experimental Psychology, 43*, 43–48.
Miyamoto, M., Slightom, J.L., and Goodman, M. (1987). Phylogenetic relations of humans and African apes from DNA sequences in the psi-nu-globin region. *Science, 230*, 369–373.
Molfese, D.L., Freeman, R.B., Jr., and Palermo, D.S. (1975). The ontogeny of brain lateralization for speech and nonspeech sounds. *Brain and Language, 2*, 356–368.
Monod, J. (1969). On symmetry and function in biological systems. In A. Engstrom and B. Strandberg (Eds.), *Symmetry and function of biological systems at the macromolecular level* (pp. 15–27). New York: Wiley.

Morgan, C.L. (1894). *Introduction to comparative psychology*. London: Walter Scott.

Morgan, E. (1982). *The aquatic ape: A theory of human evolution*. London: Souvenir Press.

Morgan, E. (1985). *The descent of woman*. London: Souvenir Press.

Morgan, M.J. (1977). Embryology and the inheritance of asymmetry. In S. Harnad, R.W. Doty, L. Goldstein, J. Jaynes, and G. Krauthamer (Eds.), *Lateralization in the nervous system* (pp. 173–194). New York: Academic Press.

Morgan, M.J., and Corballis, M.C. (1978). On the biological basis of human laterality, II: The mechanism of inheritance. *Behavioral and Brain Sciences, 1*, 270–277.

Morgnani, G. (1769). *The seats and causes of diseases investigated by anatomy*. B.A. Alexander (Ed. and Trans., London: A. Millar and T. Cadell.

Morley, M.E. (1957). *The development and disorders of speech in childhood*. Edinburgh: Churchill-Livingstone.

Morrel-Samuels, P., Herman, L., and Bever, T. (1989). *A left hemisphere advantage for gesture-language signs in the dolphin*. Atlanta: Psychonomic Society Meeting, November.

Morris, D. (1967). *The naked ape*. London: Cape.

Morris, R.D., and Hopkins, W.D. (in press). Perception of human chimeric faces by chimpanzees: Evidence for a right hemisphere advantage. *Journal of Comparative Psychology*.

Moss, F.A. (1929). *Applications of psychology*. Boston, MA: Houghton Mifflin.

Myers, J.J., and Sperry, R.W. (1985). Interhemispheric communication after section of the forebrain commissures. *Cortex, 21*, 249–260.

Nadeau, S.E. (1988). Impaired grammar with normal fluency and phonology: Implications for Broca's aphasia. *Brain, 111*, 1111–1137.

Nagylaki, T., and Levy, J. (1973). "The sound of one paw clapping" isn't sound. *Behavior Genetics, 3*, 279–292.

Napier, J.R. (1971). *The roots of mankind*. London: Allen and Unwin.

Nass, R., Baker, S., Speiser, P., Virdis, R., Balsamo, A., Cacciara, E., Loche, A., Dumic, M., and New, M. (1987). Hormones and handedness: Left hand bias in female congenital adrenal hyperplasia patients. *Neurology, 37*, 711–715.

Naylor, H. (1980). Reading disability and lateral asymmetry: An information-processing analysis. *Psychological Bulletin, 87*, 531–545.

Nebes, R.D. (1971). Superiority of the minor hemisphere in commissurotomized man for the perception of part–whole relations. *Cortex, 7*, 333–349.

Nebes, R.D. (1972). Dominance of the minor hemisphere in commissurotomized man in a test of figural unification. *Brain, 95*, 633–638.

Nebes, R.D. (1973). Perception of spatial relationships by the right and left hemispheres in commissurotomized man. *Neuropsychologia, 3*, 285–289.

Needham, R. (1973). *Right and left: Essays on dual symbolic classification*. Chicago: University of Chicago Press.

Neville, A.C. (1976). *Animal asymmetry*. London: Arnold.

Newman, H.H. (1940). *Multiple human births*. New York: Doubleday, Doran.

Newport, E.L. (1983). Task specificity in language learning? Evidence from speech perception and American Sign Language. In E. Wanner and L.R Gleitman (Eds.), *Language acquisition: The state of the art* (pp. 450–486). Cambridge: Cambridge University Press.

Nottebohm, F. (1972). Neural lateralization of vocal control in a passerine bird. II. Subsong, calls, and a theory of vocal learning. *Journal of Experimental Zoology, 179*, 25–50.

Nottebohm, F. (1977). Asymmetries in neural control of vocalization in the canary. In S. Harnad, R.W. Doty, L. Goldstein, J. Jaynes, and G. Krauthamer (Eds.), *Lateralization in the nervous system* (pp. 23–44). New York: Academic Press.

Nottebohm, F. (1989). From bird song to neurogenesis. *Scientific American, 260*, 74–79.

Ogden, J.A. (1984). Dyslexia in a right-handed patient with a posterior lesion of the right cerebral hemisphere. *Neuropsychologia, 22*, 265–280.

Ogden, J.A. (1985a). Anterior-posterior interhemispheric differences in the loci of lesions producing visual hemineglect. *Brain and Cognition, 4*, 59–75.

Ogden, J.A. (1985b). Contralesional neglect of constructed visual images in right and left brain-damaged patients. *Neuropsychologia, 23*, 273–277.

Ogden, J.A. (1986). Autotopagnosia: Occurrence in a patient without nominal aphasia and with an intact ability to point to parts of animals and objects. *Brain, 108*, 1009–1022.

Ogden, J.A. (1988). Language and memory functions after long recovery periods in left hemispherectomized subjects. *Neuropsychologia, 26*, 645–659.

Ogden, J.A. (1989). Visuospatial and other "right-hemispheric" functions after long recovery periods in left-hemispherectomized subjects. *Neuropsychologia, 27*, 765–776.

O'Grady, W., and Dobrovolsky, M. (1987). *Contemporary linguistic analysis*. Toronto: Copp Clark Pitman.

Oldfield, R.C. (1971). The assessment and analysis of handedness: The Edinburgh Inventory. *Neuropsychologia, 9*, 97–114.

Ornstein, R.E. (1967). *The psychology of consciousness*. San Francisco: Freeman.

Orton, S.T. (1937). *Reading, writing and speech problems in children*. New York: Norton.

Overman, W.H., and Doty, R.W. (1982). Hemispheric specialization displayed by man but not macaques for analysis of faces. *Neuropsychologia, 20*, 113–128.

Overstreet, R. (1938). An investigation of prenatal position and handedness. *Psychological Bulletin, 35*, 520–521.

Papert, S. (1980). The role of artificial intelligence in psychology. In M. Piattelli-Palmarini (Ed.), *Language and learning: The debate between Jean Piaget and Noam Chomsky* (pp. 90–99). Cambridge, MA: Harvard University Press.

Papousek, M., and Papousek, H. (1981). Musical elements in the infant's

vocalization: Their significance for communication, cognition, and creativity. *Advances in Infancy Research, 1*, 163–224.
Parrot, M.J. (1879). Sur le développement du cerveau chez les enfants du premier age. *Archives de Physiologie Normale et Pathologique, 2ème Série, 6*, 505–521.
Passingham, R.E. (1982). *The human primate*. San Francisco: W.H. Freeman.
Passingham, R.E., and Ettlinger, G. (1974). A comparison of cortical function in man and other primates. *International Review of Neurobiology, 16*, 233–299.
Patterson, F. (1978). Conversations with a gorilla. *National Geographic, 154*, 438–465.
Penfield, W., and Roberts, L. (1959). *Speech and brain mechanisms*. Princeton, NJ: Princeton University Press.
Peretz, I., and Morais, J. (1988). Determinants of laterality for music: Towards an information processing account. In K. Hugdahl (Ed.), *Handbook of dichotic listening: Theory, methods, and research* (pp. 323–358). New York: Wiley.
Peters, M. (1988). Footedness: Asymmetries in foot preference and skill and neuropsychological assessment of foot movement. *Psychological Bulletin, 103*, 179–192.
Peters, M., and Servos, P. (1989). Performance of subgroups of left-handers and right-handers. *Canadian Journal of Psychology, 43*, 341–358.
Petersen, M.R., Beecher, M.D., Zoloth, S.R., Moody, D.B., and Stebbings, W.C. (1978). Neural lateralization of species-specific vocalizations by Japanese macaques. *Science, 202*, 324–327.
Peterson, G.M. (1934). Mechanisms of handedness in the rat. *Comparative Psychology Monographs, 9*, 1–67.
Petrie, B.F., and Peters, M. (1980). Handedness: Left/right differences in intensity of grasp response and duration of rattle holding in infants. *Infant Behavior and Development, 3*, 215–221.
Pfeiffer, J.E. (1973). *The emergence of man*. London: Book Club Associates.
Pfeiffer, J.E. (1985). *The emergence of humankind*. New York: Harper and Row.
Phippard, D. (1977). Hemifield differences in visual perception in deaf and hearing subjects. *Neuropsychologia, 13*, 555–562.
Piaget, J. (1978). *Success and understanding*. Andover, England: Routledge and Kegan Paul.
Piattelli-Palmarini, M. (1980). *Language and learning: The debate between Jean Piaget and Noam Chomsky*. Cambridge, MA: Harvard University Press.
Piattelli-Palmarini, M. (1989). Evolution, selection and cognition: From "learning" to parameter setting in biology and the study of language. *Cognition, 31*, 1–44.
Pinker, S. (1984). Visual cognition: An introduction. *Cognition, 18*, 1–63.
Pinker, S., and Bloom, P. (1990). *Natural language and natural selection*. Cambridge, MA: Center for Cognitive Science, Massachusetts Institute of Technology.
Pinker, S., and Prince, A. (1988). On language and connectionism: Analysis

of a parallel distributed processing model of language acquisition. *Cognition, 28*, 73–193.

Poizner, H., Klima, E.S., and Bellugi, U. (1987). *What the hands reveal about the brain*. Cambridge, MA: MIT Press, Bradford.

Poole, J., and Lander, D.G. (1971). The pigeon's concept of pigeon. *Psychonomic Science, 25*, 157–158.

Popper, K., and Eccles, J.C. (1977). *The self and its brain*. Berlin: Springer-Verlag.

Porac, C., and Coren, S. (1981). *Lateral preferences and human behavior*. New York: Springer-Verlag.

Pratt, R.T.C., and Warrington, E.K. (1972). The assessment of cerebral dominance with unilateral E.C.T. *British Journal of Psychiatry, 121*, 327–328.

Premack, D. (1971). Language in chimpanzee? *Science, 172*, 808–822.

Premack, D. (1985). "Gavagai!" or the future history of the animal language controversy. *Cognition, 19*, 207–296.

Premack, D. (1986). Pangloss to Cyrano de Bergerac: "Nonsense, it's perfect!" A reply to Bickerton. *Cognition, 23*, 81–88.

Premack, D. (1988). Minds with and without language. In L. Weiskrantz (Ed.), *Thought without language* (pp. 46–65). Oxford: Clarendon Press.

Prince, M. (1906). *The dissociation of a personality: A biographical study in abnormal psychology*. London: Longmans, Green.

Puccetti, R. (1981). The case for mental duality: Evidence from split-brain data and other considerations. *Behavioural and Brain Sciences, 4*, 93–123.

Putnam, H. (1980). What is innate and why: Comments on the debate. In M. Piattelli-Palmarini (Ed.), *Language and learning: The debate between Jean Piaget and Noam Chomsky* (pp. 287–309). Cambridge, MA: Harvard University Press.

Pylyshyn, Z.W. (1973). What the mind's eye tells the mind's brain: A critique of mental imagery. *Psychological Bulletin, 80*, 1–24.

Pylyshyn, Z.W. (1980). Cognitive representation and the process–architecture distinction. *Behavioral and Brain Sciences, 3*, 154–169.

Ratcliff, R. (1979). Spatial thought, mental rotation, and the right cerebral hemisphere. *Neuropsychologia, 17*, 49–54.

Raymond, F., Lejonne, P., and Lhermitte, J. (1906). Tumeurs du corps calleux. *Encéphale, 1*, 533–565.

Renfrew, C. (1989). The origins of Indo-European languages. *Scientific American, 261(4)*, 82–90.

Reynolds, D. McQ., and Jeeves, M.A. (1978). A developmental study of hemispheric specialization for recognition of faces in normal subjects. *Cortex, 14*, 259–267.

Rich, D.A., and McKeever, W.F. (1990). An investigation of immune system disorder as a "marker" for anomalous dominance. *Brain and Cognition, 12*, 55–72.

Richards, G. (1987). *Human evolution*. New York: Routledge and Kegan Paul.

Richer, P. (1881). *Etudes cliniques sur l'hystéro-épilepsie ou grande hystérie*. Paris: Delahaye et Lacroisnier.

Riddoch, M.J., and Humphreys, G.W. (1988). Description of a left/right coding deficit in a case of constructional apraxia. *Cognitive Neuropsychology, 5*, 289–316.

Rife, D.C. (1940). Handedness with special reference to twins. *Genetics, 25*, 178–186.

Roeltgen, D. (1985). Agraphia. In K.M. Heilman and E. Valenstein (Eds.), *Clinical neuropsychology* (Vol. 2, pp. 75–96). New York: Oxford University Press.

Rogers, L.J. (1980). Lateralization in the avian brain. *Bird Behavior, 2*, 1–12.

Romanes, G.J. (1888). *Mental evolution in animals*. London: Kegan Paul.

Romer, A.S. (1962). *The vertebrate body*. London: W.B. Saunders.

Rosenberger, P.B., and Hier, D.B. (1981). Cerebral asymmetry and verbal intellectual deficits. *Annals of Neurology, 38*, 300–304.

Ross, E.D. (1981). The aprosodias: Functional-anatomic organization of the affective component of language in the right hemisphere. *Annals of Neurology, 38*, 561–589.

Rosselli, M., and Ardila, A. (1989). Calculation deficits in patients with right and left hemisphere damage. *Neuropsychologia, 27*, 607–617.

Rossi, G.F., and Rosadini, G. (1967). Experimental analysis of cerebral dominace in man. In C.H. Millikan and F. Darley (Eds.), *Brain mechanisms underlying speech and language* (pp. 167–184). New York: Grune and Stratton.

Rothbart, M.K., Taylor, S.B., and Tucker, D.M. (1989). Right-sided facial asymmetry in infant emotional expression. *Neuropsychologia, 27*, 675–687.

Rubens, A.B. (1976). Transcortical motor aphasia. In H. Whitaker and H.A. Whitaker (Eds.), *Studies in neurolinguistics* (Vol. 1, pp. 293–303). New York: Academic Press.

Rumbaugh, D. (1977). *Language learning by a chimpanzee: The LANA Project*. New York: Academic Press.

Rumelhart, D.E., and McClelland, J.L. (1986). On learning the past tenses of English verbs. In J.L. McClelland, D.E. Rumelhart, and the PDP Research Group (Eds.), *Parallel distributed processing: Explorations in the microstructure of cognition.* Vol. 2: *Psychological and biological models* (pp. 216–271). Cambridge, MA: Bradford Books/MIT Press.

Rumelhart, D.E., McClelland, J.L., and the PDP Research Group (1986). *Parallel distributed processing: Explorations in the microstructure of cognition.* Vol. 1: *Foundations*. Cambridge, MA: Bradford Books/MIT Press.

Russell, I.S. (1979). Brain size and intelligence: A comparative approach. In D.A. Oakley and H.C. Plotkin (Eds.), *Brain, behavior, and evolution* (pp. 126–153). London: Methuen.

Russo, M., and Vignolo, L.A. (1967). Visual figure–ground discrimination in patients with unilateral cerebral disease. *Cortex, 3*, 113–127.

Sackeim, H.A., Gur, R., and Saucy, M.C. (1978). Emotions are expressed more intensely on the left side of the face. *Science, 202*, 434–436.

Sacks, O. (1985). *The man who mistook his wife for a hat*. London: Duckworth.

Sagan, C. (1977). *The dragons of Eden*. New York: Random House.

Sagan, C. (1979). *Broca's brain: Reflections on the romance of science*. New York: Random House.

Salcedo, J.R., Spiegler, B.J., Gibson, E., and Magilavy, D.B. (1985). The autoimmune disease systemic lupus erythematosus is not associated with left-handedness. *Cortex, 21*, 645–647.

Sarich, V.M., and Wilson, A.C. (1967). Immunological time scale for hominid evolution. *Science, 158*, 1200–1203.

Satz, P. (1972). Pathological left-handedness: An explanatory model. *Cortex, 8*, 121–135.

Savage-Rumbaugh, E.S. (1984). Acquisition of functional symbol usage in apes and children. In H.L. Roitblat, T.G. Bever, and H.S. Terrace (Eds.), *Animal cognition* (pp. 291–310). Hillsdale, NJ: Erlbaum.

Savage-Rumbaugh, S. (1987). A new look at ape language: Comprehension of vocal speech and syntax. *Nebraska Symposium on Motivation, 35*, 201–255.

Schaffer, H.R. (1984). *The child's entry into a social world*. New York: Academic Press.

Schaller, G.B. (1963). *The mountain gorilla*. Chicago: University of Chicago Press.

Schepers, G.W.H. (1950). The brain casts of the recently discovered *Plesianthropus* skulls. In R. Broom, and G.W.H Schepers (Eds.), *Sterkfontein Ape-man, Plesianthropus* (Vol. 4, pp. 85–117). Transvaal, South Africa: Transvaal Museum Memoranda.

Schiff, B.B., and Lamon, M. (1989). Inducing emotion by unilateral contraction of facial muscles: A new look at hemispheric specialization and the experience of emotion. *Neuropsychologia, 27*, 923–935.

Schrier, A.M., Angarella, R., and Povar, M.L. (1984). Studies of concept formation by stumptailed monkeys: Concepts of humans, monkeys, and letter A. *Journal of Experimental Psychology: Animal Behavior Processes, 10*, 564–584.

Science framework for public schools. (1978). Sacramento, CA: California State Board of Education.

Searle, J.R. (1980). Minds, brains, and programs. *Behavioral and Brain Sciences, 3*, 417–461.

Searleman, A., and Fugagli, A. (1987). Suspected autoimmune disorders and left-handedness: Evidence from individuals with diabetes, Crohn's disease, and ulcerative colitis. *Neuropsychologia, 25*, 367–374.

Searleman, A., Porac, C., and Coren, S. (1989). Relationship between birth order, birth stress, and lateral preferences: A critical review. *Psychological Bulletin, 105*, 397–408.

Semenov, S.A. (1964). *Prehistoric technology*. London: Cory, McAdams, and MacKay.

Sergeant, D. (1969). Experimental investigation of absolute pitch. *Journal of Research in Musical Education, 17*, 135–143.

Sergent, J. (1986). Microgenesis of face perception. In H.D. Ellis, M.A. Jeeves, P. Newcombe, and A. Young (Eds.), *Aspects of face processing* (pp. 17–33). Dordrecht: Martinus Nijhof.

Sergent, J. (1988). Face perception and the right hemisphere. In L.

Weiskrantz (Ed.), *Thought without language* (pp. 108–131). Oxford: Clarendon Press.
Sergent, J. (1989). Image generation and the processing of generated images in the cerebral hemispheres. *Journal of Experimental Psychology: Human Perception and Performance, 15,* 170–178.
Sergent, J., and Hellige, J.B. (1986). Role of input factors in visual-field asymmetries. *Brain and Cognition, 5,* 174–199.
Seth, G. (1973). Eye–hand coordination and 'handedness:' A developmental study of visuo-motor behavior in infancy. *British Journal of Educational Psychology, 43,* 35–49.
Seyfarth, R.M., and Cheney, D.L. (1984). The natural vocalizations of non-human primates. *Trends in Neuroscience, 7,* 66–73.
Shallice, T. (1988). *From neuropsychology to mental structure.* Cambridge: Cambridge University Press.
Shankweiler, D., and Studdert-Kennedy, M. (1967). Identification of consonants and vowels presented to left and right ears. *Quarterly Journal of Experimental Psychology, 19,* 59–63.
Shepard, R.N. (1978). The mental image. *Psychological Review, 33,* 125–164.
Sherman, J.A. (1978). *Sex-related cognitive differences: An essay on theory and evidence.* Springfield, IL: Charles C. Thomas.
Shimizu, A., and Endo, M. (1983). Handedness and familial sinistrality in a Japanese student population. *Cortex, 19,* 265–272.
Sibley, C.G., and Ahlquist, J.E. (1984). The phylogeny of hominoid primates, as indicated by DNA-DNA hybridization. *Journal of Molecular Evolution, 20,* 2–15.
Simons, E.L. (1963). A critical appraisal of the tertiary primates. In J. Buettner-Janusch (Ed.), *Evolutionary and genetic biology of primates.* (Vol. 1, pp. 65–129). New York: Academic.
Simons, E.L. (1989). Human origins. *Science, 245,* 1343–1350.
Simpson, E. (1980). *Reversals: A personal account of victory over dyslexia.* London: Victor Gollancz.
Sinclair, A.R.E., Leakey, M.D., and Norton-Griffiths, M. (1986). Migration and hominid bipedalism. *Nature, 324,* 307–308.
Skelton, R.R., McHenry, H.M., and Drawhorn, G.M. (1986). Phylogenetic analysis of early hominids. *Current Anthropology, 27,* 21–43.
Skinner, B.F. (1948). *Walden two.* New York: Macmillan.
Skinner, B.F. (1957). *Verbal behavior.* New York: Appleton-Century-Crofts.
Smith, B.D., Meyers, M.B., and Kline, R. (1989). For better or for worse: Left-handedness, pathology, and talent. *Journal of Clinical and Experimental Neuropsychology, 11,* 944–958.
Smith, B.H. (1986). Dental development in *Australopithecus* and early *Homo. Nature, 323,* 327–330.
Smith, J. (1987). Left-handedness: Its association with allergic diseases. *Neuropsychologia, 25,* 665–667.
Snowdon, C.T. (1982). Linguistic and psycholinguistic approaches to primate communication. In C.T. Snowdon, C.H. Brown, and M.R. Petersen (Eds.), *Primate communication* (pp. 212–238). Cambridge: Cambridge University Press.

Snowdon, C.T. (1989). Vocal communication in New World monkeys. *Journal of Human Evolution, 18,* 611–633.

Sperry, R.W. (1974). Lateral specialization in the surgically separated hemispheres. In F.O. Schmitt and F.G. Worden (Eds.), *The neurosciences: Third study program* (pp. 5–19). Cambridge, MA: MIT Press.

Sperry, R.W. (1982). Some effects of disconnecting the cerebral hemispheres. *Science, 217,* 1223–1227.

Sperry, R.W. (1984). Consciousness, personal identity, and the divided brain. *Neuropsychologia, 22,* 661–673.

Sperry, R.W. (1988). Psychology's mentalist paradigm and the religion/science tension. *American Psychologist, 43,* 607–613.

Sperry, R.W., Zaidel, E., and Zaidel, D. (1979). Self recognition and social awareness in the deconnected right hemisphere. *Neuropsychologia, 17,* 153–166.

Springer, S., and Searleman, A. (1980). Left handedness in twins: Implications for the mechanisms underlying cerebral asymmetry of function. In J. Herron (Ed.), *Neuropsychology of left-handedness* (pp. 139–158). New York: Academic Press.

Springer, S.P., and Deutsch, G. (1985). *Left brain, right brain.* New York: Freeman.

Starkweather, C.W., Franklin, S., and Smigo, T.M. (1984). Vocal and finger reaction times in stutterers and nonstutterers. Differences in correlations. *Journal of Speech and Hearing Research, 27,* 193–196.

Steenhuis, R.E., and Bryden, M.P. (1989). Different dimensions of hand preference that relate to skilled and unskilled activities. *Cortex, 25,* 289–304.

Steklis, H.D., and Marchant, L.F. (1987). Primate handedness: Reaching and grasping for straws? *Behavioral and Brain Sciences, 10,* 284–286.

Strauss, E., and Moscovitch, M. (1981). Perception of facial expressions. *Brain and Language, 13,* 308–322.

Stringer, C.B., and Andrews, P. (1988). Genetic and fossil evidence for the origin of modern humans. *Science, 239,* 1263–1268.

Suberi, M., and McKeever, W.F. (1977). Differential right hemispheric memory storage of emotional and non-emotional faces. *Neuropsychologia, 15,* 757–768.

Susman, R.L. (1988). Hand of *Paranthropus robustus* from Member 1, Swartkrans: Fossil evidence for tool behavior. *Science, 240,* 781–784.

Tanner, N.M., and Zihlman, A.L. (1976). Women in evolution 1. Innovation and selection in human origins. *Signs: Journal of Women in Culture and Society, 1,* 585–608.

Taylor, E.S. (1976). *Beck's obstetrical practice and fetal medicine* (10th edition). Baltimore: Williams and Wilkins.

Taylor, H.G., and Solomon, J.R. (1979). Reversed laterality: A case study. *Journal of Clinical Neuropsychology, 1,* 311–322.

Taylor, J. (1958). *Selected writings of John Hughlings Jackson.* London: Staples.

Teilhard de Chardin, Pere (1959). *The phenomenon of man.* London: Collins.

Teng, E.L., Lee, P., Yang, P.C., and Chang, P.C. (1976). Handedness in a Chinese population: Biological, social, and pathological factors. *Science, 193,* 1148–1150.

Terrace, H.S. (1979). Is problem solving language? *Journal of the Experimental Analysis of Behavior, 31,* 161–175.

Terrace, H.S., Petitto, L.A., Sanders, R.J., and Bever, T.G. (1979). Can an ape create a sentence? *Science, 206,* 891–902.

Terzian, H. (1964). Behavioral and EEG effets of intracarotid sodium amytal injection. *Acta Neurochiurgia (Vienna), 12,* 230–239.

Teuber, H.-L., and Weinstein, S. (1956). Ability to discover hidden figures after cerebral lesions. *Archives of Neurology and Psychiatry, 76,* 369–379.

Thatcher, R.W., Walker, R.A., and Giudice, S. (1987). Human cerebral hemispheres develop at different rates and ages. *Science, 236,* 1110–1113.

Thigpen, C.H., and Cleckley, H. (1957). *The three faces of Eve.* New York: McGraw-Hill.

Thomas, K. (1984). *Man and the natural world.* Harmondsworth: Penguin Books.

Thompson, L.J. (1971). Language disabilities in men of eminence. *Journal of Learning Disabilities, 4,* 39–50.

Tobias, P.V. (1987). The brain of *Homo habilis*: A new level of organization in cerebral evolution. *Journal of Human Evolution, 16,* 741–761.

Tomasch, J. (1954). Size, distribution, and number of fibers in the human corpus callosum. *Anatomical Record, 119,* 7–19.

Torgerson, J. (1950). Situs inversus, asymmetry, and twinning. *American Journal of Human Genetics, 2,* 361–370.

Toth, N. (1985). Archeological evidence for preferential right-handedness in the Lower and Middle Pleistocene, and its possible implications. *Journal of Human Evolution, 14,* 607–614.

Travis, L.E. (1931). *Speech pathology.* New York: Appleton.

Treisman, A.M., and Geffen, G. (1968). Selective attention and cerebral dominance in perceiving and responding to speech messages. *Quarterly Journal of Experimental Psychology, 20,* 139–150.

Trinkaus, E. (1983). *The Shanidar Neandertals.* New York: Academic Press.

Turing, A.M. (1950). Computing machinery and intelligence. *Mind, 59,* 433–460.

Turkewitz, G. (1988). A prenatal source for the development of hemispheric specialization. In D.L. Molfese and S.J. Segalowitz (Eds.), *Brain lateralization in children* (pp. 73–81). New York: Guilford.

Turkle, S. (1984). *The second self: Computers and the human spirit.* New York: Simon and Shuster.

Vallar, G., Bisiach, E., Cerizza, M., and Rusconi, M.L. (1988). The role of the left hemisphere in decision-making. *Cortex, 24,* 399–410.

Van Riper, C. (1971). *The nature of stuttering.* New York: Prentice-Hall.

Van Strien, J., Bouma, A., and Bakker, D.J. (1987). Birth stress, autoimmune diseases, and handedness. *Journal of Clinical and Experimental Neuropsychology, 6,* 775–780.

Van Valen, L., and Sloan, R.E. (1965). The earliest primates. *Science, 150,* 743–745.

Vargha-Khadem, F., and Corballis, M.C. (1979). Cerebral asymmetry in infants. *Brain and Language, 8*, 1–9.

Vernon, M.D. (1960). *Backwardness in reading* 2nd edition). Cambridge: Cambridge University Press.

Vles, J.S.H., Grubben, C.P.M., and Hoogland, H.J. (1989). Handedness not related to fetal position. *Neuropsychologia, 27*, 1017–1018.

von Bonin, G. (1962). Anatomical asymmetries of the cerebral hemispheres. In V.B. Mountcastle (Ed.), *Hemispheric relations and cerebral dominance* (pp. 1–6). Baltimore: Johns Hopkins University Press.

von Frisch, K. (1955). *The dancing bees*. New York: Harcourt, Brace.

Wada, J.A., Clarke, R., and Hamm, A. (1975). Cerebral asymmetry in infants. *Archives of Neurology, 32*, 266–282.

Wada, J.A., and Rasmussen, T. (1960). Intracarotid injection of sodium amytal for the lateralization of cerebral speech dominance: Experimental and clinical observations. *Journal of Neurosurgery, 17*, 266–282.

Wagner, M.T., and Hannon, R. (1981). Hemispheric asymmetries in faculty and student musicians and non-musicians during melody recognition tasks. *Brain and Language, 13*, 379–388.

Walker, A., Leakey, R.E., Harris, J.M., and Brown, F.H. (1986). 2.5-Myr *Australopithecus boisei* from west of Lake Turkana, Kenya. *Nature, 322*, 517–522.

Walker, S.F. (1980). Lateralization of function in the vertebrate brain: A review. *British Journal of Psychology, 71*, 329–367.

Walker, S.F. (1983). *Animal thought*. London: Routledge and Kegan Paul.

Wallace, A.R. (1895). *Natural selection and tropical nature*. London: Macmillan.

Warrington, E.K., and James, M. (1986). Visual object recognition in patients with right hemisphere lesions: Axes or features? *Perception, 15*, 355–366.

Warrington, E.K., and James, M. (1988). Visual apperceptive agnosia: A clinico-anatomical study of three cases. *Cortex, 24*, 13–32.

Warrington, E.K., and Pratt, R.T.C. (1973). Language laterality in left handers assessed by unilateral E.C.T. *Neuropsychologia, 11*, 423–428.

Warrington, E.L., and Taylor, A.M. (1973). The contribution of the right parietal lobe to object recognition. *Cortex, 9*, 152–164.

Washburn, S.L., and Moore, R. (1980). *Ape into human: A study of human evolution*. Boston: Little, Brown.

Watson, J.B. (1913). Psychology as the behaviorist views it. *Psychological Review, 20*, 158–177.

Watson, N.V., and Kimura, D. (1989). Right-hand superiority for throwing but not for intercepting. *Neuropsychologia, 27*, 1399–1414.

Webster, W.G. (1986). Response sequence organization and production by stutterers. *Neuropsychologia, 24*, 813–821.

Weiss, M.S., and House, A.S. (1973). Perception of dichotically presented vowels. *Journal of the Acoustical Society of America, 53*, 51–58.

Werker, J.F. (1989). Becoming a native listener. *American Scientist, 77*, 54–59.

Wernicke, C. (1874). *Der Aphasische Symptomenkomplex*. Brelau: Cohn and Weigart.
West, R., Howell, P., and Cross, I. (1985). Modelling perceived musical structure. In P. Howell, I. Cross, and R. West (Eds.), *Musical structure and cognition* (pp. 21–52). New York: Academic Press.
Whitaker, H.A. (1978). Is the right leftover? *Behavioral and Brain Sciences, 2,* 323–324.
White, L., Jr. (1967). The historical roots of our ecological crisis. *Science, 155,* 1203–1207.
White, R. (1989). Visual thinking in the ice age. *Scientific American, 261(1),* 74–81.
Wiesel, T.N. (1982). Postnatal development of the visual cortex and the influence of the environment. *Nature, 299,* 583–591.
Wigan, A.C. (1844). *The duality of the mind.* London: Longman, Brown, Green, and Longmans.
Wilson, D. (1872). Righthandedness. *The Canadian Journal, No. 75,* 193–203.
Wilson, E.O. (1975). *Sociobiology.* Cambridge, MA: Belknap Press of Harvard University.
Wilson, E.O. (1978). *On human nature.* Cambridge, MA: Harvard University Press.
Wilson, P.J. (1980). *Man, the promising primate: The conditions of human evolution.* New Haven, CT: Yale University Press.
Wofsy, D. (1984). Hormones, handedness, and autoimmunity. *Immunology Today, 5,* 169–170.
Wolff, P.H., Hurwitz, I., and Moss, H. (1977). Serial organization of motor skills in left- and right-handed adults. *Perceptual and Motor Skills, 15,* 539–546.
Woo, T.L., and Pearson, K. (1927). Dextrality and sinistrality of hand and eye. *Biometrika, 19,* 165–199.
Wood, B.A., and Chamberlain, A.T. (1987). The nature and affinities of the "robust" australopithecines. *Journal of Human Evolution, 16,* 625–641.
Wood, J.P. (1966). *The snark was a boojum: A life of Lewis Carroll.* New York: Pantheon.
Woodruff, G., and Premack, D. (1979). Intentional communication in the chimpanzee: The development of deception. *Cognition, 7,* 333–362.
Woolley, H.T. (1910). The development of right handedness in a normal infant. *Psychological Review, 17,* 37–41.
Wright, R. (1972). Imitative learning of a flaked stone tool technology—the case of an orangutan. *Mankind, 8,* 296–306.
Wundt, W. (1894). *Lectures on human and animal psychology.* London: Swan Sonnenschein.
Wyke, M. (1971). The effects of brain lesions on the learning performance of a bimanual coordination task. *Cortex, 7,* 59–72.
Yerkes, R.M. (1925). *Almost human . . .* New York: Century.
Young, A.W. (1988). Functional organization of visual recognition. In L. Weiskrantz (Ed.), *Thought without language* (pp. 78–107). Oxford: Clarendon Press.

Young, G. (1977). Manual specialization in infancy: Implications for lateralization of brain function. In S.J. Segalowitz and F.A. Gruber (Eds.), *Language development and neurological theory* (pp. 289–311). New York: Academic Press.

Young, J.Z. (1962). Why so we have two brains? In V.B. Mountcastle (Ed.), *Interhemispheric relations and cerebral dominance* (pp.). Baltimore: Johns Hopkins Press.

Yunis, J.J., and Prakash, O. (1982). The origin of man: A chromosomal pictorial legacy. *Science, 215*, 1525–1530.

Zaidel, E. (1976). Auditory vocabulary of the right hemisphere following brain bisection or hemidecortication. *Cortex, 12*, 191–211.

Zaidel, E. (1981). Reading by the disconnected right hemisphere: An aphasiology perspective. In Y. Zotterman (Ed.), *The Wenner-Gren Symposium on dyslexia* (pp. 67–91). London: Plenum.

Zaidel, E. (1983). A response to Gazzaniga: Language in the right hemisphere, convergent perspectives. *American Psychologist, 38*, 542–546.

Zaidel, E. (1985). Language in the right hemisphere. In D.F. Benson and E. Zaidel (Eds.), *The dual brain: hemispheric specialization in humans* (pp. 205–231). New York: Guilford.

Zaidel, E., and Peters, A.M. (1981). Phonological encoding and ideographic reading by the disconnected right hemisphere: Two case studies. *Brain and Language, 14*, 205–234.

Zangwill, O.L. (1976). Thought and the brain. *British Journal of Psychology, 67*, 301–314.

Zatorre, R.J. (1984). Musical perception and cerebral function: A critical review. *Music Perception, 2*, 196–221.

Zatorre, R.J. (1989). Effects of temporal neocortical excisions on musical processing. *Contemporary Music Review, 4*, 255–266.

Zihlman, A.L., Cronin, J.E., Cramer, D.L., and Sarich, V.M. (1978). Pygmy chimpanzees as a possible prototype for the common ancestor of humans, chimpanzees, and gorillas. *Nature, 275*, 744–746.

Zurif, E.B. (1974). Auditory lateralization: Prosodic and syntactic factors. *Brain and language, 1*, 391–404.

Index